Human Geography

A Short Introduction

Second Edition

John Rennie Short

University of Maryland, Baltimore County

New York   Oxford
OXFORD UNIVERSITY PRESS

Oxford University Press is a department of the University of Oxford. It fur-
thers the University's objective of excellence in research, scholarship, and
education by publishing worldwide. Oxford is a registered trade mark of
Oxford University Press in the UK and certain other countries.

Published in the United States of America by Oxford University Press
198 Madison Avenue, New York, NY 10016, United States of America.

For titles covered by Section 112 of the US Higher Education
Opportunity Act, please visit www.oup.com/us/he for the latest
information about pricing and alternate formats.

Library of Congress Cataloging-in-Publication Data

Names: Short, John R., author.
Title: Human geography : a short introduction / John Rennie Short, University
of Maryland, Baltimore County.
Description: Second edition. | New York : Oxford University Press, [2018] |
Includes bibliographical references and index.
Identifiers: LCCN 2017014395 (print) | LCCN 2017023027 (ebook) | ISBN
9780190679880 | ISBN 9780190679835 (pbk.) | ISBN 9780190679927
(looseleaf)
Subjects: LCSH: Human geography.
Classification: LCC GF41 (ebook) | LCC GF41 .S488 2018 (print) | DDC
304.2—dc23
LC record available at https://lccn.loc.gov/2017014395

Printing number: 9 8 7 6 5 4 3 2

Printed in Canada by Webcom Inc.

I dedicate this book to the memory of two of my grandparents.

My paternal grandmother, Janet Adamson Craig Short (1895–1966), had a profound respect for learning and social justice, an endless devotion to my education, and an almost infinite patience with my spelling lessons. I think of her often, especially when my thoughts turn to the connections between love and teaching, devotion and education, wonder and learning.

My maternal grandfather, John Rennie (1903–1963), was a coal miner who spent all his working years in the darkness and the danger. Perhaps because he worked in such cramped and tight spaces, in one memorable holiday, just the two of us, he gave me a sense of limitless freedom.

Brief
Contents

Contents

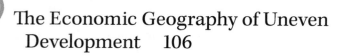

PART 3   THE ECONOMIC ORGANIZATION OF SPACE   89

Preface

The aim of this book is to introduce students to a wide range of important and exciting work in human geography. The primary audience is students in colleges and universities. I decided to write this book because many of the standard texts are too big, and increasingly too expensive, to provide the accessible and affordable base most of us need for our human geography courses. I sense a pushback by teachers and students against the overly large and expensive books available now. They have grown into, to use Henry James's description of many nineteenth-century novels, "loose and baggy monsters." There is room for a more interesting and subtle book than the standard texts. This briefer and more accessible alternative is written in a more familiar style that can be augmented by other resources.

As a former mountain climber I will use as metaphor the attempts on the big Himalayan peaks. In the 1970s, the attempts were increasingly organized as large teams with many climbers and elaborate systems of camps and base camps. Then, in the late 1970s, a number of climbers dispensed with the large teams and sought to climb alone or with one other climber. Less burdened by organizational weight, they were much more successful in reaching the summits in quick direct assaults. This book adapts a similar ethic of "light and fast" that affords more flexibility to instructors than a traditional textbook. Not an exact metaphor, to be sure, but close enough to give you a sense of the book's character and mission.

The title, *Human Geography: A Short Introduction*, employs the word "short" in two ways. First, it indicates a relatively brief introduction rather than a wide survey—although only in the word-rich world of college textbooks can a 100,000-word text be described as short; "concise" may be a more appropriate term. Second, the play on my own name is to signal that it is a book with a distinctive authorial voice. Textbooks are written at specific times in specific places by specific people, and these three basic facts color and shape the material covered and the nature of the coverage. I have drawn heavily on much of my own work conducted over the past thirty-five years. The main title announces the subject matter, while the subtitle lets the reader know that it is the world of human geography as seen by just one person. This is less an act of egotism than a reminder to the reader that the text is not revealed truth but the singular vision of just one scholar.

The aim is to be both engaging and comprehensive. The text is intended to be both student- and instructor-friendly. The structure, while providing a coherent whole, also allows sectional choices to meet the different needs, time constraints, and interests of individual instructors. Each chapter has a list of further readings and websites that instructors can employ in teaching and develop as resources in ways suitable for the size and constitution of their particular classes.

This is an ambitious book that gives readers a sense of the complex human geography of the contemporary world. It brings together a global perspective with an understanding of national concerns and the growth of select urban regions. Broad arguments are enlivened with detailed case studies. The writing style is accessible to the general reader, and the scholarship is comprehensive, so that different interpretations are presented. Six large themes dominate:

- the relationships between people, environment, and resources,
- the economic organization of space,
- the cultural organization of space,
- the global organization of space,
- the political organization of space,
- the urban organization of space.

Running through a discussion of these broad themes are case studies that include examples of specific places as well as examples of the geographical imaginations—models, ideas, and theories—that inform and shape the relationship between people and their environments.

Part 1 sets the context. The first chapter provides a brief introduction to the physical geography of the world, while the second provides an intellectual context by discussing major themes in the development of the discipline of human geography. Not all human geography courses include these topics, but I feel they have a place. To understand contemporary human geography, it is necessary to have a basic grounding in the physical geography of the planet and the intellectual history of the discipline. This section is an elective for those with the time to set the course in its broader physical and intellectual history. Some may elect to move straight to Part 2, which makes the connections between population, environment, and resources. Part 3 discusses the economic organization of space. Part 4, new to the second edition, looks at the cultural organization of space and focuses on the important geographies of population and culture. This section is important because one can't understand much that is current these days without a certain understanding of human populations, languages, and religions. Part 5 centers on the analysis of global trends and processes. Many introductory texts have only one chapter devoted to globalization, almost as an afterthought. An important part of contemporary scholarship identifies the global scale as important for understanding a wide variety of the world's most pressing issues. The book will enable readers to make sense of complex and seemingly unrelated global and regional phenomenon. Part 6 examines the political organization of space. Here I take as the starting point the idea that space embodies and contains power relations. The book concludes with Part 7, which focuses on cities, now home to the majority of the world's population. Specific chapters examine trends of urbanization, urban networks, and the internal structure of the city.

The text is constructed so that instructors can tailor the readings to suit class needs. The entire book may be used, but some may want to exclude the introductory chapter, while others may want to focus on just four of the five major topic areas. Those with a more economic interest, for example, will certainly want to include Part 3, while those with more cultural emphasis will definitely use Parts 4 and 5. Those with focus on political geography will use Part 6 and those with an interest in cities will use Part 7. Each of the sections is self-contained, so an instructor may elect to choose any four or five and leave more time for other class activities.

New to the Second Edition

A second edition allows for improvements on the first. There are four major changes.

- The first is the addition of an entire new section entitled *The Cultural Organization of Space* that includes three new chapters on the geography of population, geography of religion, and the geography of language. The most recent works in these traditional concerns of human geography are explored and explained.

- The second major change, afforded by a second edition, is a full revision of the entire text making it an even more up to date source of the latest geographical work.

- The third major change is to make the text more student-friendly. Each chapter now opens with a list of learning objectives and closes with a list of learning outcomes. Key terms for each chapter are bolded and defined in a new glossary of over 160 entries. More subheadings make it easier to navigate the text and follow the narrative flow. All these changes make the text even more accessible to teachers and students.

- The fourth change is to the look of the book. New figures, illustrations, and maps complement the revised text. Along with the new fresh design the book is visually appealing and pedagogically adaptable.

Teaching and Learning Package

This book is supported by a carefully crafted ancillary package designed to support both professors' and students' efforts in the course:

- *Digital files of all the graphics in the book.* Instructors will find all of the images from the book available to them, both as raw jpegs and pre-inserted into PowerPoint.

- The book is accompanied by a full complement of digital, interactive materials available on a companion web site (www.oup.com/us/short), via Oxford University Press' Dashboard platform (see description below), or for single sign-on use with most local course management systems. Included here are:

  ❑ *Review questions for students.* Carefully crafted by experienced instructors, these computer-graded review questions accompany each unit of the textbook. Professors can assign them for homework, or students can use them independently to check their understanding of the topics presented in the book.

  ❑ *Interactive exercises based on curated resources.* These exercises—more extensive than chapter review exercises—guide students through visualizations and animations about key geographic topics. These interactives are assignable, and include homework exercises and questions for review.

  ❑ *Test questions and testing software.* Written by experienced instructors and answerable directly from the text, these questions provide professors and instructors with a useful tool for creating and administering tests.

  ❑ *Dashboard.* A text-specific, integrated learning system designed with clear and consistent navigation. It delivers quality content and tools to track student progress in an intuitive, web-based learning environment. Dashboard features a streamlined interface that connects instructors and students with the functions they perform most, simplifying the learning experience to save time and put student progress first.

Instructors should contact their Oxford University Press representative for more information about the supplements package.

Acknowledgments

I'd like to thank the many thoughtful scholars who, during the writing process, dedicated valuable time to reviewing and offering comments on the manuscript. Their comments improved the book significantly.

Jeff Baldwin, *Sonoma State University*
Randy Bertolas, *Wayne State College*
Brian Blouet, *The College of William and Mary*
Gina M. Bryson-Prieto, *Hialeah-Miami Lakes Senior High School, Miami, FL*
Karl Byrand, *University of Wisconsin Colleges*
Dorothy Cassetta, *Carroll High School, Southlake, TX*
Thomas Chapman, *Old Dominion University*
Patrick Clancy, *Strath Haven High School, Wallingford, PA*
Timothy W. Collins, *University of Texas at El Paso*
Daniel J Dempsey, *College of the Redwoods*
Mike DeVivo, *Grand Rapids Community College*
Elizabeth Dunn, *University of Colorado Boulder*
Owen Dwyer, *Indiana University–Purdue University Indianapolis*
Kyle T. Evered, *Michigan State University*
Adrienne Goldsberry, *Michigan State University*
Richard Grant, *University of Miami*
Steven M. Graves, *California State University, Northridge*
Benjamin Harris, *Bishop Kelly High School, Boise, ID*
Ann Fletchall, *Western Carolina University*
Joshua Hagen, *Marshall University*
Stephen Healy, *Worcester State University*
Peter R. Hoffman, *Loyola Marymount University*
Valentine Udoh James, *Clarion University of Pennsylvania*
Richard E. Katz, *Roosevelt High School, Seattle, WA*
Heidi Lannon, *Santa Fe College*
Jonathan Leib, *Old Dominion University*
Donald Lyons, *University of North Texas*
Heather J. McAfee, *Clark College*
Richard Medina, *George Mason University*
Douglas C. Munski, *University of North Dakota*
Nancy J Obermeyer, *Indiana State University*
Linda F. Pittman, *Richard Bland College*
Nathan J. Probasco, *Briar Cliff University*
Darren Purcell, *University of Oklahoma*
William Rowe, *Louisiana State University*
James C. Saku, *Frostburg State University*

Arun Saldanha, *University of Minnesota*
Richard H. Schein, *University of Kentucky*
Colleen Schmidt, *Carnegie Vanguard High School, Houston, TX*
Roger M. Selya, *University of Cincinnati*
Jeremy Slack, *The University of Texas at El Paso*
Jill Stackhouse, *Bemidji State University*
Thomas Sullivan, *University of Montana*
Selima Sultana, *University of North Carolina at Greensboro*
Kate Swanson, *San Diego State University*
Rajiv Thakur, *University of Tennessee*
Scott Therkalsen, *Grossmont Community College*
Paul M. Torrens, *University of Maryland, College Park*
Amy Trauger, *University of Georgia*
Erika Trigoso, *University of Denver*
Nicholas Vaughn, *Indiana University*
Jamie Winders, *Syracuse University*
David J. Wishart, *University of Nebraska—Lincoln*
Ryan Weichelt, *University of Wisconsin—Eau Claire*
Max D. Woodworth, *The Ohio State University*

The book in your hands is the product of a collaborative effort. I am very fortunate to work with a team of great professionals at Oxford University Press. The book has benefited tremendously from their hard work and dedication and I am grateful to be associated with them: I owe a special debt to Dan Kaveney, Executive Editor, who believed in the project, gave it his full support, and shepherded it from the idea stage through to publication. Many thanks are due to: Christine Mahon, Associate Editor, for attending to the many details preparing a manuscript for production, and for assembling the supplements package; Megan Carlson, Assistant Editor, for arranging review panels and with manuscript preparation; Sarah Goggin for her excellent work on writing and assembling the supplements package; David Jurman, Marketing Manager, for his tireless work getting the word out; Marianne Paul, Production Editor, for attending to the myriad details involved in turning a manuscript in to a published work; and Michele Laseau, Art Director, for a great design and cover. I also want to tip my hat to the a tireless band of salespeople for promoting the book on campuses all across the country.

Lisa Benton-Short is a constant source of love and endless encouragement with just the right amount of sass to keep me firmly grounded.

Welcome to human geography: an endlessly fascinating and always rewarding subject!

Human Geography

PART 1

The Context

This section sets the stage. Chapter 1 gives a brief account of the planet we call home. Attention is paid to the evolution of the Earth, its emerging physical geography, and its humanization. The term "geography" derives from the Greek for "earth description." Chapter 2 looks at the evolution of earth description from its earliest roots to current concerns.

1

The Home Planet

We inhabit a tiny blue dot in a vast inky darkness. It has been in existence for 4.6 billion years, but our occupancy is more recent. Our species emerged around 200,000 years ago. Later, our immediate ancestors migrated out of Africa to populate the world. Today we live in a humanized world deeply impacted by our presence. The health of the planet is our primary responsibility.

The Big Picture

According to the aptly named **Big Bang theory**, it all started with a very big explosion that produced enough energy to expand a single point outward to infinity. From a singular point in space-time of intense heat and pressure, the universe began to expand, cooling as it spread outward. The universe is still moving outward from this specific moment and particular location. Measuring the speed of the expansion allows us to calculate the approximate moment of the "birth" of our universe, around 13.7 billion years ago. A billion years after the Big Bang (a term coined by cosmologist Fred Hoyle in 1949), galaxies first came into being from differential gravitational pull in the young universe.

The Big Bang expansion exacerbated minutely small differences in density into sites of star clusters and galaxies. For the first 7 to 9 billion years, the attraction of this matter slowed down the rate of expansion of the universe in a cosmic pull of competing forces, the explosive energy of the Big Bang dampened by the gravitational tug of the matter that it created in its wake. Then, the expansion of the universe began to accelerate as a mysterious energy source, with the foreboding name of **dark energy**, overcame gravity. Almost three-quarters of the universe is made up of this unknown force; of the rest, a little under a quarter is made up of dark matter that neither reflects nor emits light, and only around 4 percent is the matter in the universe that we can see and, as yet, understand. We live in a dark universe.

1.1 Poor Pluto

From 1930 to 2006, there were nine planets, a period during which Pluto was discovered, identified as a planet, and then dropped from the list of major planets. The American astronomer Clyde W. Tombaugh discovered the planet in 1930 from the Lowell Observatory in Arizona. Before this date it was too small—only 18 percent of the Earth's diameter—and too far to be visible. It was designated as the ninth and farthest planet from the Sun, but in 2006 it was reclassified as one of the dwarf planets, now classified as *plutoids*, and dropped from the list of major planets. While denied membership in the club of major planets, poor Pluto still makes its eccentric orbit at the edge of the solar system. In a touching moment of solidarity with the marginalized and the shunned, the supporters of the Turkish soccer club Beşiktaş carried banners that proclaimed, "We Are All Pluto."

And to add to the pervasive strangeness of it all, there is also the intriguing proposition that there was more than one Big Bang. We are living in the aftermath of a Big Bang, but perhaps there were more, possibly an infinite number as the universe expands and contracts, each Big Bang leading to a giant implosion followed in turn by another Big Bang in an endless cycle.

The universe consists of around 200 billion galaxies and 30 billion trillion stars. Our home planet is situated in the Milky Way, a galaxy composed of 100 to 400 billion stars bound together by gravity and stretching across 100,000 light years. One **light year** is the equivalent of 5.8 trillion miles; it is calculated from the speed of light in a vacuum, 186,000 miles per second, or 700 million miles per hour. One of the stars, located 24,000 light years from the center of this galaxy, is the Sun, the center of our planetary system and rightfully deserving its capitalization. Eight planets revolve around this star. From near to far they are Mercury, Venus, Earth, Mars, Jupiter, Saturn, Uranus, and Neptune. The four nearest to the Sun are relatively small and composed of rock and metal; the four farthest, the gas planets, are larger, with planetary rings of particles and cosmic dust. Jupiter and Saturn are composed of hydrogen and helium, and the two farthest planets, Uranus and Neptune, traveling the darker, colder edges of the solar system, are ice giants made up of water, ammonia, and methane.

In 2016 scientists at the California Institute of Technology presented evidence that there may be a ninth planet with a mass of five to ten times that of the Earth. This planet, only known as Planet Nine, has not been observed directly but only inferred from the orbits of surrounding objects. The same type of inductions occurred in the nineteenth century when observations of Uranus indicated that there may be another planet out there. There was—Neptune. As technology allows us to delve deeper into the darkness of the universe, perhaps one day Planet Nine may be sighted.

We are situated close enough to the Sun to get more heat and light than the farthest planets but not so close as to be burned up by the intense heat experienced by the two planets closer to the Sun. The third planet out from the Sun has enough oxygen and water to support life.

We live on a planet in motion. It revolves around the Sun once every 365.25 days. A year marks how long it takes the Earth to complete one full movement around the Sun. The Earth is tilted approximately 23 degrees from the perpendicular in its orbit. This creates the seasons, especially marked farther away from the equator, where the distance from the Sun varies more substantially when the tilt is angled away or toward the Sun. The tilt of the Earth divides this yearly cycle into seasons; closer to the poles, the seasonal effects are exaggerated, as intense cold turns into a distinct warming when the long, dark days of winter become the light-filled days of summer. At the equator the seasons are less pronounced and the daily division into light and dark more even throughout the year. At the poles the annual cycle moves from a Sun that never sets to a Sun that is barely visible over the dark horizon. At the equator the Sun is a more constant and reliable presence.

Our planet rotates on its own axis on a roughly twenty-four-hour cycle. The daily cycle is created as the Earth turns toward and then away from the Sun. Although we continue to use the terms "rising" and "setting" Sun, it is the Earth that moves. Our planet revolves around the Sun and rotates on its own axis.

The Earth does not revolve in a perfect circle. It wobbles, the angle of tilt moves, and its orbit around the Sun varies in its eccentricity (departure from circularity). These small differences in distance from the Sun may account for very long-term climate changes on the Earth, especially the rise and fall of ice ages.

Soon after the formation of the solar system—after only 30 to 50 million years—a giant **asteroid**, almost half the size of the Earth, hit our planet. The impact created the Moon, which now revolves around the Earth every 27.3 days, its pitted surface a silent witness to the destructive forces still at work in the universe. The gravitational effect of the Moon on the Earth is responsible for the **tides** that move the oceans and seas in ceaseless and regular vertical motion. The **lunar cycle** is the source of our division of time into months.

The yearly, monthly, and daily cycles that are such important rhythms of our lives are caused by the movement of our planet and the Moon. Even in a more electronic age we are affected by the beats of the cosmos. Two researchers, Scott Golder and Michael Macy, examined millions of tweets over a two-year period. Using selected words in messages to connote moods, they found that there was distinct periodicity, with more positive words in the early morning. Each new dawn offers the promise of a new beginning. Our moods also vary over the seasons. **Seasonal affective disorder (SAD)** is the tendency for more negative moods among normally healthy people during distinct seasons. Winter blues are more common in northern latitudes because of the rapid decrease in sunlight. SAD varies with latitude, with only 1.4 percent experiencing it in Florida but 9.7 percent in colder, darker New Hampshire.

Shaky Ground: Plate Tectonics

Look at any map of the world. The continents sitting in the blue seas and vast oceans look solid, firm, permanent. From a very long perspective the image is deceptive. The present distribution of landmasses is just the most recent frame in a dynamic, complex picture of continents forming, reforming, splitting up, and moving across the Earth's surface.

Earth came into being around 4.55 billion years ago. Soon afterward, around 4.51 billion years ago, the Earth was hit by a giant asteroid, turning it into a fiery ball of intense heat. Over the years, the surface cooled more than the interior. The Earth's surface, the "solid ground" of so many metaphors, is in fact a thin, brittle crust, no more than four to sixty-five miles thick, that floats precariously on a viscous mass of molten metal. The crust that we occupy is the cold top level that formed on the Earth's surface just as a skin forms when boiling milk cools. Below is the hot mantle and, even deeper, the extremely hot core. The solidified crust fractured across the large round object of the Earth, breaking up into distinct **tectonic plates**. There are nine large plates and numerous small ones (Figure 1.1). They sit atop powerful currents; below them, the mantle of molten metal heaves as hotter liquid moves up in convection currents from the boiling mass at the Earth's core while the cooler liquid closer to the surface sinks to the bottom. This continuous subterranean motion moves the plates on the surface, like bumper cars in a fairground ride that move in a restricted space. Driven by the upward convection currents deep in the Earth's mantle, they bump against, slide past, and move away from each other. Mountains are formed, trenches are created, **earthquakes** occur, and boiling magma spews out in volcanic eruptions to reveal the fiery material that shapes the Earth's geological formations. The surface that we live on is a fragile membrane across deep and powerful subterranean forces. Volcanic and earthquake activity are particularly severe at the edges where plates meet (Figure 1.2 and Figure 1.3).

The plates are always on the move. In the geological equivalent of political empires, giant landmasses rise and fall. Over 1.4 billion years ago, small masses of land collided to form the supercontinent of Rodinia. This landmass was rent asunder into different blocs including Avalonia; Baltica; Laurentia, eventually to become North America; and Gondwana, the source continent for South America, Africa, India, Antarctica, Australia, and New Zealand. When Baltica and Laurentia joined up with Gondwana around 300 million years ago, a new supercontinent, Pangaea, was created. It took up almost one entire side of the globe—the other side was mostly water—and was centered on the South Pole. This began to break up, but the large landmass of Gondwana remained. Beginning around 200 to 160 million years ago, it too began to break up, as South America broke off and drifted west, and a landmass of what was to become India began a 150-million-year journey northward to Asia, forming the Himalayas when it crashed into the Eurasian plate.

Around 60 million years ago, another landmass, what we now call Australia, detached from what is now

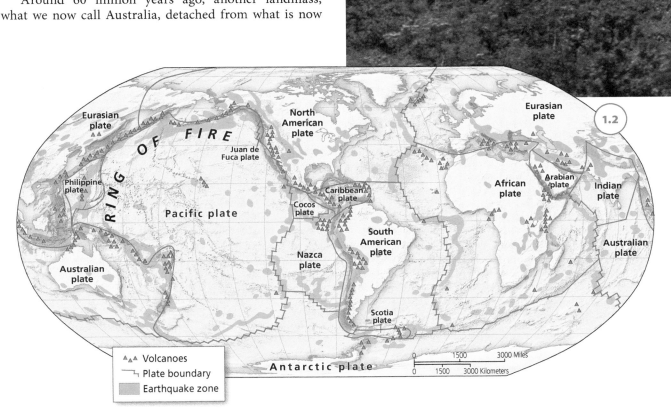

1.2 Location of seismic activity (earthquakes and volcanoes). They cluster along plate boundaries. Notice the **Ring of Fire** around the Pacific. **1.3** Along the line of mountains in Central America is a zone of active volcanic and earthquake events. This is an active volcano just outside the city of Antigua in Guatemala.

Antarctica and began a 45-million-year, 500-mile journey through various climate zones toward the warmth of a more northerly location. Australia was on the move. Slowly. It is still moving northward away from Antarctica at a rate of around 2.7 inches per year. There are many traces of the long journey. Australia shares with South America similar species of marsupials, turtles, and lungfish that are over 400 million years old, a biotic reminder of when the two continents were joined in Gondwana. And the distribution of an extinct fern, *Glossopteris*, in Antarctica, Africa, Australia, India, and South America is a biological reminder of the supercontinent.

The Indo-Australian plate eventually crunched up against the Eurasian plate around 15 million years ago. The border area between the two plates is still one of the most active volcanic zones in the world. Let us consider three events. Around 70,000 years ago, a giant volcano exploded at Mount Toba in Sumatra (now part of Indonesia). Almost a billion tons of ash and dust were thrown into the atmosphere, darkening the skies and reducing global temperatures by around 5°C for up to five years. One anthropologist, Stanley Ambrose, writes of a long volcanic winter, an instant ice age that severely reduced early human populations. The total human population in the world was probably reduced to around 10,000, so that local variations became more pronounced in human **evolution**, creating the contemporary racial differentiation. The huge, violent explosion caused **global climate change** that further impacted human populations. The extent to which the consequent **environmental stressors** may have promoted social cooperation and thus facilitated the human dispersal from Africa is a debatable but intriguing proposition.

The region saw a second event, another massive volcanic explosion, when the island of Krakatoa exploded on August 27, 1883. Not as violent as the Mount Toba explosion, it was still devastating; the resultant tsunami killed approximately 36,500 people in the region as tidal waves drowned towns and villages along the coast. The air pollution caused by all the dust and ash blotted out the Sun's heat and light and reduced the world's temperature by about 0.5°C. Simon Winchester also argues that Krakatoa influenced political developments. He suggests that the 1888 Banten Peasants' Revolt, which pitted local people against Dutch colonial rule, was partly influenced by the cataclysm that befell the Indonesians and fed into an anticolonial and Islamic fundamentalist narrative that "explained" the explosion and mass deaths as the work of a wrathful Allah signaling displeasure at colonial control and lax religious practices.

On December 26, 2004, movement along the plate caused the third event—another major disaster. The Sumatra-Andaman earthquake occurred nineteen miles below sea level off the coast of Sumatra. Along a 1,000-mile zone, plates shifted almost fifty feet and raised the surface of the seabed, displacing millions of gallons of water that caused the **tsunami**, the post-earthquake waves that suddenly overwhelmed coastal communities in fourteen countries fringing the Indian Ocean. As waves cresting 100 feet crashed into unsuspecting communities, more than 230,000 people died, 125,000 were injured, and almost 1.7 million were displaced from their homes.

Life on Earth

For half of its entire existence the Earth did not sustain life. It took a long time for the planet to cool down and its landmasses to become more stable. The beginnings of life around 2.2 billion years ago were in the modest form of blue-green algae. Evolution was slow until the development of sexual reproduction. An explosion of life took place around 545 million years ago in the Cambrian period of the early Paleozoic era as living forms with skeletons first appeared. The growth of the amount and diversity of life was not a simple upward trajectory. The same era also marked the first global mass extinction, as many species of shallow water fauna were killed off by rising sea levels.

Around 418 million years ago, the first land animals appeared. Insects and amphibians also emerged. In the Mesozoic era, from 248 to 65 million years ago, the first dinosaur appeared, as did the first mammal, a tiny shrew-like animal dwarfed by the giant reptiles around it. In the late Mesozoic there was another mass extinction of animal life. Dinosaurs became extinct, and birds lost almost 80 percent of their species, as did the marsupials. Some scientists suggest that colossal volcanic eruptions in a mountain chain in eastern India spewed huge amounts of carbon dioxide and **sulfur dioxide** into the atmosphere, poisoning the air. Other scientists point to an asteroid hitting the Earth around the same time, 65 million years ago. A crater measuring 100 miles wide and 6 miles deep was discovered in the Yucatan peninsula near the town of Chicxulub. The impact of the asteroid created global fires that sucked oxygen out of the atmosphere and caused huge shock waves that triggered volcanoes and earthquakes. The volcanic eruptions in India and the asteroid strike in the Yucatan created devastating global environmental conditions. Dust covered the Earth, leading to temperature decrease. **Acid rain** fell. Dark and coldness settled on the Earth, plants stopped growing, and animals died. More than 90 percent of all marine life died as the seas became acid baths. The dinosaurs became extinct, as they could not survive in the wasteland of permanent winter. They starved to death.

The Tertiary period, from 65 million to 1.8 million years ago, is the age of mammals. Whales appear, as do elephants, cats, and dogs. Around 50 million years ago, the first primate appears, a little lemur-like creature barely more than two pounds in weight. The primates diversified and spread. Four million years ago, the first **hominid**

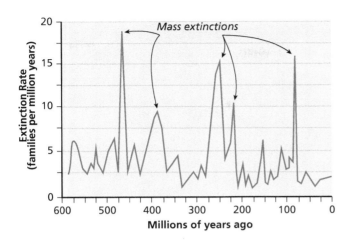

1.4 Mass extinctions

TABLE 1.1	Mass Extinction Events		
TIME PERIOD (MYA)		**CAUSES**	**CONSEQUENCES**
Ordovician–Silurian (443)		Climate change	85 percent of all sea life was wiped out
Late Devonian (359)		Impact events	75 percent of species wiped out
Permian (248)		Multiple	96 percent of species wiped out
Triassic-Jurassic (200)		Multiple	50 percent of animals wiped out
Cretaceous–Tertiary (65)		Multiple	75 percent of species wiped out

*mya = million years ago

to stand upright, *Australopithecus,* walked on the Earth. Between 2 and 3 million years ago, at least eight different hominid species emerged from this one species, correlating with a time of increased climate change. The rapidly changing conditions rewarded species that could adapt quickly, especially those that had the brain capacity to adjust to constant **environmental change**. Adaptability to diverse environmental conditions is one of the defining hallmarks of our human species.

Mass Extinctions

The fragility of life on Earth is revealed most dramatically during episodes of **mass extinctions**. Five such events are recorded. Figure 1.4 depicts five spikes in extinction events. Table 1.1 provides more detail about the causes and consequences of these events. In the first one, climate change in the form of an ice age disrupted the oceans' chemistry during the Ordovician–Silurian extinction event, killing a majority of sea life species. Impact events, either an asteroid hitting the Earth and/or giant volcanic eruptions around 65 million years ago, were the principal cause behind the more recent extinction event during the Cretaceous–Tertiary, when the dinosaurs were killed off. The catastrophic events caused a massive increase in greenhouse gases and rapid warming that made it unlivable for many terrestrial species, even the fearsome *Tyrannosaurus rex.*

These mass extinctions show that life is fragile, but also resilient, as new life forms emerge in their aftermath.

A Humanized World

Tracing human origins was long the preserve of archeologists and physical anthropologists. Digging in old sites and dating the old bones was the preferred method. Dates were always provisional, and the ambiguity of the information always provided lots of room for debate. Things began to change when scientists could use live humans to chart the past. Our bodies, it appears, are the equivalent of an archeological site, containing memories and traces of the far distant past. Plotting the distribution of the information in our bodies allows us to map possible sites of human origin and paths of human dispersal.

In the 1950s, differences in blood group protein were identified that allowed a picture of human ancestry. A more sophisticated analysis was made possible in 1987, when scientists first used the DNA in mitochondria, the rod-shaped "power plants" of cells, to construct a genetic tree. Using this method allowed scientists to examine genetic materials passed from mother to child, and led them to the startling conclusion that we are all descended from one woman who lived in Africa around 200,000 years ago.

Out of Africa

After decades of competing ideas and dueling theories, there is an emerging consensus that all humans are descended from a group of people who left Africa. Human origins lie in Africa. The first humans stepped out of Africa to populate and humanize the world (Figure 1.5). Evidence of **genetic variation** sustains this **Out of Africa** hypothesis. Peoples in central and southern Africa have more genetic variation than elsewhere in the world. The highest levels of genetic diversity in the world are among the Namibian and Khomani Bushmen of southern Africa, the Biaka Pygmies of Central Africa, and the Sandawe of East Africa. There is a decrease of genetic diversity with increasing distance from Africa, suggesting a founder's group in Africa as the basis for all subsequent human population. As smaller subgroups broke off to settle in different places, the amount of genetic diversity decreased. The amount of genetic variation declines with distance from Africa—a classic case of the **distance-decay effect**—

1.5 The peopling of the world

suggesting that humans came out of Africa to populate the Earth. It was the first wave of globalization.

The human diffusion from Africa, around 120,000 to 70,000 years ago, initially involved a tiny population of probably between 3,000 to 5,000 people. Notice the implied leeway in the range. We have gotten a general idea, not a sure fix. The movement out of Africa took different routes at different times. The oldest was a group that moved out through the Horn of Africa into the Middle East. One group then moved along a coastal route skirting the Indian Ocean. They took around 20,000 years to migrate around the Indian Ocean, taking the beachcomber route into Southeast Asia and Oceania. The migration took the form of **budding**, as small subgroups broke away to settle new places. Traveling along the coast and up rivers, they avoided the difficulties of passing through deserts and across mountains. Consider, as just one example, the settling of Australia. Around 50,000 years ago, when sea levels were lower, Australia was linked to Papua New Guinea, and there was only a relatively short distance, probably no more than sixty-five miles, between the edge of Asia and the coast of Australia. Bamboo rafts probably were constructed to make the sea crossing. Current archeological evidence dates the human presence in Australia to between 55,000 to 42,000 years ago. Most of the early Australian sites have deep charcoal deposits that suggest both early settlement and a long record of occupation. The migrants traveled quickly up the rivers and along the coast. Colonization was rapid, as there are sites scattered throughout the vast continent in the north, northwest, southwest, and southeast and as far

south as Tasmania, all dated to around 32,000 to 40,000 years ago. Central Australia was settled last, as there were no easy coastal or riverine routes to follow. People had to venture into a dry, forbidding interior. But venture they did. We have a site, the Puritjarra Cave Rock Shelter, with rock art suggesting occupancy as far back as 32,000 years ago.

Another later wave of human migration around 50,000 years ago moved into Eurasia. Giant walls of ice restricted movement north; it was still the Ice Age, also known as the Pleistocene era. However, sea levels were much lower in the Pleistocene, exposing land bridges that are now covered with seawater. The English Channel was dry land, allowing people and animals to move from Europe into what is now Britain and Ireland. On the other side of the world, the **land bridge** between the northeasternmost edge of Eurasia and the northwest edge of America allowed people and animals to make the journey across the continents around 22,000 to 25,000 years ago.

Human groups were restless, continually moving to and settling new areas. People even sailed across the vast expanses of the Pacific Ocean to settle on such distant specks of islands as Hawaii. The early humans explored the world, settled in different habitats, and impacted their environment.

These colonizers' impact was immediate. Despite early interpretations of them as docile elements in the landscape, they were very much active agents in modifying and changing the environment. Early humans first shared the Eurasian landmass with other hominids such as the Neanderthals and the Denisovans. These were cousins to humans on the evolutionary tree, emerging as separate

1.2 Wallace's Line: A Biogeographical Boundary

In the mid-nineteenth century, Alfred Russell Wallace (1828–1913) spent eight years travelling the Malay Archipelago. He travelled over 14,000 miles and collected over 100,000 samples. He was less well known than Darwin, and his letters back to members of England's scientific community indicating a theory of natural selection for the origin of species prompted some to counsel Darwin to publish his own work in 1859. In 1863 Wallace published a paper in the *Journal of the Royal Geographical Society* that identified a boundary line between two realms. This line, now known as Wallace's Line, divides the Australian and Asian biotas.

Wallace's work was part of a surge of mapping in the nineteenth-century natural sciences. Maps were, and still are, used as a conceptual framework and as a tool for presenting vast amounts of data in simple yet suggestive ways. Wallace's Line evolved from papers he wrote in 1855 and 1860 as well as the 1863 paper. Later, the zoologist and biogeographer Max Carl Wilhelm Weber drew his own line based on surveys of mammalian fauna. His line is farther to the east than Wallace's Line. The difference between the two lines is that flora found it easier to move across the "line," while the seas constituted a larger barrier to animal movement. The area between the Wallace and Weber lines can be considered a zone of transition.

The biogeographical boundaries identified by Wallace and Weber are a result of **plate tectonics**. The northward drift of the Australian plate made it crash against the Eurasian plate around 15 million years ago. Previously separate biological realms were brought into contact. The Australian plate was like a raft filled with distinctive flora and marsupial animals now abutting the Asian flora and fauna. Without a theory of plate tectonics (Alfred Wegener did not moot it until 1912, or publish it in 1915), Wallace could not discern the geological basis for the division. However, his careful mapping highlighted the benefits of fieldwork and the representation of these data in maps. Wallace's Line is the biological equivalent of plate boundaries, even though Wallace was not aware of the plates and the theory of continental drift.

Wallace's Line

REFERENCES

Camerini, J. R. (1993) Evolution, Biogeography and Maps: An Early History of Wallace's Line. *Isis* 84: 700–727.

Vetter, J. (2006) Wallace's Other Line: Human Biogeography and Field Practice in the Eastern Colonial Tropics. *Journal of the History of Biology* 39: 89–123.

Wallace, A. J. (1863) On the Physical Geography of the Malay Archipelago. *Journal of the Royal Geographical Society* 33: 217–234.

subspecies around 500,000 years ago. The Neanderthals were not as dumb as common characterizations assume. They walked upright and effectively communicated with each other. They and the Denisovans emerged in and lived throughout Eurasia, but with the entry of *Homo sapiens,* these two hominid species died out around 30,000 years ago. As the human population spread across the Eurasian landmass, we effectively annihilated our closest competition for food and environmental resources. One competitive advantage was the division of labor. Among human populations, women and young children were responsible for skill-intensive crafts such as making clothing and shelter as well as gathering food. Men were primarily responsible for hunting. The division of labor allowed access to a wider range of food sources compared to the other hominids, who, because they did not have the gender and age division of labor that allowed economic specialization, were almost entirely dependent on hunting game.

When the game disappeared, they starved and died out. The gender and age division of labor in early human populations created more specialized economic roles that allowed a more efficient use of labor and gave access to a wider food resource base.

The other hominid groups died out, but they live on in our genetic makeup. They must have been close enough to humans so that interbreeding occurred. We have a genetic record of the process. All non-Africans have between 1 and 4 percent of Neanderthal DNA. The people of New Guinea have around 5 percent of Denisovan DNA. Most of the world's non-African population carries this genetic reminder of the width of human sexual appetite and willingness to "interact" with their evolutionary cousins. The distinction between human and nonhuman hominids was more of a liminal area than a sharp and fixed boundary.

Pleistocene Overkill

The early human colonization of the Earth was also marked by ecological changes. The most pronounced is the **Pleistocene Overkill**, which refers to the extinction of **megafauna** such as the mammoth and the mastodon that occurred around 11,000 years ago. This extinction occurred on every continent and was experienced by every community of large terrestrial vertebrates at a time of humans' increased technological sophistication in hunting. While there has been some debate about the relative role that climate change played in the mass extinction, human hunting definitively contributed. In some continents both hunting and climate change were important, while in others only hunting was the primary cause. The extinction was particularly marked in North America around 20,000 years ago.

America is a test case of the **overkill**, since human entry into the region was relatively late, and there are no records of other hominids. John Alroy created a grid of 754 spatial cells of one degree of latitude and one degree of longitude. His computer simulation model for North America clearly shows that human population growth and hunting led to mass extinction 1,200 years after humans first appeared.

The larger and slower megafauna were very vulnerable to skilled hunters. The megafauna, weighing more than 220 pounds and thus slower than other animals, became especially vulnerable. In Australia, for example, many megafauna became extinct around 46,000 years ago, soon after the entry of humans into Australia. We will never again see the three-meter-high giant kangaroo, the rhino-sized wombat, or the marsupial lion. We do have a visual record from rock paintings in Arnhem Land in northern Australia, which may depict a *Palorchestes*, a large browsing animal; a *Genyornis*, similar to an emu but three times as tall; and a marsupial lion. Farther west in Kimberley,

rock paintings depict giant kangaroos and huge echidnas. The paintings are a reminder of the time when megafauna and humans shared the same space and an early example of the immense role that humans played and continue to play in shaping the global ecosystem.

Climate Change

Over the long history of the Earth, its climate has changed. Over the past 100,000 years the Earth has warmed, ice sheets have melted, and sea level has risen. In the past 1,000 years the climate warmed during the Medieval Warm Period from 1100 to 1300, and then cooled from 1300 to 1700 in what is known as the Little Ice Age. In the past 100 years our planet's average temperature has heated up by 1.5°F and is projected to rise even more in the next 100 years due to human impact, the most important being **greenhouse gas** emissions (especially **carbon dioxide**), which warm the atmosphere.

The full implications of climate change have yet to be unraveled, but a number of major trends are now identified. At the higher latitudes and altitudes, glaciers are shrinking and **permafrost** is warming and thawing. Animal life has responded in a variety of ways including shifts in geographic range (with heat-loving species able to live farther north in the Northern Hemisphere and farther south in the Southern Hemisphere), migration patterns, and species interaction. There is mounting evidence that climate change is increasing the risks of extinctions and the potential impact on human health as changing ecologies provide new opportunities and vectors for viruses and the diseases they cause.

A number of respected scientists now argue that we are in a sixth mass extinction event, caused by climate change

1.6 The sixth extinction

and human activity. There is a background rate of around two mammal extinctions per 10,000 species per 100 years, but in the last century the rate is now 110 times higher than this level. Over the past 500 years there has been a 10 percent reduction in land-based wild animals and a 13 percent reduction in the number of bird species. The rapid loss of biodiversity means that a sixth extinction event may be underway.

More extreme weather events including drought, floods, and major storms are also more likely, owing to climate change. The consequences vary with location. In North America there is greater risk of drought in the west and southwest, and a higher chance of more powerful tornadoes in the central region and more punishing storms along the east and southeast coasts. This increased likelihood is not a straight-line relationship such that each year sees more frequent extreme weather events; rather, the probability of their happening is greater, so that "100-year storms" occur once every ten years.

Some ecosystems are especially threatened. Coral reefs survive in only a very narrow temperature range and are thus highly vulnerable to warming oceans. **Climate change** is particularly marked in the high latitude due to what is known as the Arctic Amplitude, which causes greater near-surface warming than at middle and low latitudes.

Humans have responded to climate change in two main ways: adaption and mitigation. Adaption involves engineered and technological solutions such as building storm barriers, or changing behaviors such as the shift in garden preferences from lawns to desertscapes in the southwestern USA. Mitigation involves trying to reduce our **carbon footprint** and lessen the emission of greenhouse gases.

Cited References

Alroy, J. (2001) A Multispecies Overkill Simulation of the End-Pleistocene Megafaunal Extinction. *Science* 292: 1893–1896.

Ambrose, S. (1998) Late Pleistocene Human Population Bottlenecks, Volcanic Winter, and Differentiation of Modern Humans. *Journal of Human Evolution* 34: 623–651.

Ceballos, G., Ehrlich, P., Barnosky, A. D., Garcia, A., Pringle, R. M., and Palmer, T. M. (2015) Accelerated Modern Human-induced Species Loss: Entering the Sixth Mass Extinction. *Science Advances.* http://advances.sciencemag.org/content/1/5/e1400253

Golder, S. A. and Macy, M. W. (2011) Diurnal and Seasonal Moods Vary with Work, Sleep, and Daylength across Diverse Cultures. *Science* 333: 1878–1881.

Grove, M. (2011) Change and Variability in Plio-Pleistocene Climates: Modelling the Hominin Response. *Journal of Archaeological Science* 38: 3038–3047.

Kuhn, S. L. and Stiner, C. M. (2006) What's a Mother to Do? The Division of Labor among Neanderthals and Modern Humans in Eurasia. *Current Anthropology* 47: 953–980.

Urban, M. C. (2015) Accelerating Extinction Risk from Climate Change. *Science* 348: 571–573.

Winchester, S. (2003) *Krakatoa: The Day the World Exploded: August 27, 1883.* New York: Viking.

Select Guide to Further Reading

Hazen, R. M. (2012) *The Story of Earth: The First 4.5 Billion Years, from Stardust to Living Planet.* New York: Viking Penguin.

Kolbert, E. (2014) *The Sixth Extinction: An Unnatural History.* New York: Henry Holt.

Stix, G. (2008) Traces of a Distant Past. *Scientific American* 299, 1: 56–63.

Tattersall, I. (2009) Human Origins: Out of Africa. *Proceedings of National Academy of Sciences* 106: 16018–16021.

Selected Websites

Shifting plates
https://ees.as.uky.edu/sites/default/files/elearning/module04swf.swf
http://geology.com/plate-tectonics.shtml
http://www.pbslearningmedia.org/asset/ess05_int_boundaries/
http://www.nasa.gov/audience/forstudents/5-8/features/F_Earth_Has_Faults_prt.htm

An interactive atlas of human origins and dispersal
http://www.bradshawfoundation.com/journey/
http://www.businessinsider.com/prehistoric-human-migration-from-africa-animated-map-2015-5
The Biodiversity Heritage Library
http://www.biodiversitylibrary.org
Convention on Biological Diversity
http://www.cbd.int

The World Meteorological Organization is a specialized agency of the United Nations that provides information on the state and behavior of the Earth's atmosphere, its interaction with the oceans, the climate it produces, and the resulting distribution of water resources.
  http://www.wmo.int/pages/index_en.html

The Intergovernmental Panel on Climate Change (IPCC) is the leading international body for the assessment of climate change.
  http://ipcc.ch
The US EPA site is a useful source of data on climate change across the regions of the USA.
  http://www3.epa.gov/climatechange/

Interactive Map Websites

http://nhb-arcims.si.edu/ThisDynamicPlanet/index.html
https://www.learner.org/interactives/dynamicearth/plate.html
http://geology.com/plate-tectonics.shtml

http://www.pbslearningmedia.org/asset/ess05_int_boundaries/
http://www.businessinsider.com/prehistoric-human-migration-from-africa-animated-map-2015-5

Learning Outcomes

This chapter provides a basic grounding into the history of the Earth.

- The Big Bang theory comes from the observation the universe is infinitely expanding. By calculating its current rate of expansion, we are able to theorize on a specific time and location where all matter first aggregated, then exploded outward to its current state of existence, propelled by dark energy and a possible succession of multiple subsequent Big Bangs.
- The universe consists of approximately 200 billion galaxies and 30 billion trillion stars. Our own galaxy, the Milky Way, consists of 100 to 400 billion stars, is 100,000 light years across, and contains our Sun, which is surrounded by eight planets.
- The Earth is at a distance from the Sun that allows it to support life, with a tilted axis that allows its 365.25-day orbital cycle to be defined by the seasons we now know. Seasonal effects in any specific location are defined by one's location between the equator and the poles. The Earth's daily rotation produces the "rising" and "setting" Sun.
- The Moon, created by an asteroid's impact with the Earth, is responsible for the ebbs and flows of tides as well as our division of time into months.
- Researchers have found that the movement of our planet and the lunar cycle have profound effects on human moods, such as seasonal affective disorder (SAD) and "winter blues," which are more commonly found during periods of decreasing light in northern latitudes.
- The Earth's crust was formed around 4.51 billion years ago. It is a brittle layer, ranging from four to sixty-five miles thick, floating upon the liquid molten metal of the planet's mantle, below which is the much hotter inner core.
- The crust is composed of nine large plates and numerous smaller ones. They sit atop the continuous currents of the mantle, which push the plates against, past, and away from each other. Such movement forms many geological features, such as mountains and trenches, and results in what we refer to as natural disasters such as earthquakes and volcanic eruptions.

- Over the last 1.4 billion years, the Earth's continents formed as plates separated from a single mass called Pangaea, with some later colliding with each other, but leaving clues as to their origins and movement.
- There are numerous examples of the effects of plate tectonics on human movement and development, including the impact that ice ages have had on human dispersal and social cooperation, as well as the disastrous effects of earthquakes, volcanoes, and tsunamis.
- Although life on Earth began around 2.2 billion years ago, the evolution of animals and humans began much later, perhaps as recently as 200,000 years ago for humans, and was marred by periods of climate change and geological disaster.
- At least five major mass extinctions have occurred.
- Most scientists now conclude that the first humans diffused from Africa, settling as new subgroups and spreading around the planet over the next tens of thousands of years.
- *Homo sapiens* effectively annihilated other related species around 30,000 years ago while competing for food and other resources. The division of labor among these early humans helped lead to greater adaptation and our lasting survival.
- Human development helped contribute to ecological changes by, for example, aiding in the extinction of megafauna as humans became more sophisticated and advanced hunters.
- Climate change has become more pronounced in the last 100 years, leading to warming of land and sea, more extreme weather events, changing patterns of disease, and more extinctions.
- We may be in the sixth phase of mass extinction.
- Humans' two major responses to climate change have been adaption and mitigation.

The Nature of Geography

In this chapter I will place the contemporary study of human geography in a deeper historical context. Human geography of today is shaped by the debates of the past. I will identify five particular themes that resonate down through the years: coordinating absolute space, the emergence of geography from **cosmography**, the discussion of relative space, environment and society, and the links between geography and society. A discussion of these themes is a revealing entry point into the history of human geography. I will also explore the contemporary concerns of human geography.

Mapping Absolute Space

A significant and recurring theme in geography is the attempt to accurately represent the world we live in through maps. It is an important element of the geographical imagination. One of the earliest geographers who sought to measure and coordinate the world was Claudius **Ptolemy**. He was a Greek-Egyptian who worked in the great library of Alexandria. He lived sometime between 90 CE and 168 CE and spent his adult life in Alexandria, a Greek city founded in 331 BCE and named after Alexander the Great. The city was one of the wonders of the classical world, and its population grew to almost half a million. It featured palaces and large public buildings, and at the heart of its intellectual life was the library, a major center of scholarship with over 700,000 volumes. Euclid and Archimedes worked at the library, and Eratosthenes was, for a time, the chief librarian.

In one of his major books, *Geography*, Ptolemy defined **latitude** and **longitude**, discussed methods of mapmaking, and posed the problem of how to represent the round world on a flat surface. He proposed two map projections, a **conic projection** and a partial conic projection, on which to represent the habitable world. Conic projections map the Earth on a cone

2.1 Measuring the World

The measuring of the world is a significant part of the early history of geography. By the fifth century BCE, the Greeks knew that the world was a sphere. Eratosthenes (ca. 276 BCE–ca. 194 BCE), who, like Ptolemy, worked at the great library of Alexandria, calculated the circumference of the Earth. Assuming a spherical Earth, he calculated the circumference with only a 1–2 percent difference from the correct figure of 24,901 miles.

Ptolemy's measurement of the circumference was much less accurate, around 18,000 miles. The wide diffusion of Ptolemy's writings meant that this figure was taken as fact during the early modern period. Columbus probably had this much smaller value in mind when he was planning to sail to the East Indies; Ptolemy's shorter value made the voyage that much less intimidating.

In 1669–1670, the Frenchman Jean Picard tried to accurately measure the Earth by extrapolating from the distance of one degree of latitude measured in the French countryside. However, similar surveys of a degree of latitude undertaken in Peru and Scandinavia at the same time had different values. The reason was not an instrumentation error but because the Earth is oblate, flattened at the poles. We live on a less than perfect sphere in all its misshapen complexity.

2.1 Ptolemy's map of the world in a conic projection, ca. 1300

and then show it as flattened with lines of latitude as concentric circles (Figure 2.1). He compiled lists of the latitude and longitude of places in the world known to him, basically a world centered on the Mediterranean.

In the European Dark Ages, Ptolemy's influence quickly disappeared. His work, however, was kept alive in the Arab world. Ptolemy's work was translated by the Arab cosmologists of the Abbasid court in Baghdad. Al-Battani (ca. 880) restated the *Almagest*, Ptolemy's work on astronomy, with improved measurements, and *Geography* was revised by al-Khwarizmi around 820 in his *Face of the Earth*, which included a map of the world. Al-Idrisi (ca. 1100–1166) was a practicing geographer who traveled through the Arab world. He constructed a celestial sphere and a map of the world. He was part of a larger geographical school of scholarship in the Arab world, involving book knowledge as well as geographical fieldwork and travel, that included the *Book of Countries* by Al-Yaqubi (ca. 891), Al-Balkhis's *Figures of the Climates* (921), Al-Muqaddasi's *The Best Description for an Understanding of All Provinces* (ca. 985), and Al-Bakri's *Book of Roads and Kingdoms* (ca. 1050).

In the early 1400s, Byzantine scholars brought their Greek manuscripts, including manuscripts of Ptolemy's writings, to Italy. The first printed copy of *Geography* appeared in Vicenza in 1475. Other editions appeared in Rome (1477), Florence (1480), and Ulm (1482). The Ulm edition, with its richly colored woodblock maps featuring deep blue seas and bright yellow borders, is arguably one of the most beautiful. The printing of Ptolemy's *Geography* stimulated new knowledge and innovative cartographic techniques. After 1508, editions of Ptolemy's *Geography* included maps that incorporated the discoveries of the New World. The first printed map of the New World appeared in an edition of *Geography*. Ptolemy's continuing legacy includes orientating maps with north at the top, the grid of latitude and longitude, using gazetteers in maps and atlases, and employing a variety of map projections. Ptolemy introduced us to the idea and practice of a coordinated world.

It is important to remember, however, that Ptolemy's view of the world was centered on the West. It was not the only view. Figure 2.2 is one of the oldest extant world maps from East Asia, created around 1389, roughly the same time as the world's map shown in Figure 2.1. Notice how China sits in the center of the map, the pivot of the world's

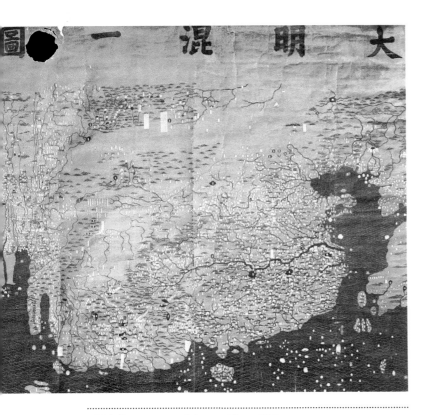

2.2 Korean map of the world, ca. 1389

landmass, with a hazily outlined Europe located on the far western periphery.

The Shift from Cosmography to Geography

Geography was part of a much wider cosmographical understanding of an interconnected cosmos that also included astronomy, astrology, and magic. The history of geography in the past five hundred years is the dismemberment of this cosmological enterprise.

It begins in the Renaissance, where there is evidence of both continuances and ruptures. There was an esoteric side to the Renaissance that included alchemy, astrology, and magic. These were not marginal concerns but rather central to the life and work of influential Renaissance scholars. John Dee (1527–1608) was a mathematician, mapmaker, and astrologist. His writings combine an interest in exploration and travel, magic and alchemy, mathematics and geography. Dee, like Ptolemy, had a concern with a cosmography that linked the Earth and the heavens, people and their wider environment.

We sometimes assume that the Scientific Revolution marked a major change: before was superstition; after was

rational, enlightened thought. Recent scholarship has questioned the traditional assumption of such a radical break.

The shift from magus to scientist and from cosmography to geography was neither sudden nor abrupt. The new scientific order bears the mark of the old cosmologies. These connections are clear when we look at some of the early cosmographer/geographers. The great mapmaker Gerard Mercator (1512–1594) had a religious dimension to his work. Born Gerard Kremer into a modest household in the small town of Rupelmonde near Antwerp, he went to Louvain in 1530 and Latinized his surname to Mercator. He was always interested in theology; he wrote religious tracts, and his first map was a map of the Holy Land. In 1544 he was accused of heresy, one of forty-three accused in Louvain. Two of the accused were burned at the stake, one was beheaded, and a woman was buried alive. Supported by friends and colleagues at the university, he was released. Mercator was a friend of John Dee; they corresponded, and Dee visited him in 1547. Mercator is best known today for his maps. In 1585 he published volumes covering France, Belgium, and Germany. In 1589 he published twenty-two maps, including those of Italy, Slovakia, and Greece. He died in 1594, but his son Rumold collected all the maps in one volume, entitled it *Atlas,* and published it in 1595. We continue to use the term "**atlas**" to describe a collection of printed maps.

Perhaps the last cosmographer and first modern geographer was Baron Friedrich Wilhelm Heinrich Alexander von Humboldt (1769–1859; see Figure 2.3). Born in Berlin,

2.3 Portrait of von Humboldt, 1843

2.2 Thinking about Maps

Maps and mapmaking are an integral part of the history of geography. The history of cartography was long dominated by a narrative that stressed mapping as a steady rise in greater knowledge and increased understanding of the world. It was a tale of the increasing scientific rationality of mapmaking from a dim and distant premodern past to an increasingly enlightened present. Maps were milestones along this journey. This discourse was undermined by the postmodern turn, which emphasized maps as social constructions, stories with a purpose, full of erasures and silences as well as inscriptions and disclosures. Maps have since been deconstructed according to their ideological basis, political undertones, and social contexts. They are now no longer seen as uncomplicated pictures of the world. They reflect power relations and embody the knowledge and ignorance, articulations and silences, of the wider social world. The accuracy or provenance of maps is no longer the only consideration in the new history of cartography. It is important to uncover their narrative context—their truths as well as their lies. Mapping is not innocent science—it is a political act.

REFERENCES

Harley, B. (2001) *The New Nature of Maps*. Baltimore and London: Johns Hopkins University Press.

2.4 Isothermal chart, 1823

he was privately tutored as a young boy and then studied at universities in Frankfurt, Göttingen, Hamburg, and Freiburg. He traveled widely. He arrived in Paris shortly before the storming of the Bastille; he wrote later that it "stirred his soul." He had a lifelong social concern with social reform and improvement. A recurring theme of his intellectual curiosity was to identify the "life force." His belief in universal harmony echoed concerns of John Dee. He was also deeply empirical; measurement and numbers were important to him.

After 1796, when he inherited money after his mother's death, he was able to pursue his dream of traveling, exploring, and writing about the world. He traveled to France and Spain—a "measuring expedition," he called it—so he could practice surveying techniques. He traveled to South America in 1799 and stayed until 1804. He passed through Cuba, Colombia, Ecuador, Mexico, and Venezuela, recording, measuring, and explaining. He gathered 60,000 plant specimens, made maps, and amassed a range of data from climate to linguistics. His careful mapping of plants in different places demonstrated the effect of altitude and latitude on plant communities.

Humboldt was an influential figure. His fame and reputation stimulated the adoption of measurement and observation in various expeditions and surveys throughout the world. He stimulated geographical measurement and observation. Figure 2.4 is an 1823 **isothermal map** of the world, as the text notes, "drawn from the account of Humboldt and others." Humboldt traveled throughout his life, in later years venturing east to Siberia and covering almost 9,000 miles at a time when transport was rudimentary. His travel journal consisted of thirty-four volumes.

His best-known work, *Cosmos*, was published in four volumes from 1845 to 1862. It deals with responses to the physical environment and explores the world of perception and feelings. There is extensive treatment of poetic descriptions of nature, from the Greeks to modern travelers, and more standard scientific works that discuss astronomy, geology, and physical geography. Humboldt thought of calling his work *Cosmography* but stayed with *Cosmos* because it implied order. On the pivot of intellectual change, he shared some of the concerns of the older cosmographical tradition while inaugurating modern geography.

Mapping Relative Space

Absolute space is the grid of latitude and longitude. It is now easy to note that as I type these lines, I am located at 38.87 degrees north, 77.00 degrees west. We have come very far in giving an accurate spatial fix. In absolute terms, the Earth is accurately mapped, gridded, and coordinated. However, this locational fix says little about the neighborhood where I live. What kind of people live here? What are the housing conditions and neighborhood characteristics? Absolute location tells us nothing about the character of places. Much of contemporary human geography is concerned with the **relative space** of social connections, political arrangements, and economic conditions. We can consider the development of relative space by sampling some varied endeavors: the history of the mappings of crime, health, and social deprivation.

Crime

The term "**moral statistics**" first appears in an essay by André-Michel Guerry in 1833. It was used to refer to crime, pauperism, and a wide range of social phenomena. Such statistics were an important part of nineteenth-century thematic mapping. Maps of crime in France first appeared in 1829, when Guerry used data from 1825 to 1827 to plot, for each of the *départements* in the country, the incidence of crimes against persons, crimes against property, and educational instruction. Mapping in the early nineteenth century predated the development of statistical techniques. Maps and mapping were thus an important way to identify and suggest causal connections. Belgian statistician Adolphe Quetelet also used maps to suggest connections. His carefully gradated shading maps of France, the Low Countries, and parts of the Rhineland, published in 1831, show statistics on crimes against property and people. Quetelet was an influential figure in Europe, and his work and maps were translated into other languages. In the United States, Chicago was an early laboratory of social analysis and spatial mapping. A 1942 study by Shaw and McKay, for example, maps the covariation of various social conditions with reported incidences of juvenile delinquency.

Today the mapping of crime is an important element in policing. Keith Harries summarizes the now vast material on this practice. Maps of crime allow a spatial visualization vital to patrol officers, investigators, police managers, policy makers, and community organizations. Improvements in

GIS (geographic information systems) now allow data to be mapped, presented, and correlated. Maps are used to identify areas of high crime activity (so-called hot spots), and mapping the timing and spacing of crimes allows a more efficient use of police resources. **Crime maps** of local areas are regularly posted by police departments.

Mapping is also used helps to solve crimes. **Geographic profiling**, like psychological profiling, has become an important part of law enforcement. Geographic profiling assumes that criminal activity is place-specific and that criminals like to use known areas that are not too close to their home location. Criminals do not travel very far from their anchor points to commit crime. The dates and places of committed crimes can thus be mapped to create a probability surface of clues to the location of criminals.

The background to the mapping of these "moral" statistics is not only the concern to know where social deviance occurs but also to look for the covariance of other factors so that the deviance can be understood, controlled, and negated. Mapping moral statistics is not innocent of wider political considerations. "Deviance," for example, is a politically freighted term whose definition tells us about those with power, their concerns and perceived threats. Contemporary human geographers no longer concern themselves just with mapping crime. A large body of work is now concerned with exploring the socio-spatial nature of crime and the policing of space. Phil Hubbard, for example, looks at the geographies of prostitution, while Katherine Beckett and Steve Herbert study how the spatial strategies of policing in Seattle banish certain types of people, especially the poor and indigent, and enforce zones of exclusion in the city.

Public Health and Disease

The first maps of the incidence of disease, at least in the United States, were by Valentine Seaman (1770–1817), who identified individual cases of yellow fever along the New York City waterfront. Mapping the spread of disease became more common in the nineteenth century. Public anxiety over the health of industrial cities in Britain led to the creation of the Poor Law Commission, which prompted the Ordnance Survey, the government mapping agency, to produce detailed maps of towns and cities.

One of the most famous sets of medical maps was drawn by John Snow, a doctor working in mid-nineteenth-century London. He was convinced that cholera was communicated by contaminated water. The 1855 second edition of his work *On the Mode of Communication of Cholera* contained two maps. The first showed the areas of London served by two different water companies. These companies used different sources for their water. While the area served by the company using fresh water had death rates due to cholera of 5 per 1,000, the one that drew polluted water from the River Thames had rates of 71 per 1,000. Snow's second map plotted the distribution of cholera cases at a more finely grained level and showed that they clustered around particular pumps. People using the water pump in Broad Street, poisoned by a leaking cesspool, were more likely to come down with cholera than those using water from unaffected pumps. The maps highlighted the fact that cholera was caused by contaminated water and not, as the prevailing view at the time believed, contact with polluted air.

Social geographers have long mapped diseases and epidemics, and the interest in mapping disease continues. Two human geographers, Andy Cliff and Peter Haggett, along with various colleagues, have produced a large body of work plotting the spread of diseases. They have produced studies of disease diffusion through islands as well as case studies of individual diseases such as AIDS, influenza, and measles. Geographer Andrew Lawson, for example, works on **disease mapping** and analyzes how diseases cluster. In one paper with colleagues he showed how minority and low-income children in South Carolina were more likely to live in areas with lead-contaminated soils. The work confirmed the existence of **environmental racism** and reaffirmed the connection between lead exposure in children and increased incidence of intellectual disabilities.

There is more to the contemporary **social geography** of health and disease than mapping. More recent work examines the geographies of health inequalities, health care provision, health, and well-being.

Deprivation and Inequality

London in the 1880s was a place of turmoil: an economic depression in the middle of the decade increased unemployment and heightened social unrest. Charles Booth, a wealthy ship owner, established a social survey, one of the first. His multivolume *Life and Labour of the People in London* (1892–1903) quantified social problems in the metropolis. Booth's use of **social statistics** to measure and map urban inequality inaugurated an important strand of urban research in which empirical methods were tied to welfare reform objectives. This strand of urban investigation became a common feature of urban studies in the early twentieth century.

Jane Addams was a social activist (Figure 2.5). In 1889 she founded Hull House on Chicago's Near West Side. She worked there until her death in 1935. She and the other residents of the settlement provided services for the neighborhood, such as kindergarten and daycare facilities for children of working mothers, an employment bureau, an art gallery, libraries, and music and art classes. Hull House surveys of the local areas, similar to Booth's survey, led to the construction of maps of household income levels and ethnicity (Figure 2.6). One interesting feature of Hull House was the important role of women. Eight out of ten of the contributors to the 1895 volume *Hull-House Maps and Papers* were women. The Hull House residents and their supporters forged a powerful reform movement that launched the Immigrants' Protective League, the Juvenile Protective Association, the first juvenile court in the nation, and a Juvenile Psychopathic Clinic.

2.5 Jane Addams, 1914. **2.6** Wage map published in *Hull-House Maps and Papers* in 1895. Courtesy of the Newberry Library, Chicago, Illinois.

They lobbied the Illinois legislature to enact protective legislation for women and children and to pass, in 1903, a strong child labor law and an accompanying compulsory education law. The Keating–Owen Child Labor Act of 1916 was the legislative result of their efforts.

Jane Addams was tireless in her dedication. She wrote many popular books and numerous articles, maintained speaking engagements around the world, and played an important role in many local and national organizations, such as the Consumers League, the National Conference of Charities and Corrections (later the National Conference of Social Work), Campfire Girls, the National Playground Association, the National Child Labor Committee, the National Association for the Advancement of Colored People, and the American Civil Liberties Union. She was awarded the Nobel Peace Prize in 1931.

The concern with urban social difference continues to be an important part of urban social geography. To take just one example from thousands: Alasdair Rae highlights areas of deprivation in the English city of Sheffield as he maps health, education, skills, and crime, and plots the trajectory of change. While the maps and techniques have become more sophisticated, there remains at the core of human geography a consistent interest in documenting, mapping, and explaining the socio-spatial patterns of inequality.

Environment and Society

One goal of early-twentieth-century geography was to study the causal effects of environment on society. Ellsworth Huntington (1876–1947), a professor of geography at Yale University, argued that certain climatic conditions are "especially favorable to human progress." His **environmental determinism** was a form of racism. In his 1924 book *Civilization and Climate* he argued, "We know that the denizens of the torrid zone are slow and backward, and we almost universally agree that this is connected with the damp, steady heat" (p. 2). Climate, for Huntington, created culture.

Geography, like all intellectual disciplines, reflects and embodies current beliefs and ideologies. In the late nineteenth and early twentieth centuries, geography was often used to justify imperial adventures as well as to back up belief systems that ranked humans and societies. At the top were the white Protestant people and societies of Northwest Europe and North America. At the bottom were the black and brown peoples of the world. Racism and imperialism were an integral part of geography.

Writing at approximately the same time as Huntington, Ellen Churchill Semple (1863–1932), who taught geography at the University of Chicago and Clark University, developed a more probabilistic view of the role of environment in human progress. She was concerned with examining the relationship less deterministically and with less emphasis on racial differences and cultural hierarchies. Her writing needs to be placed in context. At a time when technological triumphalism was dominant, she was suggesting a counter-narrative that humans are connected to the land. In her classic work *Influences of Geographic Environment* (first published in 1911), she writes on page 2: "Man has been so noisy about the way he has 'conquered Nature,' and Nature has been so silent in her persistent influence over man, that the geographic factor in the equation of human development has been overlooked." Semple's book, a product of its time in its use of such terms as "savage," is still worth reading for both its eloquent style and its sensitivity to the relationship between people and environment.

Modern geography has long moved past issues of environmental determinism. However, a socially sensitive environmentalism still has some value. At the macro level, for example, climate change plays an important role in the rise and fall of civilizations. Mesopotamian civilization collapsed around 3,400 years ago due in part to a severe 200-year drought. The Mayan civilization of Central America collapsed almost 1,100 years ago, again due to a combination of social and environmental factors. The environmental context is an important one for looking at the big picture of societies' rise and fall. Echoing the basic theme of Huntington, but with a very different emphasis, Christian Parenti considers in his book *Tropic of Chaos* the area between the Tropics of Cancer and Capricorn. The **tropical zone** is one where issues of climate change and especially water availability are partial causes behind social conflict and political ruptures. Parenti draws a connection between climate change and violence through preexisting conditions, such as the legacy of Cold War militarism and the destructive consequences of neoliberal economics. He describes a catastrophic convergence of societies littered with cheap weapons and fractured by the lack of social connectivities now worsened by the wrenching effects of climate change. In an even broader perspective, Jared Diamond identifies the links behind social collapse and environmental degradation. This is not the simple determinism of earlier geographers but a more nuanced account that shows how social change is embedded and embodied in environmental relations.

Geography and Society

The history of an academic discipline is tightly bound up in broader social developments. Three themes spring to mind when discussing the evolving relationship between geography and society: the connection with wider intellectual debates, the varied relationships to political power, and contemporary concerns.

Intellectual Debates

One historian of geography, David Livingston, describes the discipline as a contested tradition. He argues for a history of geography that does not assume an unchanging metaphysical core but rather a series of situated geographies, ones that shape and are shaped in turn by the time and place of their making. To write a global geography in London in 1910 or Washington, DC, in 1945 is a very different enterprise from a global geography written in twenty-first-century Beijing. A comparison of the different worldviews embodied in Figures 2.1 and 2.2 is instructive. Practices and procedures vary, and this variation is not incidental to the evolution of geographic writing but rather fundamental. What interests are advanced, what assumptions are made, and what underlying model of how the world hangs together all depend on the time and place. Historians of the discipline, for example, are often appalled at the **casual racism** prevalent in the work of geographers writing in decades past. In other words, there is a historical geography to the evolution of **human geography**.

The situated and compromised nature of knowledge production is best summarized in the title and subtitle of Steven Shapin's 2010 book on the history of science, *Never Pure: Historical Studies of Science as if It Was Produced by People with Bodies, Situated in Time, Space, Culture, and Society, and Struggling for Credibility and Authority.* The history of geographic thought is intimately bound up in the historical geography of ideas.

Geography and Politics

The development of human geography is intimately connected to the interests of powerful groups, whether they are private and corporate or public and governmental. John Dee was an important intellectual figure in Renaissance England. He cast horoscopes for Queen Elizabeth I, promoted the expansion of English mercantile interest, and "proved" English claims in the New World. The production of geographical knowledge is intimately connected to national economic interest. Fast-forward over three hundred years: geography, according to Sir Halford Mackinder (1861–1947), was an aid to **statecraft**. What he meant by that was that geography could help frame geopolitical strategies for the nation state; in his particular case, Britain and the British Empire. He developed a **heartland theory** to argue that whoever ruled the central part of the Eurasian landmass, the heartland, ruled the world. The idea was first formulated in 1904 and appeared in book form in 1919. At the time he was a fervent anti-Bolshevik, and so his argument can be seen in one light as a rationale for providing a cordon around Bolshevik Russia. His ideas influenced Nazi strategic thinking and US Cold War strategy.

Isaiah Bowman (1878–1950), a contemporary of Mackinder, was director of the American Geographical Society for over twenty years. He was also the chief territorial advisor to President Wilson and to the US State Department during the First World War. He had a strong belief in US expansionism based on penetration of foreign markets by US capitalism. He provided a **geographical imagination** to ideas of US global superiority and dominance. His book *The New World*, written in 1921, is still a very well written and accessible account of geopolitical issues and problems facing the world in the aftermath of the First World War and in the wake of the fragmentation of prewar empires.

Dee, Mackinder, and Bowman are at one end of the applied geography scale, their work and ideas directly shaping and influencing national interests. At the other end are more antiestablishment figures. Peter Kropotkin (1842–1921) provides an interesting contrast. He was born into the privileged life of pre-Revolutionary Russian aristocratic society; his father owned vast tracts of land and over 1,200 serfs. In 1864 he led a geographical expedition to Siberia, and on his return to St. Petersburg he became secretary to the Imperial Russian Geographical Society. His interests widened to include a more radical vision of society: an **anarcho-communist** society based on mutual aid and voluntary associations. In 1875 he was imprisoned, but because of his social status was allowed to finish a geographical report on the last Ice Age, based on the fieldwork he had conducted earlier in Finland and Sweden. Students who are late with their work might like to think of Kropotkin, who finished his geography report even while in jail. He managed to leave Russia and lived in Switzerland and England, but returned in 1917 after the February Revolution overthrew the czar. He spoke out against the authoritarian socialism inaugurated with the October Revolution and the Bolshevik accession to power. Since 1957 a Moscow subway station has borne his name.

A more intellectual radicalism shapes the work of some contemporary human geographers. David Harvey (b. 1935) is a human geographer whose PhD work was on hops production in the English county of Kent. His first book, *Explanation in Geography* (1969), was an argument for the adoption of the deductive approach in geography. The context for Harvey's next book, *Social Justice and the City* (1973), was the wider world of the Vietnam War, the persistence of poverty in the richest countries, intractable racial divisions, and the decline of the long postwar economic boom, exposing inequalities and social discontent. There was a growing dissatisfaction in the academy with traditional methods of scholarship. In *Social Justice*, Harvey reflects upon and informs the debates surrounding these societal and academic issues. *Social Justice*, then, is really two books. The first part, "Liberal Formulations," focuses geographical inquiry on socially relevant topics. The second, "Social Formulations," marks a distinct **epistemological break**. The concept of the city is radically altered between parts one and two. Rather than being an independent object of inquiry, it is an important element that mediates and expresses social processes.

In subsequent books, Harvey continues to provide radical critiques of capitalism. *The Enigma of Capital* (2010) outlines the tendency of crisis in capitalism, while *Rebel Cities* (2012) argues for the importance of the city and the neighborhood as sites of political struggle. The title of his 2014 book, *Seventeen Contradictions and the End of Capitalism*, summarizes his focus on exposing the internal working of the capitalist system and highlighting its inherent long-term instability.

Contemporary Debates

Human geography has undergone profound change in the past fifty years. A number of "turns" can be identified. The turn of **quantitative geography**, associated with such people as Peter Haggett and Brian Berry in the 1960s, was concerned with measurement, calibration, the creation of spatial models, and the testing of hypotheses. It was not just a fascination with numbers. Waldo Tobler, for example, sought to represent flows in maps and to add a much needed temporal dimension to cartographic representation and the geographical imagination. He is also credited with devising **Tobler's Law**, which states, "Everything is related to everything else, but near things are more related to each other." The quantifiers also excavated previously neglected scholars such as the 1930s German geographer Walter Christaller, whose work we will consider in chapter 17. This excavation of previously ignored work is a consistent theme of all the different approaches, and leads to a constant recreation of the history of human geography.

But the quantitative turn, which built on a long tradition of quantitative geography, was not a revolution that came to dominate the entire discipline. It is more appropriate to think of it as a seismic shudder with multiple epicenters: Cambridge and Bristol in Britain and the University of Washington in the United States. In many places the seismic shift was scarcely recorded, and if it was, it was dismissed as of minor, ephemeral significance. And here we come to another trait of contemporary human geography: no turn has involved all of the discipline, because it is just too big and too varied.

The quantifiers had an uphill task against the entrenched forces of reaction. Old men (and most of them were men) with established reputations rarely take much heed of young scholars in a hurry who make very negative statements about the work of their elders. It was in one sense a generational struggle for power and status. But scarcely had the quantifiers established a beachhead in the discipline, through creating mandatory courses on statistical techniques and establishing their own publishing outlets, before counter-critiques and new turns emerged from the early 1970s. We can identify five from the many.

First, the notion of dispassionate, objective observers was criticized for its lack of social relevance and its connections with the corporate world. In effect, the "scientific" endeavor was neither socially neutral nor politically innocent. The emphasis shifted from identifying the spatial structure of society—a world of surfaces, edges, nodes, hierarchies, and flows—to the analysis of the social organization of space. This involved a more radicalized human geography.

Second, there was a criticism of the quantifiers from a humanist perspective. The **humanist critique** pointed to the lack of concern with personal geographies, the fact that statistical averaging ignored the whole realm of feeling, perception, and intersubjectivity. The work of Peter Jackson and Susan Smith showed what a radically informed **cultural geography** could look like. In the United States, cultural geography was revived through a concern with landscape as a socially contested, materially produced cultural artifact. Geographers like Jim Duncan and Denis Cosgrove reinvigorated old concerns of landscape with the new interest in social theory.

Third, in the 1980s feminists appeared as a significant voice in the struggle for the hearts and minds of human geography. The **feminist critique** drew attention to the bias in the academic practice of geography and geographical writings. It was a discipline dominated by men and a discourse that rarely connected with the reality of gendered spaces and gendered lives. There is now a considerable body of work on the gendered separation of productive and reproductive spheres, issues of domestic labor, work–home separation, and the overall living and working experience of women. The feminist perspective also shows how gender interacts with race, class, and sexuality to produce socio-spatial patterns of inequality and marginalization. Geographers such as Doreen Massey, Linda McDowell, and Gill Valentine, to name just a few feminist scholars, have widened my own angle of vision.

Fourth, the adoption of **postmodernism** in human geography was a significant development. Postmodern geographers such as Michael Dear, Gillian Rose, and Ed Soja criticized the notion of objective universal constructs. Knowledge was not revealed to socially objective viewers but socially constructed by the perspective of the viewer. The postmodern turn shifted the debates by introducing notions of hybridity, alternative rather than dominant narratives, and a multiplicity of ways of knowing.

Fifth, there is now a greater concern with the marginal and the marginalized. For much of the history of geography in the past two hundred years, the subject was written about by a relatively small sample of people—generally men in positions of power and authority in the richer countries of the world. Geography was linked to the existing global and national power structure. In recent years, however, there has been a growing body of work that we can call **subaltern geographies**. More geography is being written from outside the circle of usual suspects. More female scholars and scholars from developing countries are beginning to shift the angle of vision. There are

2.3 The Spatial Turn

The "spatial turn" is the name given to a greater emphasis on space and place in the social sciences and humanities. It was promoted by two trends. The first is an increasing recognition that space and place play important roles in the issues and concerns important to social scientists and humanity scholars. Of course it is at the heart of the geographical enterprise, so we have been practicing the spatial turn long before it became fashionable in other disciplines. A spatial turn is noted in anthropology, architecture, history, psychology, religion, literature, sociology, and history. Topics such as maps and mapping, territoriality, landscape, spatial change, and the transformations of place are the heart of new ventures in other disciples that have found a shared interest in describing and explaining the culturally inscribed meanings of place and power relations of space.

The second trend is the growing use of GIS in the social sciences and humanities—for example, mapping in the digital humanities. One research project generated a 1.5-million-word corpus from 80 texts written about the Lake District in England between 1622 and 1900. Places that were described with terms such as "beautiful," "majestic," and "sublime" were mapped. The analysis showed the influence of the Romantic sensibilities, how "sublime" is used to refer to wide sweeps of landscapes while "majestic" is more specific. The mappings reveal the different ways used at different times to describe and evaluate the landscape of the Lake District.

REFERENCES

Gregory, I. and Donaldson, C. (2016) Geographical Text Analysis: Digital Cartographies of Lake District Literature. In: *Literary Mapping in The Digital Age. Digital Research in the Arts and Humanities*. London: Routledge, pp. 67–78.

Warf, B. and Arias, S. (eds) (2008) *The Spatial Turn: Interdisciplinary Perspectives*. London: Routledge.

also more geographical studies of marginalized groups: women, children, migrant workers, peasants, slum dwellers, sexual minorities, and so forth. Geography is enriched by these developments.

Today human geography is a rich area of intellectual inquiry. The discipline continues to be contested in a lively fashion.

The Concerns of Human Geography

Geography has some enduring concerns. Geographers are interested in the reasons behind and the consequences of the spatial distribution of human activities and human environmental interactions. Geographers look at space and place as containers, stages, sites, and outcomes of human activities and environmental interactions. They are always asking two major and related questions: where and why?

Fundamental Concepts

Space and place are fundamental concepts in human geography. Space is a more generalized place such as the nation state, the region, the city, or the neighborhood, while places include the USA, southeast England, Shanghai, and Bukchon, South Korea.

Understanding how space becomes place and how places are incorporated into space is at the heart of the human geography enterprise. Culture turns space into places. Place is space made more intimate by culture. But places are in turn incorporated into space, as when the USA is linked to the global economy by flows of people, goods, and ideas. Space and place are more than mere background; they are stages for performances, given meaning and significance by social actions, political conflict, and economic processes. Space and place are active moments, not static categories.

Geographers are concerned with the spatial organization of society and the social organization of space: how spaces and places are produced, used, consumed, destroyed, represented, managed, and controlled. They are concerned with spaces of hope and spaces of exclusion, the reconstructions as well as the destructions of place.

Spatial Categories

There is a range of spatial categories that hold special meaning for geographers: they include the nation state, the region, landscape, the city, and the neighborhood, home. Other spatial categories include the frontier, boundary, territory, and turf. Geographers look at the physical and discursive construction of these categories, their changing usage and meaning. The categories are employed not as reified spatial abstractions, but as social constructions full of cultural meaning and political significance. They are not unchanging verities but the outcome and site of negotiation and conflict.

There is spatial imagination at work in the study of human geography, and so a wide range of spatial categories are employed, including location, which is a point or place on the Earth's surface. Location can be used in absolute terms, such as latitude and longitude, as well as in relative terms, such as describing a country or region as rich or poor.

Distributional categories employed include density, concentration, and pattern. Typical questions: What is the density of population in a certain place and why? Why are certain economic activities concentered in certain locations? Is there a pattern to events such as the distribution of towns and cities?

Movement across space and between places is another important topic. Geographers are interested in how the friction of distance influences things. The time taken to travel is improved by transport improvements, leading to what is known as space-time convergence, the collapse of the time taken to travel the same distance.

The term "diffusion" is used to describe the spread of things, ideas, people, or practices across space. Diffusion occurs in different ways. It can occur through relocation, as when people often take their language and religion with them when they move, diffusing it more widely; through contagion, as when diseases spread through contact; or through hierarchical diffusion, as when ideas flow from large to small cities.

Geographers study different types of regions, which we can define for the moment as demarcated areas of the Earth's surface. Regions are defined by different criteria and so we refer to cultural regions and economic regions. Hollywood, the Rust Belt, and the Asian Tigers are all examples of some of the regional definitions used in economic geography. Human geography employs different types of regions at different spatial scales.

The last few paragraphs are very dense. However, think of them as mere pointers that introduce you to spatial categories that are more carefully defined and strategically employed throughout the rest of this book.

Methods

Geographers work at a variety of scales, from the global and national to the more intimate spaces of our personal interaction. Human geography has a large bandwidth and so employs a variety of techniques. There is an enduring concern with spatial representation and mapping. Figure 2.7 maps the Human Development Index (HDI). This is a composite index that combines measures of a long and healthy life, being knowledgeable, and having a decent standard of living. From this map, what conclusion can you make about the global distribution of human development? In which countries would you choose to live, and why?

In recent years, mapping has been dramatically altered by the rise of **Geographic Information Systems (GIS)**, systems that store geographic data and allow their manipulation and representation. GIS is used by governments and companies to store, access, and display data. Geographers also use remote sensing, which involves the collection of data taken from aircraft, satellites, and more recently drones. For example, a recent study by geographer Do-Hyung Kim and colleagues used satellite imagery to examine tropical forests

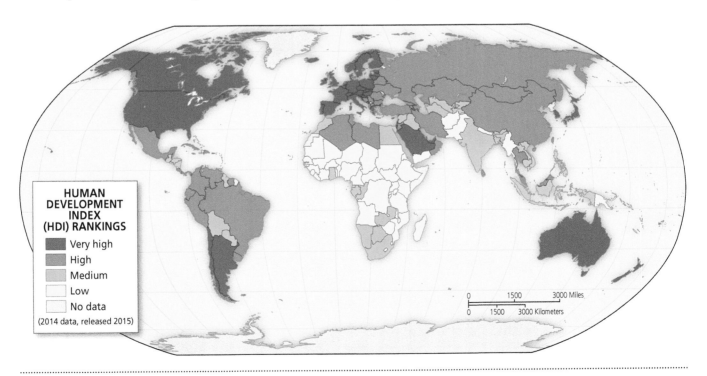

2.7 Human Development Index.

in 34 counties. They discovered that the rate of loss was higher than then suggested by the United Nations, which had been using self-reported data. While UN data showed deceleration in forest loss, their detailed study showed a 62 percent acceleration in deforestation in the humid tropics from the 1990s to the early 2000s.

Remote sensing allows us to utilize data that would be difficult to map, such as ice sheet shrinkage or the emergence of slums at the edge of cities in the developing world. Figure 2.8 shows a space-based perspective of the Kansas City metro area taken in September 2014 by astronauts on the International Space Station.

Geographic profiling is now an often-used technique. Initially used in crime studies to identify patterns of crime and locate criminals, it is now used to identify disease outbreaks and control invasive species. It was used to solve an art mystery: who was the insurgent artist Banksy? It was also used to help narrow down the location of Osama bin Laden while he was in hiding. Assumptions included a distance decay function: he would not move far from his last sighting in Afghanistan, he would be in Pakistan because it was a similar culture, and he would be in a town or city in a dwelling large enough to house his retinue with high walls to provide privacy and security. His precise location

in Abbottabad was confirmed by following a courier. DNA evidence from people in the house provided **ground truthing**.

A range of techniques is employed in contemporary human geography. Quantitative analysis relies on data and statistical analysis, while qualitative techniques include storytelling and the use of focus groups and participant observation. The combination of the qualitative and quantitative provides sympathetic and grounded accounts. In my own case study of Aboriginal art production and racial tensions in Alice Springs, Australia, I mapped peripheral settlements, and studied government data and city records, but also visited the city over many years to interview people, cultivate connections, and immerse myself in the world of art production and sale.

Human geographers look at space and society from a variety of scales, encompassing approaches that range from GIS, remote sensing, and statistical analysis techniques to qualitative ethnographic studies. They intersect with other researchers to consider such important topics as the flexible identities of gender, ethnicity, and nationality; the measurement, explanation, and understanding of globalization; the connections between space and place; the recurring question of how we represent the world around

2.8 Remotely sensed image of Kansas City metro area

2.4   GIS, Remote Sensing, and the Democratization of Mapping

While it is often used in the worlds of national security and as part of corporate research, GIS can also be used to map social inequalities. Alasdair Rae uses GIS to identify areas of deprivation in Scotland in an interactive map that employs Google Maps. His website is Stats, Maps n Pix.

There also has been a democratization of mapping. OpenStreetMap is a collaborative effort with 2 million registered users who generate a free map of the world. The community plays an important role in **crisis mapping**. During the 2010 Haiti earthquake, volunteers used satellite data to provide a real-time digital atlas of the road system, vital information in a country desperate to get aid supplies to affected communities. The project was so successful that now the OpenStreetMap community regularly teams up with aid organizations in the wake of disasters to provide urgent and much-needed spatial data.

Open source software and global volunteer networks are now an important part of crisis mapping. GIS and social media are being used to provide real-time georeferenced data in places experiencing violence or disruption. Patrick Meier provided examples from Haiti, Japan, Kenya, Russia, Somalia, and Syria.

There are also more focused examples. Geography students at the University of Montana used satellite images to construct a detailed topographic map of a tiger reserve in Nepal, which allowed people on the ground to reconcile their position on the map and track down poachers.

GIS and remote sensing can also be used in more open and democratic way, especially for societies that are closed and undemocratic. Curtis Melvin used Google Earth to a create map of North Korea that highlighted missile-storage facilities, mass graves, labor camps, and the palaces of the elite. He went on to work on a crowdsourced atlas of the secretive state, opening up one of the world's most isolated and brutal countries to greater public scrutiny.

REFERENCES AND WEBSITES

Meier, P. (2012) Crisis Mapping in Action: How Open Source Software and Global Volunteer Networks Are Changing the World, One Map at a Time. *Journal of Map and Geography Libraries: Advances in Geospatial Information, Collections and Archives* 8: 89–100.

Chaney, R. (2015, December 16) Nepali Tiger Poachers Fear University of Montana Mapmakers. *Missoulian,* http://missoulian.com/lifestyles/recreation/nepali-tiger-poachers-fear-university-of-montana-mapmakers/article_40c44f52-bbe2-5c87-80f0-ab5668c58ad0.html

https://www.openstreetmap.org/#map=5/51.500/-0.100
https://sites.google.com/site/scotdep2012/home
http://www.nkeconwatch.com/north-korea-uncovered-google-earth/
http://www.statsmapsnpix.com
http://www.38northdigitalatlas.org

us; and the pressing concerns of sustainability, environmental transformation, and climate change. At its very best, the discipline exhibits an environmental sensitivity with a social awareness. And like the two young people in Vienna shown in Figure 2.9, it looks at the world as a constant source of fascination and interest.

An editorial opinion piece in the influential British newspaper *The Guardian* summarizes the attractions of the discipline:

> [G]eography is a subject for our times. It is inherently multidisciplinary in a world that increasingly values people who have the skills needed to work across the physical and social sciences. Geographers get to learn data analysis. . . . They learn geographic information systems. They can turn maps from a two-dimensional representation of a country's physical contours into a tool that illustrates social attributes or attitudes: not just where people live, but how, what they think and how they vote. They learn about the physics of climate change, or the interaction of weather events

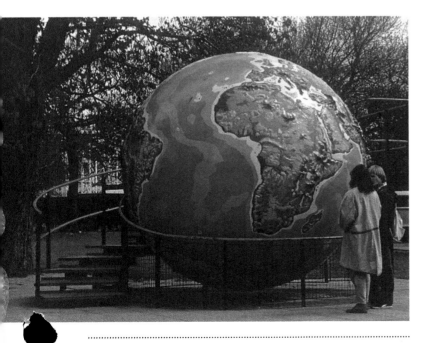

2.9 Two young people studying a globe in a park in Vienna, Austria *and flood risk, or the way people's behaviour is influenced by the space around them. All these are not just intrinsically interesting and valuable. They also encourage ways of seeing and thinking that make geographers eminently employable.*

As you read these lines, you are now an active participant and creator in this fascinating discourse. At it very best, the study of geography exercises the spatial imagination and cultivates a profound sense of place. Welcome to the most exciting and engaging of subjects.

Cited References

Aelion, C. M., Davis, H. T., Lawson, A. B., Cai, B., and McDermott, S. (2013) Associations between Soil Lead Concentration and Populations by Race/Ethnicity and Income-to-poverty Ratio in Urban and Rural Areas. *Environmental Geochemical Health* 35: 1–12.

Balch, E. (1895) *Hull House Maps and Papers.* http://www.jstor.org/stable/2276290

Beckett, K. and Herbert, S. (2009) *Banished: The New Social Control in Urban America.* New York: Oxford University Press.

Booth, C. (1903) *Life and Labour of the People in London.* London: Macmillan.

Bowman, I. (1921) *The New World: Problems in Political Geography.* Yonkers: World Book Company.

Diamond, J. (2005) *Collapse: How Societies Choose to Fail or Succeed.* New York: Penguin.

Editorial Opinion (2015, August 13) The Guardian View on Geography; It's the Must Have A-level. http://www.theguardian.com/commentisfree/2015/aug/13/the-guardian-view-on-geography-its-the-must-have-a-level?CMP=share_btn_tw

Kim, D. H., Sexton, J. O., and Townshend, J. R. (2015) Accelerated Deforestation in the Humid Tropics from the 1990s to the 2000s. *Geophysical Research Letters* 42: 3495–3501.

Harries, K. (1999) *Mapping Crime: Principle and Practice.* Washington, DC: Department of Justice. NCJ 178919.

Harvey, D. (1973) *Social Justice and the City.* London: Arnold.

Harvey, D. (2010) *The Enigma of Capital.* London: Profile.

Harvey, D. (2012) *Rebel Cities: From the Right to the City to the Urban Revolution.* London: Verso.

Harvey, D. (2014) *Seventeen Contradictions and the End of Capitalism.* Oxford: Oxford University Press.

Hauge, M. V., Stevenson, M. D., Rossmo, D. K., and Le Comber, S. C. (2016) Tagging Banksy: Using Geographic Profiling to Investigate a Modern Art Mystery. *Journal of Spatial Science* 61: 185–190.

Hubbard, P. (1999) *Sex and the City: Geographies of Prostitution in the Urban West.* Aldershot: Ashgate.

Humboldt, A. (1897) *Cosmos: A Sketch of a Physical Description of the Universe.* New York: Harper.

Huntington, E. (1924) *Civilization and Climate.* New Haven: Yale University Press.

Livingstone, D. N. (1993) *The Geographical Tradition: Episodes in the History of a Contested Enterprise.* Oxford: Blackwell.

Parenti, C. (2011) *Tropic of Chaos: Climate Change and the New Geography of Violence.* New York: Nation.

Rae, A. (2011). Deprivation in Sheffield. http://www.sheffield.ac.uk/polopoly_fs/1.137936!/file/ajr_sheffield_deprivation_nov_2011.pdf

Shapin, S. (2010) *Never Pure: Historical Studies of Science as if It Was Produced by People with Bodies, Situated in Time, Space, Culture, and Society, and Struggling for Credibility and Authority.* Baltimore, MD: Johns Hopkins University Press.

Semple, E. C. (1911) *Influences of Geographic Environment on the Basis of Ratzel's System of Anthropogeography.* New York: Holt. https://archive.org/details/influencesofgeog00semp

Short, J. R. (2012) Representing Country in the Creative Postcolonial City. *Annals of Association of American Geographers* 102: 129–150.

Snow, J. (1855) *On the Mode of Communication of Cholera.* London: John Churchill. http://www.ph.ucla.edu/epi/snow/snowbook.html

Select Guide to Further Reading

On maps and mapping:

Krygier, J. and Wood, D. (2011, 2nd ed) *Making Maps.* New York: Guilford.

Short. J. R. (2003) *The World through Maps.* Toronto: Firefly.

Tyner, J. (2015) *The World of Maps.* New York: Guilford.

For contemporary discussion of the geographies of health:

Anthamatten, P. and Hazen, H. (2011) *An Introduction to the Geography of Health.* New York: Routledge.

Brown, T., McLafferty, S., and Moon, G. (2010) *A Companion to Health and Medical Geography.* Malden: Wiley-Blackwell.

Cliff, A. D., Haggett, P., and Raynor, M. S. (2004) *World Atlas of Epidemic Diseases.* London: Hodder Arnold.

Koch, T. (2011) *Disease Maps: Epidemics on the Ground.* Chicago: University of Chicago Press.

Two sample papers on crime:

Andresen, M. A. and Brantingham, P. L. (2012). Visualizing the Directional Bias in Property Crime Incidents for Five Canadian Municipalities. *The Canadian Geographer/Le Geographe Canadien* 57: 31–42.

Rossmo, D. K. and Harries, K. (2011) The Geospatial Structure of Terrorist Cells. *Justice Quarterly* 28: 221–248.

A useful entry point into the history of an intellectual discipline is to read the biographies and autobiographies of its practitioners. Here is a very small sample of a range of human geographers:

French, P. J. (1987) *John Dee: The World of an Elizabethan Magus*. London: Routledge.

Gould, P. (1999) *Becoming a Geographer*. Syracuse, NY: Syracuse University Press.

Gould, P. and Pitts, F. (eds) (2002) *Geographical Voices*. Syracuse, NY: Syracuse University Press.

Haggett, P. (1990) *The Geographer's Art*. Oxford: Blackwell.

Hubbard, P. and Kitchin, R. (eds) (2011) *Key Thinkers on Space and Place*. London: Sage.

Kearns. G. (2009) *Geopolitics and Empire: The Legacy of Halford Mackinder*. Oxford: Oxford University Press.

Kropotkin, P. (1927) *Memoirs of a Revolutionist*. London: Smith Elder. First published 1899. http://theanarchistlibrary.org/library/petr-kropotkin-memoirs-of-a-revolutionist

Moss, P. (ed) (2000) *Placing Autobiography in Geography*. Syracuse, NY: Syracuse University Press.

Rupke, N. A. (2008) *Alexander von Humboldt*. Chicago: University of Chicago Press.

Smith, N. (2004) *American Empire: Roosevelt's Geographer and the Prelude to Globalization*. Berkeley and Los Angeles: University of California Press.

On human geography, ancient and modern:

Association of American Geographers (2017) *International Encyclopedia of Geography*. Chichester, UK: Wiley-Blackwell.

Agnew, J. A. and Livingstone, D. N. (eds) (2011) *The Sage Handbook of Geographical Knowledge*. Thousand Oaks, CA: Sage.

Cox, K. (2014) *Making Human Geography*. New York: Guilford.

Cresswell, T. (2013) *Geographic Thought: A Critical Introduction*. Chichester, UK: Wiley-Blackwell.

Gregory, D., Johnston, R., Pratt, G., Watts, M. and Whatmore, S. (eds) (2011) *The Dictionary of Human Geography*. Chichester, UK: Wiley-Blackwell.

Martin, G. J. (2005) *All Possible Worlds: A History of Geographical Ideas*. 4th ed. Oxford: Oxford University Press.

Thomson, J. O. (2013) *History of Ancient Geography*. Cambridge: Cambridge University Press.

On GIS:

Bolstadt, P. (2016) *GIS Fundamentals*. Ann Arbor: XanEdu.

Nyerges, T. L., Couclelis, H. and McMaster, R. (eds) *The Sage Handbook of GIS and Society*. London: Sage.

Selected Websites

On GIS
https://www.youtube.com/watch?v=j5WmvTxQF5w
https://www.youtube.com/watch?v=bLnJpkqjlwo
Humanitarian Open Street Mapping
https://hotosm.org
Just one of the many urban police websites that map crime
http://oakland.crimespotting.org/map
A website run by geographers that maps and analyzes user-generated geocoded data; it uses social media to produce some very cool maps
http://www.floatingsheep.org/

Websites of selected geographical societies and organizations:
American Association of Geographers: http://www.aag.org
American Geographical Society: http://americangeo.org
International Geographical Union: http://igu-online.org
National Geographic magazine: http://www.nationalgeographic.com
Royal Geographic Society: http://www.rgs.org/HomePage.htm

Learning Outcomes

- Human geography is shaped by the debates of the past, with five particular themes that have resonated through the years: the coordination of absolute space, the emergence of geography from cosmography, the discussion of relative space, environment and society, and the links between geography and society.
- Ptolemy was one of the earliest known mapmakers.
- The shift from cosmography to geography was neither sudden nor abrupt. Humboldt was on the cusp of the change.
- There is a long and enduring interest in human geography in mapping and explaining the socio-spatial patterns of crime, public health, and inequality.
- Geography, like all intellectual disciplines, reflects and embodies current beliefs and ideologies. In the late nineteenth and early twentieth centuries, for example, geography was often used to justify imperial adventures as well as to back up belief systems that ranked humans and societies.
- Modern geography has moved past issues of environmental determinism to a more sensitive environmental understanding.
- Human geographers use a variety of techniques and different scales to understand a wide range of topics.
- Human geography today has developed in response to various critiques of its lack of relevance, lack of humanism, and gender and racial biases.
- Human geography now incorporates an environmental sensitivity and a social awareness.

People, Resources, and Environment

Part 2 takes up the general themes of population and resources. Chapter 3 outlines the basic trends of growth and decline and examines the dynamics of the demographic transition. The relations between population and food supply are detailed in chapter 4 and those between population and resources in chapter 5. Chapter 6 examines people–environment relations through exploring environmental impacts on society and human influences on environment change.

OUTLINE

Population Dynamics

LEARNING OBJECTIVES

Describe the historical factors that trigger major population changes.

Describe the main trends of world population change with associated economic and cultural impacts.

Identify the four phases of the demographic transition.

Explain the social and economic concerns that accompany each stage of the demographic transition.

Relate current global cultural and political occurrences to contemporary population geographies.

Human population levels have risen and fallen in line with climate change and availability of food. In the past two hundred years there has been a steady increase in the world's population. Population dynamics have changed. The demographic transition is a shift from high birth and death rates, through declining death rates and high birth rates, to declining birth and death rates. Different parts of the world are at different stages of this transition. Some have high growth rates, while others are experiencing net population decline.

The world population is now close to 7.5 billion. It grew from a very small base. The number of humans poised to leave Africa to populate the world probably ranged from 3,000 to 5,000 souls. This tiny group was the basis for the spread and growth of human population around the globe. Global population remained small and steady until 4,000 years ago, when there was a slow, steady rise, then an explosive upward growth beginning around 1800 (Figure 3.1). Population growth accelerated in the past century. It took thirty years, from 1930 to 1959, for an extra billion people to be added. It took only another thirteen years, from 1998 to 2011, for yet another billion to be added.

Throughout most of human history, the world's population was relatively small. Specific bottlenecks occurred when the human population came precariously close to extinction. Around 70,000 years ago, the eruption of the volcano Toba in Sumatra caused a volcanic winter of a long stretch of colder, cloudier weather. The total human population may have been reduced to only thousands. The result was a very small pool of genetic diversity. Today's huge population is genetically very close. We share a common ancestry, and despite the cultural production of racial and ethnic differences, we are biologically just part of one large extended family.

Despite such setbacks, humans prospered. They were adaptable, smart, and mobile. They transformed their surroundings, trapping animals, creating gardens, and efficiently developing gender divisions of labor between male hunting and female gathering and foraging. Men and women accumulated a long-term, deep, and sophisticated knowledge of the environment.

Population Declines

The long-term trend is one of global population growth. The increase in the **world population** masks significant examples of population decline. While wars and social conflict can reduce population in specific regions at specific times, longer-term declines are mainly due to disease and climate change.

3.1 World population growth

Disease

In Athens between 430 and 427 BCE, almost a third of the population of the city-state died from a disease as yet undetermined, likely a form of typhus. Yet disease has sometimes reached pandemic proportions. A **pandemic** is an epidemic on a wide geographic scale. We can consider three examples.

The first was the **bubonic plague** that swept in a series of waves across the world from 541 to 747 CE. It was especially pronounced in the Byzantine Empire. The first wave occurred from 541 to 544 and is named after the ruling emperor of the time, Justinian. It spread quickly along trade routes. The total death toll is estimated at 25 million people. In Constantinople, at the height of the first wave, almost 5,000 people were dying every day, ultimately halving the city's population. The disease spread as far west as England and as far east as Persia. The mass death created social confusion, caused economic dislocations, and prompted political upheaval. Villages were abandoned. Arable land lay fallow, and labor was in short supply. The lack of labor stimulated technological improvements such as improved plow design and new crop rotations. The pandemic impeded the growth of the Byzantine Empire and led to what some describe as the birth of Europe, as there was a localization of economies and political systems rather than unification under an empire. It was the pivot of large-scale social change, including the shift from antiquity to the Middle Ages and the opening stage of the rise of Western Europe to continental and then global dominance.

A second pandemic, a recurrence of bubonic plague, swept throughout Asia, North Africa, and Europe: the "**Black Death**," first mentioned in written records in 1346, by 1353 had killed between 30 and 40 percent of the population, with some estimates of up to 60 percent mortality in Europe. The population of Europe probably declined from 80 million to 30 million. The effects of such a rapid population loss were immense. Forests regrew throughout Europe. Societies dominated by low wages, high rents, and high prices were quickly transformed into ones of high wages, low rents, and low prices. The lack of population forced the development of more efficient, less labor-intensive technologies. Landowners switched from the strenuous **arable agriculture** of grain production to the less taxing **pastoral farming** of producing meat, wool, and dairy products. The switch "from corn to horn" also gave more employment opportunities to women, who were traditionally employed in the pastoral sector. There were also social upheavals. In England the 1381 **Peasants' Revolt** was the organized resistance to attempts by authorities to reduce wage levels to pre–Black Death levels. The wider significance of the Black Death is debatable. There are those who argue for continuities before and after, while others see a more profound shift from the medieval world to the modern world—a world where religious devotion was undermined for some, strengthened for others; a world where minorities, especially Jews, could become scapegoats for disasters; a world where the authorities were made responsible for public health; a world where the relationship between rich and poor was dramatically revealed as based less on unchanging

3.1 The Reproductive Revolution

The demographic transition is less a natural fact than a social revolution. It is a result of changes in public health and nutrition that extended the average life span. It is also the consequence of a reproductive revolution. The reproductive capacity of women is potentially very high. In theory, each fertile woman can have up to, and in some cases have had more than, fifteen children. Before the demographic transition, the reproductive efficiency was low, with few women living to puberty, high **infant mortality rates**, and low life expectancy. Fewer than one in ten girls reached the age of puberty. The result was a relatively low reproductive capacity. Each child born had limited chances of reaching adulthood. Emphasis was thus placed on each and every woman having as many children as possible. This largely defined the role of women and is the fundamental basis of **patriarchy**, the belief system and practice that men are the authority figures and women are restricted to home, hearth, and the bearing and rearing of children. Patriarchy—although mediated through different cultural and religious filters—is an attribute of low reproductive efficiency.

After the demographic transition, more women survived into their fertile years, and infant mortality rates fell. This higher reproductive capacity means that women can have fewer children. As the risk of each child dying declines, women do not need to have so many children to ensure that some survive into adulthood. The increased life expectancy also extends more women's lives beyond childbearing ages. The reproductive revolution calls into question the basis of patriarchy. McInnes and Diaz argue that the origin of other social changes, including the deregulation and privatization of sexuality and the rise of the importance of individual identity, also ultimately resides in the increased efficiency of reproduction. Their case is suggestive and intriguing, yet like all ambitious arguments it is open to debate and discussion, criticism, improvement, and refinement.

REFERENCE

McInnes, J. and Diaz, J. P. (2009) The Reproductive Revolution. *Sociological Review* 57: 262–284.

custom and tradition than on the brute facts of labor supply and demand.

Perhaps the most dramatic relationship between disease and population declines involves the **Columbian Encounter** that took place when Europeans discovered and colonized the Americas. The impact on the indigenous population was devastating. The pre-Columbian population of the Americas is difficult to gauge precisely, but estimates put the figure between 50 million and 100 million. The indigenous population was descended from peoples who had come from Eurasia before the development of agriculture. There was thus a major difference between the Eurasian and the American human populations. In Eurasia the agricultural revolution involved a closer association between animals and people. Over time diseases that passed from animals to humans, such as influenza and measles, were less fatal, as those fatally susceptible died out, leaving behind a more resilient human population. This process did not occur in the New World, making its indigenous inhabitants fatally susceptible to everyday diseases from the Old. Diseases such as influenza, measles, and smallpox proved fatal to the people of the Americas. Soon after contact, indigenous populations collapsed as fatal diseases spread quickly with devastating effect. "Great was the stench of the dead," noted a 1571 report of an epidemic in Guatemala. In the northwest region of Guatemala, the population fell from 260,000 in 1520 to 47,000 in 1575. Across the entire continent, from 1492 to 1640, the indigenous population was reduced by close to 90 percent. Later European colonists arrived in a land largely emptied of its original inhabitants. The wilderness that nineteenth-century observers noted—the vast forests, empty plains, and abundant animal life—was not a pristine landscape but an environment of much reduced human impact. Geographer William Denevan deconstructs the "**pristine myth**" that North America was a vast wilderness before the coming of the Europeans. The later colonists encountered not what they thought of and described as a "natural" wilderness but what was in fact a regenerated wilderness, the product of a dramatic decline in the indigenous population and its environmental impact. The American landscape of the eighteenth and nineteenth centuries was not an unchanging wilderness but the result of the population collapse of the sixteenth and seventeenth centuries. The Columbian Encounter did not reveal the American wilderness but created it.

Societies develop equilibrium when demographic trends stay the same or exhibit only slight increase or decrease. Change, even large change, if it develops slowly, can be easily incorporated into social institutions. But when population levels fall drastically, the resultant ruptures and tears in social norms and social relationships produce dramatic changes and long-term unfolding consequences. All three pandemics had dramatic effects. The rise of Western Europe, the creation of modernity, and the decline of American indigenous people all partly originate in rapid population loss.

Climate Change

The development of agriculture made human population more dependent on **agricultural productivity**. A decline in agricultural output could increase mortality and lead to fewer births. The decline in food supply occurred especially when climate change reduced the length of the growing season. Patrick Galloway noted a connection between population change and fluctuations in climate change for the mid-latitude regions of the world. Across these regions there was a warming trend, peaking around 1200 CE, followed by cooling until 1450; then warming until 1600, followed by a rapid cooling until around 1690. These changing temperatures caused a decline in agricultural productivity—colder weather reduced harvests and increased the risk of hunger and starvation—that suppressed population growth. In England, for example, during the cooling period of 1600 to 1650, population levels declined because, with lower food supplies, women had fewer children.

Harry Lee and colleagues, testing Galloway's model, examined the data for China over the past thousand years. They found that five major population contractions occurred during cooler temperatures. Long-term cooling led to falling harvest yields, fewer births, and increasing mortality rates.

Historian Geoffrey Parker extended the analysis to the global level. He pulled together a vast array of material to consider the political implications of global climate change in the seventeenth century. From around 1640 to 1690 the climate was appreciably colder, at a time when most people were dependent on agriculture. This **Little Ice Age** led to harvest failures, food shortages, and hunger. A third of the world's population died and the unfolding consequences rocked the political foundations. From China to England there was an unparalleled outbreak of revolts and revolutions as people rebelled against and resisted an existing order unable to cope with the crisis.

The Demographic Transition

Until 1800, global demographics were characterized by short life expectancy, high fertility, a young population, and slow population growth. In 1800 the average lifespan was only twenty-seven years, most fertile women gave birth to six children, one in three people were less than fifteen years old, and the annual population growth rate was around 0.51 percent. People's lives were short, women spent much of their time having and raising children, and there were very few old people.

Around 1800, changes in mortality and fertility, known as the **demographic transition**, altered this global pattern.

The transition has two dominant trends. First, there was the decline in **mortality rates**, initially in the richer parts of the world, such as Western Europe and North America, brought about by improved personal hygiene, public health measures, and increased affluence, which led to better diets. These all combined to reduce death rates and lengthen **life expectancy**. More people lived longer. A benign cycle was created in which increased affluence led to more food being available, which led to longer and more productive lives, which in turn led to increased economic development. In 1950 the average life expectancy in Japan was 63.5 years; by 2015 it had reached 84.7. Lower-income countries have taken longer to increase life expectancies. Zimbabwe, for example, still has an average life expectancy of only 55.6 years. Malaria, largely eradicated elsewhere, annually sickens more than 200 million people and kills close to 800,000 children in Africa, where it is the leading cause of death for those under five years old. There are still global disparities in health care provision that have led to marked spatial differences in child mortality and overall life expectancy.

Second, the increase in life expectancy was soon followed by a decline in **fertility,** at first only in the richer countries and then later more widespread. In 1800 there were an average of 6 births for every woman; by 2015 the figure was 2.3. This is very close to the **replacement rate**, or the number of births needed to keep the population stable. The replacement rate is 2.1 in richer countries and 2.3 in poorer countries. The difference is the result of increased juvenile mortality rates in poorer countries.

Many countries moved toward lower birth rates between 1890 and 1920. Today, in rich countries like Japan, the birth rate is 1.4. In the United States, it is 1.9. Many factors were involved, including access to more reliable birth control methods, changes in household economies such that children were not so much workers as dependents, and increased investment in an individual child's caring and rearing. Family size declined. In the lower-income countries the fertility rates began to drop later, beginning in the 1960s. They are still higher than in the richer countries—6.3 births per woman in Mali, for example—but across the world, there is a transition toward lower birth rates.

This demographic transition has transformed the human population. Life is now longer as life expectancy increases and mortality rates decline. There is also a steady graying of the population. In 1950 only 4 percent of the total global population was over sixty-five; by 2000 this had increased to almost 6 percent. By 2100 it is estimated to be 21 percent. The world's population is aging, a trend especially marked in the richer countries of the world.

These demographic changes have cultural consequences. Consider, for example, gender roles and representations. Masculinity was often constructed around the image of the strong, virile body. But what of masculinity

The demography of Ireland is an interesting one. The graph of its demographic history, like most of the rest of the world, shows a steady increase from 1700 with acceleration from 1800 until it peaks at over 6 million in 1840. After the Great Famine of 1845 to 1849, the population plummeted as 1 million died and 2 million more left the country, leveling off at around 4 million. The country was slow to pass through the demographic transition, with 3.5 births per woman even into the early 1980s. High levels of emigration to Australia, Britain, and the United States kept the population level down. Then two things happened. First, after 1979 contraception was readily available, and the birth rate fell sharply. The demographic dividend of increasing working-age population was a factor in accelerated economic growth, which reached almost 6 percent annually in the 1990s. Ireland became the Celtic Tiger of rapid economic and demographic growth. Second, as the economy grew stronger, providing more employment opportunities, fewer young people left the country's shores, and indeed there was increased immigration, with Poland being a rich source. So while birth rates fell, the population increased. There was also a marked increase in female participation rates.

Since 2008, Ireland's economy has been badly hit by the global recession and risky investments by banks that were underwritten by the government, which in turn led to fiscal crisis, deep cuts in public spending, and a decline in economic growth. Immigration has fallen off, emigration has increased, and Ireland has joined Portugal, Greece, and Italy as ground zero for the European debt crisis.

REFERENCE

Bloom, D. E. and Canning, D. (2003) Contraception and the Celtic Tiger. *Economic and Social Review* 34: 229–247.

when men live long past their peak physical strength and virility? Similarly, as birth rates drop, women are freed from constant childrearing. As their lives extend beyond their fertility, we need new models of what it means to be woman. Family size declines, so that large families are now replaced by smaller, nuclear families. The family is less a large extended group of people than a tight nexus of a few individuals. Family life takes on a new hue from how it was long imagined and lived. And what of the elderly? When they were a small group, they were revered and often subsidized. But when this group grows in size and lives long past its economic productivity, they can become a fiscal liability as well as an important resource of accumulated knowledge and wealth. With a rapidly aging population, **intergenerational inequities** and conflicts can become more pronounced, just as intergenerational transfers of wealth can become more important.

Phases of the Demographic Transition

There is a chronology and a geography to the demographic transition. In the richer countries, the transition is almost complete. Life expectancy has increased, and birth rates have dropped. In poorer countries, while life expectancy has increased, the birth rate has declined more slowly. The result is a complex pattern of different demographic regimes in countries at different phases of the transition.

Figure 3.2 highlights five distinct phases of the demographic transition. Figure 3.3 depicts the population pyramid of four individual countries at different stages in this transition. The pyramids are the graphic representation of the national population along the vertical axis in five-year age cohorts, with males to the left and females to the right of the vertical line through the middle of the pyramid.

The First Phase

The first phase is one of high birth and death rates and stable slow population increase. This was the pattern across most of the world prior to 1800.

The Second Phase

The second phase is associated with high birth rates, falling death rates, and—because death rates are falling faster than birth rates—a very rapid increase in population. Countries with this profile include Afghanistan, Uganda, and Zambia. In Uganda, for example, with a population of 38.8 million, population growth is around 3.4 percent, driven by the high average birth rate of 5.9 births for every woman. The population pyramid depicted in Figure 3.3 shows the typical pattern of a large youthful base, because of the high birth rates, tapering off quickly because of the relatively low life expectancies. Birth rates are high

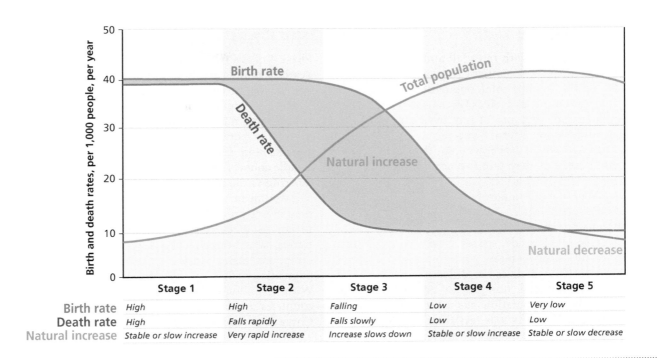

3.2 Phases of the demographic transition

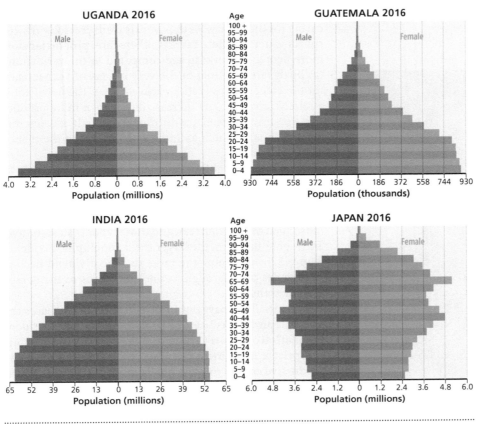

3.3 Population pyramids

because there is little effective contraception and cultural mores inhibit its use. High infant and child mortality rates may encourage parents to have more children so that at least one or more survive into adulthood. The net result is a very young population with high growth rates. Almost half of the population of Uganda is younger than fifteen. There are also considerable **gender inequalities** in income and job opportunities, since a woman's role is largely defined as giving birth to many children. The challenge for Uganda, as for all second-phase countries (which now constitute 9 percent of the world's population), is how to slow down the rate of growth to provide enough jobs and opportunities for a young population coming on to the job market and to extend the economic opportunities for women.

The Third Phase

In phase 3, birth and death rates fall and population increase begins to slow. Countries in phase 3 constitute 7 percent of the world's total population and include middle-income countries such as Ghana, Guatemala, and Iraq. In Guatemala, because of improved public health, life expectancy is now seventy-one years. The birth rate has declined to three births per woman. The decline is due to increased use of modern family planning methods and the desire for smaller families. Almost half of all women in Guatemala use modern family planning methods. The population growth rate is still high at 2.6 percent, and more than half of the population is aged below nineteen, so Guatemala, like Uganda, shares the problems of finding opportunities for its many young people. A distinctive feature in Guatemala is the difference between the indigenous Mayan and non-Mayan populations. The indigenous Mayan population has a birth rate of over five per woman, whereas it is less than two for richer, non-Mayan women. National population statistics can hide marked differences in income, race, and ethnicity (Figure 3.4). In the case of Guatemala, there is a difference between the indigenous Mayan population, which is more engaged in peasant agriculture, where children are important contributors to family income, and the non-Mayan, which tends to be more affluent and less reliant on children's labor.

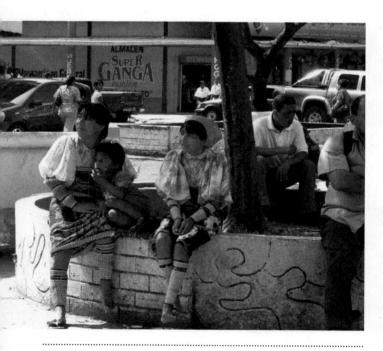

3.4 Throughout Central and South America the birth rates for indigenous peoples is much higher than for nonindigenous peoples. An Embera family in Colón, Panama.

The Fourth Phase

Phase 4 of the transition is associated with low birth and death rates and a stable or only a very slow increase in population. Countries in phase 4 now constitute 38 percent of the world's population and are characterized by sharply falling birth rates of below three per woman. Figure 3.3 shows the population pyramid of India. Compared to Uganda and Guatemala, there is a thickening in the middle cohorts, from ages twenty to fifty, and less marked tapering as the population ages. India has seen a remarkable decline in birth rates from over 5 births per woman in the 1950s to around 2.4 by 2015. There are marked regional variations. Birth rates per woman in the southern state of Kerala have dropped below 2, while they are still high in the poorer states of the north. In Uttar Pradesh, with a population of 200 million, the rate is 4 births per woman. With 1.3 billion people, India is one of the most populous countries in the world. In ten years' time, it may overtake China as the most populous country on the planet.

The Demographic Dividend

For countries at phase 3 moving into phase 4, there is the possibility of a **demographic dividend**. The dividend occurs when birth rates fall substantially, requiring less investment in the very young, and before the population starts to age dramatically, requiring more money spent on caring for the elderly. As the cohorts from previous growth spurts enter the job market, there are proportionately more people of working age compared to the very young (as birth rates decline) or the very old (as life expectancy for the older groups is still relatively low). The net effect is a relative and absolute increase of younger, more productive workers. Between one-third and one-half of economic growth in states such as India may be due to the demographic dividend.

One way to highlight the demographic dividend is to estimate the **dependency ratio**. One measure is calculated by dividing the working age population, those aged fifteen to sixty-five, by the nonworking age population, those aged zero to fourteen and those older than sixty-five. The higher the ratio, the larger the demographic dividend. Figure 3.5 shows how the ratio and the dividend decline as societies pass through to the later stages of the demographic transition.

The dividend pays out in several ways. It increases the labor supply and reduces the relative size of the **dependent population** of the very young and the very old. More public and private investment can thus be devoted to increasing productive capacity as less public money is spent supporting the very young and very old. Working-age adults tend to save more, and this can create a larger pool of capital available for investment. The human capital of the country is increased, because with many fewer children, women

3.3 Brazil and the Demographic Dividend

At the time of its first official census in 1872, the population of Brazil was only 10 million. In 2016 it is estimated at 209 million. After a century of rapid demographic growth, the country is undergoing a major demographic transition. Birth rates are dropping; the number of births per woman declined from 6.3 in 1960 to 1.8 in 2016. This remarkable fall is a product of many things, including rapid urbanization, the secularization of Brazilians, and the increasing use of contraception. The religious codes of the Catholic Church no longer exercise much influence. Popular soap operas with their emphasis on modernity play a hugely influential part in creating new gender roles and generating alternative models of family life and size. Better access to reproductive health and increasing female participation are part cause and part effect of the rapid fall in birth rates. It is not only the affluent who want to limit family size. The causes and effects feed off each other. Economic growth makes it possible for poorer families to have fewer children, which means that more can be spent on each child's education, which in turn means children are better educated and more able to find employment.

Brazil has experienced a demographic dividend since 1970 as the rapid fall in fertility has reduced the absolute and relative proportion of dependents. From 1970 to 2015, more than half of all economic growth could be explained by the dividend, which will be in operation until 2025. On the basis of this economic growth and increased democratization, recent decades have witnessed the creation of relatively generous social welfare programs of social security and a public education system. However, as the increasing life expectancy grays the population, the number of beneficiaries to workers will increase. Mortality rates have also fallen dramatically to the extent that life expectancy has increased from 53.7 in 1960 to 73.8 in 2016. While the social welfare programs reduce inequality and promote economic growth, they also will be a significant source of political debate as the population ages. How to balance old-age security and income equality with sustaining economy growth will become a significant policy issue for Brazil and the many other countries moving into the latter stages of the transition.

are able to join the workforce, and because family income is spread across fewer children, more can be invested in each child's education.

Part of the economic growth of countries such as the USA in the 1950s and 1960s and, later, South Korea was based on this demographic dividend. Even more recently, the rapid economic growth of countries such as Brazil and China in the past twenty years is based on the demographic engine of the absolute and relative increase in its working population. In these two countries, the proportion of the population aged between fifteen and sixty-five—generally considered the most economically active age range—is 69 and 74 percent, respectively. In poorer countries such as Afghanistan and Cameroon, the respective figures are 52 and 54 percent. Both these countries have also seen significant emigration as people search for economic opportunities beyond the national borders.

The opportunity for this demographic dividend is relatively short, because increasing life expectancy soon ages the population, quickly reducing the ratio of workers to nonworkers. There is a short window of

time when there is a rising share of people of working age and the demographic profile is weighted toward the most productive. During this time, holding everything else constant, average per capita income rises, both reflecting and prompting enhanced economic growth. A full dividend is only possible if female **participation rates** are high. If societies continue to restrict the employment opportunities of women, the full benefits of the dividend are unrealized. One

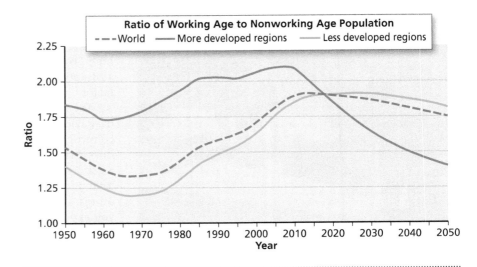

3.5 Ratio of working to nonworking age

reason behind China's spectacular growth is not simply the demographic dividend but its full implementation. Female participation rates are very high, at 94 percent for women aged twenty-five to thirty-four. The dividend is fully realized if the workforce is educated and skilled and the benefits of the growth are spread across the population rather than garnered by only a few. The tigers of rapid economic growth, such as Japan and South Korea from 1960 to 1990 and China since 1990, capitalized on their demographic dividend by increasing employment opportunities fast enough to maintain labor productivity growth.

The demographic dividend may provide a supply of relatively cheap labor, but only for a limited period. The **Lewis Turning Point**, named after economist William Arthur Lewis, occurs when the supply of cheap labor runs out. The increased demand for labor pushes up wages, and benefits are increased. Wages rise, profits may fall, and the economy loses its cheap labor-cost advantages. A vital question for China, with its below-replacement birth rate, is whether it is reaching or passing through a Lewis Turning Point.

From 1980 to 2010, China's spectacular growth was in part a function of the high proportion of working-age citizens between 20 and 65 years old. However, with declining birth rates and extended life spans, from 2010 to 2013 the proportion of people aged 65 and over will triple from 8 to 24 percent. By 2040 China will have more people aged over 65 than the USA, and the decades of rapid growth may end as the demographic dividend fades.

The Youth Bulge

In the early and middle phases of the demographic transition, a **youth bulge** occurs (Figure 3.6). This is the rapid increase in the number of people aged between fifteen and twenty-four. It occurs when a rapid reduction in child mortality occurs before a rapid falloff in fertility. In ten years' time, the increase in the number of children under fourteen becomes the youth bulge of fifteen to twenty-four. When this bulge constitutes more than 20 percent of the population, it has been linked to an increase in political instability and the possibility of increased political violence. Two political geographers, Gary Fuller and Forrest Pitts, looked at the emergence of this youth cohort in South Korea in the 1970s and 1980s and its correlation with the rise of political unrest.

A number of social scientists have sought to test the connection between youth bulges and **social conflict** in more detail. Using data from 1998 to 2005 covering 127 countries, Alfred Marcus and colleagues found that youth bulges are correlated with violent conflict. They also found that violence did fall off when the youth bulge was followed by a decline in the youth cohort. Henrik Urdal also tested this idea against empirical data. He found that an increase in the youth bulge of 1 percent increases the likelihood of domestic armed conflict in a society by 7 percent points. Countries with particularly large youth bulges are three times more likely to experience serious armed conflict. Economic stagnation plays an important role in turning a youth bulge into a tinderbox of armed conflict. In a more detailed study, Urdal looked at fifty-five large urban centers in Asia and sub-Saharan Africa for the 1960–2006 period. The male youth population bulge did not increase levels of social disorder. A more important factor was youth exclusion from economic opportunity, even in the absence of extraordinarily large urban youth-population bulges. It is not simply the number of youth but their range of opportunity for active and meaningful participation in economic and political life that influences political conflict.

Across the Middle East and North Africa, the proportion of fifteen- to twenty-four-year-olds is 21 percent of the total population and 34 percent of the working-age population. The unemployment rate among this group is a very high 26 percent. The Arab Spring had many causes, but a disenfranchised youth bulge was certainly a significant factor.

The relative and absolute size of specific cohorts does have wider social implications, even for the more affluent countries. The **Easterlin hypothesis** is named after economist Richard Easterlin, who suggested that the fortunes of a cohort depend on the size of that cohort relative to the total population. There are luckier times to be born than others. If you were born in the United States in the 1930s, you experienced a demographic trough—only 18.7 babies were born for every 1,000 people, compared to 29.5 in 1915 or 24.1 in 1950. Holding everything else constant, you had smaller class sizes, easier access to college, and more employment opportunities. While we imagine our economic

3.6 A youth bulge can occur at the middle stages of the demographic transition. Young people in Tunis, Tunisia.

success or failure to be a result of personal characteristics, it is also a matter of demographic luck.

Demographic troughs also have wider social effects. Crime rates, for example, showed a marked decline across several countries, including Brazil, Canada, and the United States, as the percentage of the population aged fifteen to twenty-four declined. This occurred in Canada and the United States in the 1990s and in Brazil in the early 2000s. Very different types of societies at different times all show the falloff in crime rates as the percentage of the fifteen-to-twenty-four age group declined. Demography is behind the fall in crime rates, as the cohort more likely to commit crime diminishes in numbers and relative weight.

The Fifth Phase

The final stage of the demographic transition is marked by very low birth and death rates and stable or decreasing population. As countries become more affluent they tend to move to the fifth phase.

In fact it is more accurate to see the transition as a continuum along which countries are moving, sometimes slowly, sometimes very quickly. Rather than a strict categorization, we can think of the phases as way stations where countries may be only temporary residents. Take the case of the most populous nation in the world, China, which is rapidly shifting from phase 4 to phase 5. Since the Communist takeover, death rates have declined markedly and birth rates have been tightly controlled by a one-child policy. The birth rate is now 1.4 per woman, below replacement and much less than the USA. Population growth has declined to 0.7 percent per annum, and there are now problems with labor shortages. The working population, those aged between 15 and 65, has declined each year since 2012. Fear of long-term labor shortages forced the government in 2015 to lift its decades-old one-child policy.

Countries in this phase constitute 46 percent of the world's population. The population pyramids of these countries, like those of Japan shown in Figure 3.3, have a narrower base because of low birth rates and less obvious tapering because of longer life expectancies. There are fewer young people compared to the increasing proportion of old people. Japan has one of the highest life expectancies and lowest birth rates in the world. The world's average life expectancy is seventy, while in Japan it is eighty-four. With only 1.2 births per woman, Japan is falling below replacement levels. A similar trend can be noted for other East Asian countries, including South Korea (1.2 births per women), Taiwan (1.1), Singapore (1.2), and Hong Kong (1.1) (Figure 3.7). In Japan the result is a declining population, a trend that is particularly marked in rural areas of the country. Much of rural and small-town Japan has been effectively depopulated as young people move to the larger cities for jobs and services. Across the nation, the population is aging. Almost one in four Japanese are

3.7 In richer societies with low birth rates, children are often the recipients of significant public and private investment. Schoolchildren in South Korea.

over sixty-five. Each year 1.3 million die in Japan but only 1 million babies are born. With limited immigration the population total has started to fall, so that by 2100 it is expected to lose 34 percent of its total population as the total drops from 126 million in 2015 to 83 million.

One major reason behind the decrease in fertility in Japan, and throughout East Asia and elsewhere, is that women of reproductive age are not getting married and not having children. Because of job discrimination, only about one-third of Japanese women remain in the workforce after having a child, compared to two-thirds of women in the United States. Because of the cultural mores against out-of-wedlock children and the heavy pressure to look after their husbands, for most Japanese women the choice is a stark one: stay single and have a career and financial independence, or get married and devote oneself to husband and child. The percent of Japanese women who have opted to remain single has doubled over the past twenty years. Over the coming decades, the proportion of working-age Japanese will continue to dwindle. This situation is the opposite of the demographic dividend: it is the demographic deficit of a declining workforce and an increasingly dependent population. The Japanese workforce will decline by 70 percent by 2050. In countries with a similar age profile, immigration from outside the country fills some of the job vacancies. In Japan, there are strict restrictions on foreign immigration. Less than 2 percent of the population is foreign-born, compared to the United States, where the figure is 12 percent, or Australia, where it is 22 percent. Japan will experience a major crisis when its anti-immigration posture crashes against the reality of its demographic deficit. Another solution to this impending

3.4 Russia's Changing Population

In 2010 the Russian government offered the equivalent of $9,000 in cash to have a second child, evidence of a pronounced demographic shift. At the end of the Soviet era, the population of Russia was around 148 million, having risen steadily from a 1950 population of around 102 million. It fell to around 138 million in 2011 and then recovered to 143 million on 2015. What is behind this remarkable collapse and more recent resurgence?

Russia experienced increased mortality rates due to a decline in life expectancy for males; it was sixty-five in 1989 but fell to a staggering fifty-nine by 2005. Life expectancy for females is seventy-two, the disparity a direct effect of heavier alcoholism rates among men compared to women. The decline of the old Soviet system with its extensive welfare programs and its replacement, especially in the earlier years of reform, by a harsh market-driven approach initially led to a decline in general public health provision, increased unemployment, and lower living standards. This economic shock therapy led very quickly to a decline in general social welfare and an increase in mortality. The number of deaths increased from 1.58 million in 1989 to 2.36 million in 2003 as the death rate rose from 10.7 to 16.4 per thousand. There was also a decline in birth rates from 1.8 in 1989 to a historic low of 1.1 in 1999, well below replacement level. There were 2.1 million births in 1989 but only 1.4 million in 2003. More Russian women, with very easy access to abortion, simply limited the number of children they were willing to have. Economic uncertainty and social dislocation made having children a more difficult choice, and with fewer males there were not enough partners with whom to start or raise a family.

The population decline occurred because more people died and fewer children were born. Immigration was tightly controlled and kept low, so the foreign-born did not contribute very much to population increase.

There is now some indication of demographic recovery. The male life expectancy rate has recovered; it is now back to sixty-five, still low in comparison with Western Europe and North America but improving. Birth rates increased from 1.1 in 1999 to 1.7 in 2015. The figures reflect increasing living standards as a commodity boom, especially in gas and oil, lead to more personal affluence and greater public spending on health, which in turn lead to higher life expectancy and more births. The case of Russia highlights how population growth and decline embody wider economic and political changes.

crisis would be to make it easier for women with children to enter the workforce. Again, however, this policy goes against traditional cultural norms. In 2011 a government think tank in South Korea, the Korea Development Institute, made the case for reducing the social taboos in the country associated with out-of-wedlock pregnancies, as one way to counter the declining fertility. With a fertility rate of 1.2 births per women, South Korea ranks 217th out of 222 countries in fertility. As the number of single women increases, easing legal safeguards of children born outside of marriage is one way to counter a falling population.

At current rates, the world's over-sixty-five population is likely to double in fifty years. Tod Fishman uses the phrase "shock of grey" to refer to the reality of rapidly aging societies. The old will become an increasingly large part of the population in both absolute and relative terms. As people live longer, what it is to be old is redefined. The over-sixty-five cohort is growing so large and so quickly that new categories are being developed: the young-old, aged sixty-five to seventy-four; the old-old, aged seventy-five to eighty-four; and the oldest, aged eighty-five and above. We can also reconceptualize old age in a more positive light as an age of personal fulfillment rather than one of degeneration and dependency. People previously considered old now live longer and more engaged lives. And with more elderly, there are new market niches, such as retirement communities. Some residential developments in the United States, for example, have minimum age requirements as the affluent elderly sequester themselves into homogenous communities.

Another form of demographic dividend is the increase in the number of older persons who are able to share their life assets with the younger generation. This is an important phenomenon in countries with strong extended family ties on one hand and limited social welfare programs on the other. Intergenerational transfers of wealth and income from an older generation to a younger generation are an important part of economic growth in East and Southeast Asia and Latin America. In South Africa, Monde Makiwane conducted fieldwork on these transfers and concluded, "Most of the elderly in South Africa use their meager old age pensions to support unemployed children and orphaned and vulnerable children."

This fifth phase of the demographic transition has developed in some richer countries where it is associated with very low birth and death rates, more pronounced aging, more reliance on immigration to fill job vacancies (especially at the lower wage levels), and a disconnect between marriage and children. More children are born out of wedlock and more couples do not have children.

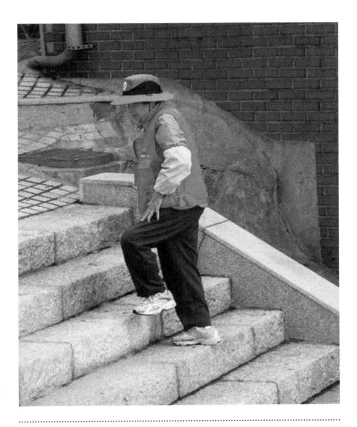

3.8 The elderly will double in population over the next fifty years. An elderly lady in Inchon, South Korea.

TABLE 3.1	Demographic Differences		
COUNTRY	**POP AGED < 15 (%)**	**MEDIAN AGE**	**WORKING POPULATION (%)**
Uganda	48	15	50
Guatemala	35	21	55
India	31	27	64
Japan	13	61	46

Median age is the age that divides the country's population into two equal halves. Working population is defined as the percentage of the population aged greater than 15 and less than 64 .

examined population pyramids. Some of their basic demographic differences are summarized in Table 3.1. In terms of the relative distribution of the population, note how countries at the earlier stages of the transition have a higher proportion of those aged below 15 and a corresponding young median age. In contrast, a country such as Japan, at the later stages of the transition, has fewer young people and more older people. Countries such as India are in the middle of the demographic dividend with the highest percentage of people of working age.

Problems and Opportunities of the Demographic Transition

At a very fundamental level, the demographic transition is an enormously life-enhancing phenomenon. People now live longer, healthier lives. The crushing tragedy of infant mortality and the lost opportunity of short lives, if not eradicated, are much reduced. People get to enjoy the gift of life for longer and in a healthier state, and they lead more productive lives.

Each stage of the demographic transition creates both problems and opportunities. In phase 2 and early in 3, as death rates decline, more people live longer, healthier lives. However, the still high birth rate means that a society at this stage has to cope with a very youthful population with a limited number of people in the most economically productive cohorts. There is the potential instability of a youth bulge.

In the middle phases of the demographic dividend, there is an increasing proportion of the economically active population. This demographic dividend can turn into a demographic time bomb, however, if economic

A variety of household living arrangements occur as traditional marriage is no longer such a dominant model of household formation. These trends are especially marked in the countries of Northwest and Eastern Europe. In Norway and Sweden, 54 and 55 percent of all births, respectively, are to unmarried women. The figures reflect increasing cohabiting between couples rather than marriage and more single-parent households, in part the result of choices of women and their families. These choices may have some negative economic consequences in some countries. In the United States, almost 41 percent of all births are to unmarried women. One in three households that experience child poverty are single-parent, female-headed households, and only 6.4 percent are from married, two-parent families. In many northern European countries, in contrast, children born out of wedlock are generally born to stable, cohabiting couples or to those with access to more generous social welfare and public health programs.

Summary of Transition

By way of a summary, let us return to the four countries that we considered earlier in the chapter when we

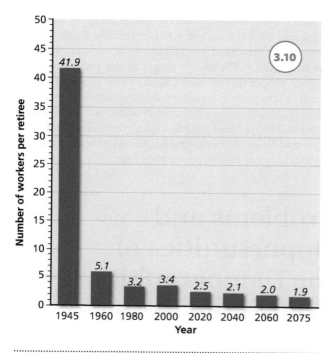

3.9 As the elderly become a larger proportion of the population, retirement communities like this will become more important features of the urban landscape. **3.10** Number of workers to retirees in US Social Security system

people—the median ages in Egypt and Tunisia are twenty-five and thirty-one, respectively, compared to forty-six in Germany—there is the possibility, in association with anemic economic growth and sclerotic political systems, of social unrest. Other things being equal, younger people more easily take to the streets than older people. And if the demography is weighted toward the young while the economy is geared toward the more elderly, the possibility of unrest rises.

In the later phases of the transition, birth rates drop and life expectancies stretch out the length of the average life. There are more people with the advantage of a long life experience. A society with more of these people should be wiser, more able to take the sage, long-term perspective. Again opportunity and crises are possibilities at the later stages of the transition. On the one hand, there is the possibility of a second demographic dividend as the elderly, able to amass a long lifetime of assets, pass on their wealth to their children and younger family members. On the other hand, there may be little dividend if the elderly hang on to their assets and continue to receive and demand public benefits.

At the later stages of the demographic transitions, there is an increase in the elderly nonworking population. We can follow one of the consequences by looking at Social Security in the United States, a social insurance fund that current workers pay into to support retired workers. When it was first introduced, there were many more workers paying into the system than recipients receiving benefits. In 1950, for example, the number of workers per beneficiary was 16.4. In 1960 the number of workers per beneficiary was 5.1, by 2013 it had fallen to 2.8, and it is likely to fall to below 2 by the end of the century. Figure 3.10 highlights the dramatic shift. In 1950 48.2 million workers supported 2.9 million beneficiaries, but by 2013 the respective numbers were 163 million workers and 57 million beneficiaries. As more retirees live longer, they are supported by fewer workers. Over the longer term, this may be fiscally unsustainable. In social democracies with generous social insurance programs, the aging of the population creates a fiscal crisis for the welfare state. It can lead to generational inequalities, as fewer workers have to support more retirees.

Generational inequity really kicks in if those paying for the elderly are unlikely to see the same benefits. In the United States, for example, older adults are advantaged because they have publicly provided pensions and healthcare that are unlikely be so generous to future generations. The net worth of a typical US household member aged over sixty-five is forty-seven times that of one younger than thirty-five. The disparity was ten to one just twenty-five years ago. Over a quarter of a century, there has been a relative shift of wealth from the young to the elderly and the creation of an intergenerational divide in life chances and benefits. Around the world, in countries with social

growth does not keep up with the number of people coming into the job market. An increase in the working-age population can produce a demographic dividend but also raises the possibility of social unrest if the army of young people fails to find gainful and rewarding employment. It is too simple to assign the Arab Spring, for example, only to the large number of young people in the societies without job prospects—there are other countries with similar conditions that did not experience social uprisings—but it was an important demographic component in the content of long-term totalitarian regimes, marked inequality, and stagnant economic opportunities for the majority. When there are a large number of young welfare programs, the aging of the population creates fiscal tensions and the possibility of intergenerational conflicts. Class, race, and gender have long been identified as sources of difference, advantage, and disadvantage. We need to add to that list both age and the cohort's position in the demographic transition.

The shift toward a more elderly population can have major implications for continued economic growth, as without younger workers the number of dependents will exceed the number of productive workers and there will be increasing pressure on generous social welfare systems. In some rich countries, there is now a greater reliance on foreign immigrants to fill job vacancies. However, this can create tensions as the issue of citizenship, for example, becomes more problematic with more "foreign" workers. As welfare costs mount, there is increasing concern with noncitizen access to generous welfare arrangements. The economic need for more immigration can sometime clash with a cultural debate about national identity and emerging political debates on citizenship.

There is a distinct global geography to the demographic transition, and it interconnects with levels of economic development and rates of growth. We can identify three types of countries:

1. Graying countries, such as Japan and Russia, with aging populations.
2. Demographic and economic growth countries, such as Brazil and Thailand, with rapidly growing populations but developing economies, with the possibility of a positive demographic dividend.
3. Rapid demographic growth with limited economic growth countries, such as Afghanistan.

Youth revolts, crime levels, culture wars about the family, discussions about welfare reform and national identity: there is a demographic context to some of these interesting social changes. Demographic changes are at the heart of some of most profound cultural transformations and intense political debates occurring around the world.

Cited References

Denevan, W. (ed.) (1992) *The Native Population of the Americas in 1492*. Madison: University of Wisconsin Press.

Fishman, T. C. (2010) *Shock of Grey*. New York: Scribner.

Fuller, G. and Pitts, F. R. (1990) Youth Cohorts and Political Unrest in South Korea. *Political Geography Quarterly* 9: 9–22.

Galloway, P. R. (1986) Long-Term Fluctuations in Climate and Population in the Preindustrial Era. *Population and Development Review* 12: 1–24.

Lee, H. F. and Zhang, D. D. (2010) Changes in Climate and Secular Population Cycles in China, 1000 CE to 1911. *Climate Research* 42: 235–246.

Lee, H. F., Fok., L., and Zhang, D. D. (2008) Climate Change and Chinese Population Growth Dynamics over the Last Millennium. *Climate Change* 88: 131–156.

Makiwane. M. (2011) Intergenerational Relations in Africa with a Focus on South Africa. Paper presented at the United Nations Expert Group Meeting on Adolescents, Youth and Employment. Population Division, United Nations Secretariat. http://www.un.org/esa/population/meetings/egm-adolescents/p09_makiwane.pdf. Accessed November 7, 2011.

Marcus, A. A., Islam, M., and Moloney, J. (2008) Youth Bulges, Busts, and Doing Business in Violence-Prone Nations. *Business and Politics* 10 doi: 10.2202/1469-3569.1227. http://www.bepress.com/bap/vol10/iss3/art4. Accessed November 8, 2011.

Parker, G. (2013) *Global Crisis: War, Climate Change and Catastrophe in The Seventeenth Century*. New Haven and London: Yale University Press.

Urdal, H. (2004) *The Devil in the Demographics: The Effect of Youth Bulges on Domestic Armed Conflict 1950–2000*. World Bank: Social Development Paper No. 14. Washington, DC.

Urdal, H. and Hoelscher, K. (2009) *Urban Youth Bulges and Urban Social Disorder*. Washington, DC: World Bank.

Select Guide to Further Reading

Benedictow. O. J. (2004) *The Black Death, 1346–1353: The Complete History*. Woodbridge: Boydell.

Bongaarts, J. (2009) Human Population Growth and the Demographic Transition. *Philosophical Transactions of the Royal Society B: Biological Sciences*. 364: 2985–2990.

Cai, F. (2010) Demographic Transition, Demographic Dividend, and Lewis Turning Point in China. *China Economic Review* 3: 107–119.

Demeny, P. (2011) Population Policy and the Demographic Transition: Performance, Prospects, and Options.

Population and Development Review 37 (supplement): 249–274.

Dyson, T. (2010) *Population and Development: The Demographic Transition.* New York: Zed Books.

Goldscheider, F., Bernhardt, E., and Lappegård, T. (2015) The Gender Revolution: A Framework for Understanding Changing Family and Demographic Behavior. *Population and Development Review* 41: 207–239.

Laslett, P. (1989) *A Fresh Map of Life: The Emergence of the Third Age.* London: Weidenfeld and Nicholson.

Lee, R. (2003) The Demographic Transition: Three Centuries of Fundamental Change. *Journal of Economic Perspectives* 17: 167–190.

Lee, R. and Reher, D. (2011) Introduction: The Landscape of Demographic Transition and Its Aftermath. *Population and Development Review* 3: 1–7.

Lesthaeghe, R. (2014) The Second Demographic Transition: A Concise Overview of Its Development. *Proceedings of the National Academy of Sciences* 111: 18112–18115.

Little, L. L. (ed) (2006) *Plague and the End of Antiquity: The Pandemic of 541–750.* Cambridge: Cambridge University Press.

Lovell, W. G. (1992) "Heavy Shadows and Black Night": Disease and Depopulation in Colonial Spanish America. *Annals of Association of American Geographers* 82: 426–443.

Nordås, R. and Davenport, C. (2013) Fight the Youth: Youth Bulges and State Repression. *American Journal of Political Science* 57: 926–940.

Pablos-Mendez, A., Radloff, S., Khajavi, K., and Dunst, S. A. (2015) "The Demographic Stretch of the Arc of Life: Social and Cultural Changes That Follow the Demographic Transition." *Global Health: Science and Practice* 3: 341–351.

Pearce, F. (2013) Why the "Youth Bulge" Can Make or Break a Country. *New Scientist*, 219: 42–45.

Rosen, W. (2007) *Justinian's Flea: Plague, Empire and the Birth of Europe.* New York: Viking.

Strulik, H. and Vollmer, S. (2015) The Fertility Transition around the World. *Journal of Population Economics* 28: 31–44.

Thompson, V. and Roberge, M. C. (2015). An Alternative Visualization of the Demographic Transition Model. *Journal of Geography* 114: 254–259.

Tuchman, B. W. (1978) *The Distant Mirror: The Calamitous 14th Century.* New York: Knopf.

Websites

Check out the world's current population at
http://www.worldometers.info/world-population/

Data for demographic research
http://www.icpsr.umich.edu/icpsrweb/DSDR/

Global demography of aging
http://www.hsph.harvard.edu/pgda/

Interactive world population map
http://www.prb.org/Publications/Datasheets/2011/world-population-data-sheet/world-map.aspx#/map/population

Minnesota Population Center
www.popdata.org

Spread of the Black Death in Europe 1346–1353
http://en.wikipedia.org/wiki/File:Blackdeath2.gif

Population Reference Bureau
http://www.prb.org/

United Nations Population Data
http://www.un.org/esa/population/
http://www.un.org/popin/data.html

World Populations
www.WorldPopulationHistory.org
www.Worldof7Billion.org

Interactive Map Websites

http://ourworldindata.org/data/population-growth-vital-statistics/fertility-rates/

http://data.worldbank.org/indicator/SP.DYN.TFRT.IN/countries?display=map

Learning Outcomes

- The Earth began with a relatively small human population. Although it is estimated that no more than 5,000 individuals left Africa to populate the Earth, populations exploded in the nineteenth and twentieth centuries, reaching 7.5 billion inhabitants today.

- Three examples of major pandemics with far-reaching social effects include the bubonic plague, which impeded growth but led to technological advancements and the birth of Europe during the sixth through eighth centuries; the "Black Death," a recurrence of the bubonic plague that swept through Asia, North Africa, and Europe in the fourteenth century and killed up to half of the existing population; and the Columbian Encounter that took place when Europeans colonized the Americas and may have reduced indigenous populations in the Americas by up to 90 percent.

- Population levels are influenced by long-term climate change that impacts agricultural yields. The Little Ice Age of the seventeenth century caused a decline in population, increased hunger, and political crises.

- The demographic transition that began around 1800 resulted in a decline in mortality rates, improved overall health, and increased life expectancies and economic productivity.
- Among the factors that led to lower birth rates in the late nineteenth and early twentieth centuries were access to more reliable birth control methods and changes in household economies such that children were not so much workers as dependents.
- The major demographic change of today is a shift toward an older population, especially in richer countries. In 1950 only 4 percent of the total global population was aged more than sixty-five; by 2000 this had increased to almost 6 percent. By 2100 it is estimated to be 21 percent. At current rates, the world's population aged over sixty-five is likely to double in fifty years.
- With an aging population, intergenerational inequities and conflicts can become more pronounced.
- During the demographic transition, richer countries experienced increased life expectancies and lower birth rates, while poorer countries also increased their populations' life expectancies but saw a slower decline in birth rates. The result is a complex pattern of different demographic regimes in countries at different phases of the transition.
- The four distinct phases of demographic transition include: (1) high birth rate and fluctuating death rate (about 9 percent of the world's population today); (2) declining death and birth rates (7 percent); (3) birth rates declining to below three per woman (38 percent); and (4) low birth rates and high life expectancy (46 percent).
- The demographic dividend, which results in a net effect of more younger and productive workers, pays off for countries as the labor supply increases relative to those who are dependent, more money is invested in production, and more capital is made available for investment.
- The strong economic performance and growth in the last 20 years in countries such as Brazil, China, and India is based on relative and absolute increases in working population.

- A full demographic dividend is possible only if female participation rates are high. If societies restrict women's employment opportunities, the full benefits of the dividend are unrealized.
- In the early and middle stages of the demographic transition, a youth bulge—the rapid increase in the number of people between the ages of 15 and 24—can appear. If this cohort increases beyond 20 percent of the population, social conflict may occur.
- Another demographic dividend from an older population is the increase in assets that are able to be passed on to younger generations, especially in countries with stronger family ties and limited social welfare programs.
- The fourth stage of demographic transition is associated with very low fertility levels, more pronounced aging, greater reliance on immigration to fill job vacancies, and a disconnect between marriage and children.
- Each stage of the demographic transition creates both problems and opportunities. Whereas the first state sees declining death rates, high birth levels lead to a higher proportion of unproductive members. The middle phases increase the proportion of economically productive individuals, if there are enough opportunities for working-age individuals. When birth rates fall and life expectancies rise in later stages, experience prevails and transferable assets can accumulate, but a more elderly population may hoard wealth and demand more public benefits.
- The economic need for increased immigration can sometimes clash with a cultural debate about national identity and emerging political debates on citizenship.
- A distinct global geography is created by the demographic transition, resulting in three types of countries: aging countries, demographic and economic growth countries, and demographic growth countries with limited economic growth opportunities.
- Demographic changes are at the heart of some of most profound cultural transformations and intense political debates occurring around the world.

Population and Food

4

LEARNING OBJECTIVES

Explain the historical context of Malthus's theory.

Understand Boserup's contribution.

Summarize the ways that the global food supply has kept up with world population size.

Discuss the lack of access to food, threats to food security, and limits to food production.

Reflect on the corporate industrialized food production system and associated concerns over food quality, health, and the environment.

Consider the unevenness of contemporary food production and consumption patterns.

In the last chapter we looked at population numbers and their change over time and across space. Absolute numbers are significant, but they take on extra meaning when we look at them relative to resources. In this chapter, we will consider how population connects with one of the most important resources, food.

Population and Agriculture

Before the advent of agriculture, the total world population did not exceed 10 million people, fluctuating between 5 million to 15 million and spread thinly across the surface of the globe. **Settled agriculture** allowed an increase in population because the food resource base was increased and made more reliable. With a specialized division of labor, effective utilization of crops and animals, and centralized political control, population could increase beyond the limits of **hunting-gathering societies**. By 2,000 years ago, the population had increased to 200 million. Improvements in agricultural technology, such as the use of the plow and hybrid crops, enabled more food to be produced, sometimes by extending the amount of arable land and sometimes by increasing the productivity of existing agricultural land. The areas of greatest population density had rich, fertile soils and reliable sources of water as well as centralized political control.

Malthusian Melancholy

The relationship between food and population has long interested social observers. One has left a permanent intellectual legacy. Thomas Robert Malthus (1766–1834) was one of eight children born to a prosperous middle-class family in Surrey, England. He is best known for his work *An*

Essay on the Principle of Population. He worked on six successive editions of this text from 1798 to 1826.

The historical context is important. Writing in the convulsive times of the French Revolution, Malthus was one of the many English reactionaries who were wary of radical change, distrustful of social progress, and skeptical of massive improvement and the idea of upward progress. His basic argument is that while population increases at a geometric rate, food supply grows only at an arithmetic rate. The end result is that there are more people than the available food supply. When things are good, people have more children, but eventually this population growth exceeds the food supply needed to feed them. Families respond to this dire situation by having fewer children, and so population decreases until things are back in equilibrium, when the cycle starts all over again. Human history is one of population increase overshooting the available food supply. Misery, poverty, and famine, the so-called **Malthusian checks**, are "natural" elements that bring population growth back into alignment with the food supply.

This simple model has policy implications. As a social conservative, Malthus was against social policy that disrupted the existing social order and its workings. He was highly critical of policies aimed to help the poor. Social welfare provision, Malthus believed, allowed the poor to have children and, for him and his followers, that simply made things worse. He believed that the number of children born to the poor should be limited.

His ideas influenced contemporary British politicians, who introduced the census in 1801 as a means of counting the population; and the New Poor Law of 1834, which launched workhouses for the poor, so familiar to the readers of Charles Dickens. His ideas filtered out to social theorists such as Herbert Spencer (1820–1903), who espoused **social Darwinism**, based on the idea of the survival of the fittest. Malthus's works were roundly criticized in the radical working-class journals of the day. In 1820 William Godwin calculated that, using China as his model, the world's population could comfortably reach 9 billion. He was writing at a time when the population was little more than 1 billion and few at the time could even imagine today's population of 7.5 billion.

Malthus's work was socially specific, written at a time of intense political debates and just as the demographic transition and the agricultural and industrial revolutions were taking off. Yet his ideas have proven very influential, echoing down through the subsequent years.

One continuing legacy is a pessimism regarding the ability to expand population beyond a fixed base because the food supply is considered relatively fixed or capable of only slow increase. Malthus presents a specter of population increasing faster than the supply of food to feed it. He was writing just as the agricultural revolution was transforming agricultural productivity. Since then, in the past two hundred years, agricultural productivity has increased remarkably, creating a much larger food supply base. In one important way, Malthus's gloomy predictions proved incorrect. The global population has increased to over 7 billion, and yet global hunger is not inevitable, even with a larger population. According to the Food and Agriculture Organization (FAO), the proportion of people in developing countries with food intakes below 2,200 calories per day fell from 57 percent in 1965 to just 10 percent in 1998. For those with access to food, the price has steadily declined by 60 percent in the last 40 years. The FAO also estimate that the number of undernourished people in the world fell from over a billion in 1992 to 792 million in 2015, even as the population increased from 5.4 to 7.5 billion.

We can also make a distinction between "undernourished," lacking enough calories to sustain a healthy state, and "malnutrition," a diet based on poor quality food lacking necessary vitamins and minerals to sustain good health. Later, we will discuss the problems of malnutrition, a growing burden in many countries that nevertheless have solved the issue of undernourishment.

The Danish economist Ester Boserup (1910–1999) reversed the Malthusian argument. Rather than agricultural technology determining population, it is population that determines agricultural technology. Necessity, in this case population pressure, is the mother of invention. In the history of agriculture, when population levels and densities are low, shifting agriculture can be employed, but when population levels and densities increase, field use becomes more permanent, which spurs agricultural innovations. These include irrigation, development of hybrid plants,

4.1 Grocery store in Budapest, Hungary. Local, national, and international food chains converge on this one store in Budapest to ensure fresh fruit and vegetables in January.

weed control, specialization in specific crops, and fertilizer use. Agriculture intensifies as population pressure mounts.

In one sense, at least, Malthus's prophecy of doom has failed to materialize, while Boserup's more optimistic view is sustained. Population has increased, yet so has global food supply (Figures 4.1 and 4.2). All the evidence suggests that world agricultural production can grow even more, in line with increased demand. Charles Godfray and colleagues looked at the ability of the global food production system to meet the demand of 9 billion people by 2050. They suggest that it is possible through a variety of methods:

- Closing the **yield gap**—that is, the difference between possible and actual productivity. In Southeast Asia, for example, there is a 60 percent difference between average and maximum rice yields. Yield gaps are larger in failed or dysfunctional states or in poor countries with limited access to the world markets or to easy and cheap credit. Political stability and world trade can increase yields by making it profitable for farmers to invest more.

- Increasing production limits. The **Green Revolution** successfully increased agricultural production throughout the world from the 1960s and 1970s by developing hybrid varieties of maize, rice, and wheat and increasing use of fertilizers and pesticides. Genetically modified crops can increase productivity. However, there is mounting public disquiet about the long-term implications of engineering our food supply and the

unfolding consequences of genetically modified elements in the food chain.

- Reducing waste. Around 35 percent of food in the world is wasted. The reason varies from lack of cold storage to "supersized" portions. Food is so cheap in some places that there is little incentive to avoid waste. Each year in the USA, over 70 billion pounds of food is wasted, up 20 percent from a decade ago.

- Changing diet. As societies get richer they tend to consume more meat. This involves growing grain to feed to livestock. Reducing the consumption of meat allows grassland for animals to be turned into arable farming. Reducing meat consumption also has a positive health benefit, since it reduces heart disease rates.

- Expanding aquaculture. Fish farming increases the amount of relatively cheap protein. There are, however, issues of environmental pollution and genetic contamination with an increase in aquaculture.

Despite the tremendous increases in agricultural productivity that have rendered Malthus's original predictions obsolete, problems remain. Let us consider four of them: hunger, sustainability, nutrition, and ethics.

Hunger, Famine, and Food Insecurity

The declining levels of undernutrition are testament to the ability of agricultural production to surpass the Malthusian limits. Yet hunger remains. Figure 4.3 shows the distribution of a global hunger. Notice the extreme concentration in tropical Africa.

The term "food security" is employed to refer to food's availability. **Food security** is when people have relatively easy access to safe and nutritious food. Food insecurity occurs when this does not occur, either on a permanent or transitory basis; it leads to hunger and undernourishment, making people more susceptible to disease and illness. Worldwide, it is estimated that there are approximately one billion without enough calories and a further one billion who are malnourished in a form of hidden hunger because they do not consume enough nutrients for a healthy life. Almost 250 million preschool children lack vitamin A, and 500 million women have an iron deficiency that causes 60,000 deaths each year during pregnancy.

We can make a distinction between the demand for food—a universal need, as we all have to to eat to remain alive—and the effective demand for food, which is the ability to pay for food. The issue is not the relationship between demand and supply but between effective

4.2 Food preparation and sale in Grenada. Women constitute the vast majority of food preparers and sellers at this weekly market in Grenada's capital city, St. George's, as they do in much of the world.

4.1 Overfishing

The oceans of the world have long been an important food source. Along the coast and deep in the ocean, fish stocks are protein-rich sources of food and nutrition. However, as industrial-style fishing with large boats and fish-detecting sonar now sweep the oceans, the supply of fish is endangered. Wilf Swartz and colleagues looked at fish catches from 1950 to 2005 and mapped the levels of exploitation. **Overfishing** in heavily exploited areas is highlighted in red. The global increase in the distribution of severe exploitation is vividly demonstrated in the comparison of the two maps. The global depletion of fish supply is one reason behind the growth of fish farming. Some wild species, such as the bluefin tuna, are on the verge of extinction due to overfishing. As the fleets go into deeper and deeper water, deep-sea fish that mature and reproduce slowly (such as the orange roughy, with a lifespan of 149 years) are particularly vulnerable. With nonselective fishing and a weak regulatory system, deep-sea fishing will soon exhaust the fish stock. Wild fish supplies peaked in the 1980s, and their global decline will put extra pressure on farming to replace the protein provided by the world's fisheries.

REFERENCES

Norse, E. A., Brooke, S., Cheung, W. W. L., Clark, M. R., Ekeland, I., Froese, R., Gjerde, K. M., Haedrich, R. L., Heppell, S. S., Morato, T., Morgan, L. E., Pauly, F., Sumaila, R., and Watson, R. (2012) Sustainability of Deep-Sea Fisheries. *Marine Policy* 36: 307–320.

Pauly, D. (2010) *Five Easy Pieces: How Fishing Impacts Marine Ecosystems.* Washington, DC: Island Press.

Swartz, W., Sala, E., Tracey, S., Watson, R., and Pauly, D. (2010) The Spatial Expansion and Ecological Footprint of Fisheries: 1950 to Present. *PLoS ONE* 5(12): e15143. doi:10.1371/journal.pone.0015143.

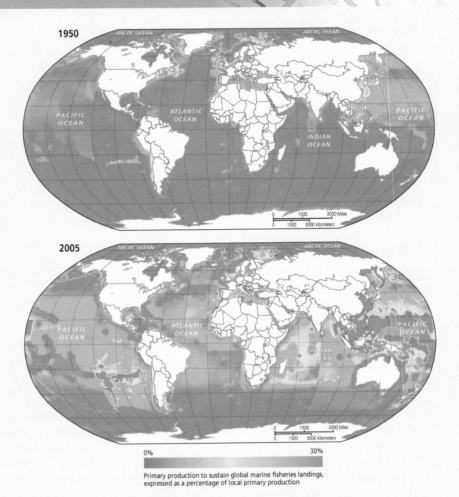

1950

2005

0%   30%

Primary production to sustain global marine fisheries landings, expressed as a percentage of local primary production

Top Fisheries exploitation, 1950 and 2005. **Bottom** On large commercial fishing boats in Gulf of Thailand, most of the fishermen are indentured Burmese workers.

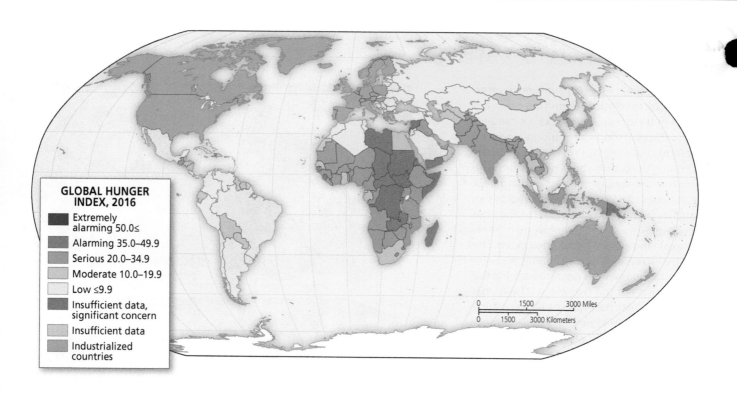

GLOBAL HUNGER
INDEX, 2016

Extremely
alarming 50.0≤

Alarming 35.0–49.9

Serious 20.0–34.9

Moderate 10.0–19.9

Low ≤9.9

Insufficient data,
significant concern

Insufficient data

Industrialized
countries

4.3 Global Hunger Index

demand and supply. Hunger occurs not only because there is not enough food but because people are too poor to buy food.

Lack of access to food is also a major cause of hunger. Extreme hunger such as **famine** regularly affects millions. Amartya Sen argues that famines are less the result of lack of food and more the collapse of the system that links the supply and demand for food. He drew upon the experience of the Bengal famine of 1943, in which 3 million people died. The famine did not occur because there was no food, but because wages did not keep up with growing food prices. Food price inflation, poor food distribution, hoarding, and poor government responses caused the collapse.

Three things cause famine: the redirection of food, the destruction of the productive capacity to grow food, and total neglect of the starving. Famines reflect lack of power more than lack of food. Historical geographer David Nally looked in detail at the famine in Ireland that began when the **Potato Blight** created multiple harvest failures beginning in 1845. More than a million people died, and 2 million left the land. He looked at the effect of the Irish Poor Law system developed prior to the famine, in 1838. Under one piece of the legislation, the Gregory Clause, tenant farmers on more than a quarter-acre holding could not receive public relief without leaving the land and entering the workhouse. Once tenants left, their dwellings were knocked down and the landholdings were consolidated.

The famine provided an opportunity for the British authorities to clear "excess" population from the land and rationalize the landholding system. A colonial state used the famine to advance its agenda of agricultural rationalization and social improvement. Nally notes, "For those holding political and economic power, famine became the function of new regulatory and corrective mechanisms that unleashed the terror of the possible."

Famine occurs when the usual markets collapse, food is taken away, or the hungry are wantonly ignored. The great famine in China that occurred from 1959 to 1962 was a result of a forced collectivization of agriculture. Family farms were turned into people's communes, and impossibly high quotas were established. Grain production plummeted, but in order to fill their quotas, local officials continued to ship the dwindling grain supply to the cities. It is estimated that around 35 million people died in a famine that was entirely human-made. Heavy police control meant that people were unable to leave their villages. People were reduced to eating tree bark.

Famines continue to be present on the world stage. They rarely appear in democracies, even poor democracies. Their connection with deeply authoritarian regimes is exemplified in the recent famine in North Korea, when between 1 and 3 million people, out of a total population of 22 million, died due to a collapse of the food distribution system and government incompetence. The country

4.2 Food Deserts

The rich world of developed economies is a place of relative food plenty. Yet even there, access to food varies. At the most extreme end are "**food deserts**," parts of a city or region with very limited access to cheap, nutritious food. The US Department of Agriculture identifies food deserts at the level of census tracts, areas containing a population of approximately 5,000 (https://www.ers.usda.gov/data/fooddesert/). Almost 24 million people in the United States live more than a mile from a supermarket. Almost 2 million households live more than a mile from a supermarket and are without access to a vehicle. In total, 5.7 percent of all US households had problems getting access to the food that they wanted. Food deserts are most prevalent in areas with low to median incomes and the highest proportion of African Americans. There are fewer food deserts in areas with predominantly white middle- and upper-income households. And groceries in food deserts cost more. Many studies show a correlation between food deserts and lower consumption of fruit and vegetables and higher levels of obesity.

REFERENCES

Dutko, P., Ver Ploeg, M., and Farrigan, T. L. (2012) *Characteristics and Influential Factors of Food Deserts*. US Department of Agriculture, Economic Research Service.

Ghosh-Dastidar, B., Cohen, D., Hunter, G., Zenk, S. N., Huang, C., Beckman, R., and Dubowitz, T. (2014) Distance to Store, Food Prices, and Obesity in Urban Food Deserts. *American Journal of Preventive Medicine* 47: 587–595.

Gordon, C., Purciel-Hill, M., Ghai, N. R., Kaufman, L., Graham, R., and Van Wye, G. (2011) Measuring Food Deserts in New York City's Low-Income Neighborhoods. *Health Place* 17: 696–700.

Shannon, J. (2014) Food Deserts: Governing Obesity in the Neoliberal City. *Progress in Human Geography* 38: 248–266.

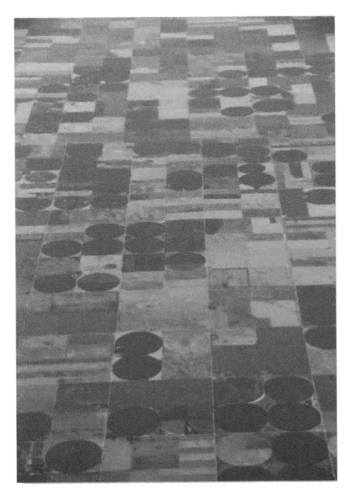

suffers from chronic food insecurity, with almost half of all children malnourished.

The Horn of Africa is a scene of recurrent famines. Almost 12 million people are currently at risk from famine. In 1984–1985, more than 1 million died in a famine in Ethiopia caused by conflict, drought, and economic mismanagement. Between 2010 and 2012, more than 260,000 people died in Somalia, half of them under the age of five. Persistent hunger in sub-Saharan Africa is one of the major failures of the global food system. Many causes are at work. Consider the case of Niger, which is one of the poorest countries of the world, with at least 40 percent of the population chronically undernourished. Food shortages affect more than 7 million people. Household vulnerability is made more severe by land degradation that is reducing the supply of farmland and forcing more people onto marginal lands. Low income, erratic rainfall, and the low status afforded to women are just some of the many reasons behind the recurring famine and food shortages. The growing reliance on food aid paradoxically depresses local agriculture production even further. The result is a population with precarious access to food too

4.4 Irrigation in Midwest United States. Central pivot irrigation systems ensure high agricultural productivity in the drier parts of the United States. However, sources of fresh water are being used up, questioning the long-term sustainability of such practices.

easily pushed over into hunger and famine by even small changes in food supply, food prices, and climate. In these circumstances, drought does not cause hunger as much as it pushes an already vulnerable population into famine.

Limits to Food Supply

The enormous gains in agricultural productivity have passed the limits imagined by Malthus. Agricultural production has increased food supply enough to dispel the original Malthusian prediction. Some see no inherent problem in feeding the world's population. And yet there are some who see a global food system precariously balanced. Whereas Charles Godfray and colleagues look to increased gains to be made, Carleton Schade and David Pimentel point to problems. A comparison of the data and arguments in the two papers makes for a compelling case study of a situation's alternative readings, and is a reminder that arguments vary even when using the same facts. Godfray et al. highlight the ability of the global food system to produce even more food, but Schade and Pimentel point to systemic problems. They suggest that the rapid increase in food supply was the result of available land, water, and energy, a benign climate, and improving crop yields, conditions unlikely to be repeated. They estimate that if the population increases to 9.2 billion there is likely to be enough food for only 6–8 billion. Food insecurity and famine are likely to afflict anywhere between 1 and 3 billion people. The problems they identify include declining land and water availability.

Increasing population reduces the amount of land available for farming and degrades some of the most productive soil. They suggest a **land deficit** of between 0.1 and 0.9 billion hectares. Irrigated land is the most productive land on the planet (Figure 4.4). However, we are losing this land to **salinization** and erosion. The production of food requires large quantities of fresh water. We already have reached the ceiling of surface supply, and the supply of water from aquifers is fast running out. Climate change will also affect the global agricultural system through generating more extreme and damaging weather events, such as flooding. While the higher-latitude countries such as Canada and Russia may experience an extension of their growing season with global warming, other regions may endure harm. In general, the growth of population combined with their enlarging the human ecological footprint is having an environmental impact. We have lost or are losing forest cover, wild fish stocks in the oceans are being depleted, and global reserves of fresh water in river basins are diminishing. More than 1.4 billion people live in river basins where water sources are being depleted.

The two contrasting papers embody the different perspectives. Whereas Godfray and colleagues point to the slack in the system, Schade and Pimentel point to its limits. They suggest that we have reached if not already passed peak water supply, peak land availability, and peak crop yields. The future that they see is one of Malthusian limits reimposed. Adherents of either side can look to current trends to bolster their claims. On the one hand, levels of food insecurity are being reduced. On the other, there is evidence of unstable food prices (Figure 4.5). They increased rapidly in 2006, then fell as the world economic

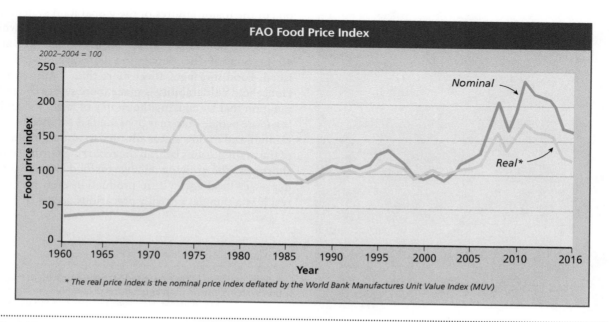

4.5 World food prices

crisis hit in 2008. Since then prices have remained historically high but are falling. Only time will tell which of these trends, declining insecurity or increasing costs, is temporary and which is persistent.

Questioning the Food Production System

The Malthusian limits were bypassed because of the creation of a very efficient **industrialized agricultural system** that generated large amounts of relatively cheap food. Only seven countries produce more food than they need—the United States, Canada, Australia, Thailand, Argentina, Ukraine and Vietnam—but a global trading system has widened the availability of food. In recent years this system has come under greater scrutiny.

The original Malthus–Boserup debate was concerned only with the quantity of food, and in this regard food supply has largely kept up with population increases. Boserup triumphs over Malthus. Gains in productivity and the emergence of a global food system have ensured large amounts of cheap food for much of the world's population. Yet there are emerging questions regarding the quality of food and the wider implications of its nutritional and environmental impacts.

The rise of **processed foods** goes hand in hand with the increase of obesity. Much processed food is loaded with sugar and fats. The World Health Organization (WHO) now estimates that 1.9 billion adults are **overweight** and 600 million are obese, as defined by the **body mass index** (BMI), which is a person's weight in kilograms divided by the square of their height in meters. Overweight people are defined as having a BMI greater than 25, and **obesity** is defined as a BMI greater than 30. It is a crude measure, as it does not distinguish between muscle and fat tissue, but until we have better metrics, it is a useful indicator. The obesity epidemic is associated with a population that is sedentary, sitting and driving rather than walking and exercising, and with diets that contain high-energy nutrients such as fat, starch, and sugar. Increasing obesity leads to cardiovascular disease and diabetes. The obesity epidemic is particularly prevalent among young children; the WHO estimates that 42 million children under the age of five are now overweight.

Figure 4.6 highlights the geographic distribution of obesity for women. The map for men is similar although with a lower incidence of obesity in South America, South Africa, the Middle East, Eastern Europe, and Russia.

There are high levels of obesity in affluent countries such as the United States. Obesity is also found increasingly in middle-income countries such as Mexico, where one in three of adults are defined as clinically obese. Malnutrition

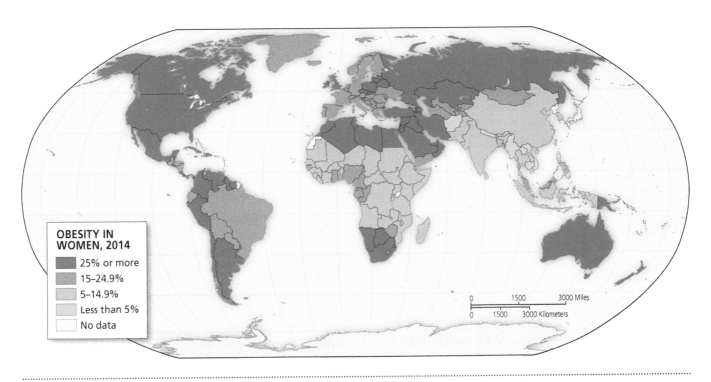

OBESITY IN WOMEN, 2014
- 25% or more
- 15–24.9%
- 5–14.9%
- Less than 5%
- No data

4.6 Obesity

4.3 Overpopulation Reexamined

In 1968 the Stanford biologist Paul Ehrlich wrote *The Population Bomb,* predicting the imminent death of millions if population growth was not contained. **Overpopulation** was, and still is, a common assumption. Since growth rates are highest in the developing world, the debate often hinges on the belief that population controls need to be introduced into these high-growth-rate countries. But when we look at the issue in more detail, a number of important points need to be made. First, it is not simply the number of people. When we factor in the impact of different populations, the problem becomes not too many poor people but too many people in the affluent world with their heavy ecological imprint. With only 5 percent of the world's population, the United States is responsible for 23.3 percent of the global total of carbon dioxide emissions. The average US citizen generates a much higher environmental impact and heavier ecological footprint than citizens of other countries, especially of lower-income countries. It is not that there are too many people, it is that there are too many with a heavy ecological imprint. Ehrlich saw the problem as too many other people, but in terms of environmental impact there are too many like him, affluent people living in the developed world, rather than too many poor people in the less-developed world.

It is perhaps useful to consider environmental impact as more than just a function of population. One useful general equation is

$$I = PAT$$

where I is the environmental impact, P is total population, A is a measure of that population's affluence, and T is the technological capacity available to that population. The environmental impact of a population is a function of its affluence and technology. It is not just population totals that have an environmental impact. Thus while the United States has only 5 percent of the world's population, because it is a rich country (A) with advanced technology (T) it consumes 20 percent of the world's energy. The greater technology and affluence give the United States a larger environmental impact (I) than its population alone (P) would suggest.

Second, there is a politics to population control. It reflects power rather than demographic analysis. The birth control measures of the last fifty years are invariably targeted at the poor and the weak, not the rich and the strong. The World Bank used to adopt fertility targets for countries, but none for the rich countries of the world. Debates about population control reflect power, not scientific neutrality.

Third, there is an adage from farming communities, "each mouth comes with two hands." It refers to the fact that people are a resource, not just a problem. The ultimate resource is the inventiveness of humans: the more humans, the larger the creative pool of talent.

It is not simply the number of people that is the issue; it is their relationship to the land, their use of resources, and the economic system at the heart of the people–environment relationship. To talk about only reducing population is to pass the blame on to the poorest and weakest, when what is really required is a fundamental reassessment of our relationship with the environment and the nature of our economic system.

REFERENCES

Connelly, M. (2008) *Fatal Misconceptions: The Struggle to Control World Population.* Cambridge, MA: Belknap Press of Harvard University Press.

Ehrlich, P. (1968) *The Population Bomb.* New York: Ballantine.

Harvey, D. (1974) Population, Resources and the Ideology of Science. *Economic Geography* 50: 256–277.

Simon, J. L. (1981) *The Ultimate Resource.* Princeton, NJ: Princeton University Press.

Weisman, A. (2013) *Countdown: Our Last, Best Hope for a Future on Earth.* New York: Little, Brown.

plays a role. The bodies of malnourished young children learn to retain fat, and in later life, with more income and food choices, the children of malnourishment can become obese adults. This process, along with a decline in physical exercise and the adoption of a more sedentary lifestyle, explains the rapidly rising obesity levels of countries transitioning from low to middle income in one generation.

Obesity has complex roots. Individuals can limit their energy intake and exercise more. Yet it is more than just a case of individual choice. Jamie Pearce and Karen Witten describe environments that create higher risk of obesity as **obesogenic**. The food industry could, and in some cases is starting to, reduce the fat, sugar, and salt content of the food it produces to ensure more nutritious choices. We could make our cities friendlier to walking and cycling rather than making them convenient only for sedentary motorists.

A food movement is emerging that shifts the debate from the basic supply of food to a wider concern with nutrition. It is a debate not just about the supply of food but also the quality of the food. In many countries government subsidizes the production of crops with little concern for nutritional value or environmental impact. In the United States, around 90 million acres are devoted to growing corn. A huge corn surplus is created by the combination of efficient farming and generous government subsidies. The corn surplus is turned into corn syrup that is added to the food supply, which in turn adds extra calories to the average consumer. The rise of obesity in the United States is directly linked to the corn surplus and the heavy use of corn syrup in processed foods.

Food surpluses are also the result of heavy chemical usage. The main form of pollution in much of the United States, for example, comes not from factories but from farms and agribusiness, with the runoff of pesticides and animal wastes entering streams, rivers, groundwater, and eventually the oceans. Every year between 10,000 to 20,000 farm workers suffer acute pesticide poisoning.

The rise of a **corporate industrialized food production system** due to economies of scale and large investments has produced vast quantities of cheap food, but we are now beginning to question the wider costs of this system and its impact on public health and environmental quality. Another new food movement is emerging that is concerned more with health, nutrition, sustainability, and access. It is a disparate movement with a wide variety of goals and concerns, including improving the plight of agricultural workers, raising issues of animal rights in factory farming, advocating the slow food movement, promoting local food producers, and encouraging inner-city vegetable gardens. Even in the USA large food corporations such as McDonald's and General Foods are advertising a switch to more natural and healthy ingredients. Fundamentally this movement is concerned with the nature of the food that we produce, particularly questioning the encouragement of grain-fed meat production, assessing the wider costs of cheap fast food, highlighting the darker side of the food chain, and debating the sustainability of the entire global food system.

This system, while perhaps producing a sufficient supply of food, does little to advocate a fair or even healthy supply of food. While obesity is emerging as a major health issue, food insecurity continues to be a problem for many. Parts of the world are overfed while others are undernourished, and yet other parts are both overfed and malnourished.

We are what we eat, and for those who can afford it there are choices to be made about what we eat and how we eat it. A non-vegetarian diet requires 2.9 times more water, 2.5 times more primary energy, 13 times more fertilizer, and 1.4 times more pesticides than a vegetarian diet. When we move beyond worrying about the supply of food, we begin to question the nutritional quality of the food we buy and eat, its environmental impact, its ethical implications, and its wider imprint on the world we live in.

What to make of the gloomy prognostication of Malthus? The constraints of food supply have been transcended by incredible gains in agricultural productivity and a global supply chain of food that links importers and exporters of food. Population has continued to increase long past the Malthusian limits. And yet if the basic limits of food supply no longer apply with the brutal force predicted by Malthus, problems remain. Out of 7.5 billion people, 1 billion are hungry (undernourished), 800 million are malnourished, and a further billion are obese. It is not only the supply of food that is the problem, although there are some who see food supply peaking and then declining as other resource inputs such as land and water become more problematic. Questions are also being raised about the distribution, quality, and price of food and the long-term sustainability of maintaining enough good-quality food for all of the world's population.

Cited References

Boserup, E. (1965) *The Conditions of Agricultural Growth: The Economics of Agrarian Change under Population Pressure.* London: Allen and Unwin.

Godfray, H. C. J., Beddington, J. R., Crute, I. R., Haddad, L., Lawrence, D., Muir, J. F., Pretty, J., Robinson, S., Thomas, S. M., and Toulmin, C. (2010) Food Security: The Challenge of Feeding 9 Billion People. *Science* 327: 812–818.

Keneally, T. (2010) *Three Famines: Starvation and Politics.* North Sydney: Knopf.

Nally, D. (2008) "That Coming Storm": The Irish Poor Law, Colonial Biopolitics, and the Great Famine. *Annals of the Association of American Geographers* 98: 714–741.

Pearce, J. and Witten, K. (eds) (2010) *Geographies of Obesity: Environmental Understandings of the Obesity Epidemic.* Burlington: Ashgate.

Schade, C. and Pimentel, D. (2010) Population Crash: Prospects for Famine in the Twenty-First Century. *Environment, Development and Sustainability* 12: 245–262.

Sen, A. (1981) *Poverty and Famines.* Oxford: Clarendon Press.

Selected Guide to Further Reading

Bassett, T. J. and Winter-Nelson, A. (2010) *The Atlas of World Hunger*. Chicago: University of Chicago Press.

Burgess, M. G., Polasky, S., and Tilman, D. (2013) Predicting Overfishing and Extinction Threats in Multispecies Fisheries. *Proceedings of the National Academy of Sciences* 110: 15943–15948.

Carolan, M. (2013) *Reclaiming Food Security*. New York: Routledge.

Cohen, D. A. (2013) *A Big Fat Crisis: The Hidden Forces behind the Obesity Epidemic—And How We Can End It*. New York: Nation.

Guthman, J. (2013) Too Much Food and Too Little Sidewalk? Problematizing the Obesogenic Environment Thesis. *Environment and Planning A* 45: 142–158.

Hallett, L. and McDermott, D. (2011) Quantifying the Extent and Cost of Food Deserts in Lawrence, Kansas, USA. *Applied Geography* 31: 1210–1215.

Hilborn, R. (2012) *Overfishing: What Everyone Needs to Know*. Oxford: Oxford University Press.

James, P. (2013) *Population Malthus: His Life and Times*. London: Routledge.

Jinseng, J. (2012) *Tombstone: The Great Chinese Famine 1958–1962*. New York: Farrar, Straus & Giroux.

Kiple, K. F. and Ornelas, K. C. (eds) (2000) *The Cambridge World History of Food*. Cambridge: Cambridge University Press.

Marlow, H. J., Hayes, W. K., Soret, S., Carter, R. L., Schwab, E. R., and Sabaté, J. (2009) Diet and the Environment: Does What You Eat Matter? *American Journal of Clinical Nutrition* 89: 1699–1703.

Morrison, N. O. (2011) Mapping Spatial Variation in Food Consumption. *Applied Geography* 31: 1262–1267.

McMillan, T. (2012) *The American Way of Eating*. New York: Scribner.

Patel, R. (2008) *Stuffed and Starved: The Hidden Battle for the World Food System*. Brooklyn: Melville House.

Pilcher, J. M. (ed) (2012) *The Oxford Handbook of Food History*. New York: Oxford University Press.

Pimentel, D. (2012) World Overpopulation. *Environment, Development and Sustainability* 14: 151–152.

Pollan, M. (2008) *In Defense of Food*. New York: Penguin.

Schlosser, E. (2004) *Fast Food Nation: The Dark Side of the All-American Meal*. New York: Harper.

Toth, G. and Szigeti, C. (2016) The Historical Ecological Footprint: From Over-population to Over-consumption. *Ecological Indicators* 60: 283–291.

Woodham-Smith, C. (1962) *The Great Hunger, Ireland 1845–9*. London: Hamish Hamilton.

Selected Websites

Data on food, food security, and many other issues are available at the Food and Agriculture Organization (FAO) and the World Health Organization (WHO)
http://www.fao.org/statistics/en/
http://www.who.int/en/
Global Hunger Index
http://www.ifpri.org/topic/global-hunger-index

Issues of world population are discussed at:
http://www.worldpopulationbalance.org/
http://www.worldwatch.org/
A number of nongovernment agencies are concerned with food security:
http://www.oxfam.org
http://www.stophungernow.org/

Interactive Map Website

http://ghi.ifpri.org/

Learning Outcomes

The agricultural revolution led to the growth in the size and density of human settlements.

- Thomas Malthus argued that the rate of population growth was higher than the increase in food supply. His theories led to the development of programs to reduce and control populations.
- Ester Boserup theorized that increased populations would be followed by an increase in food production technology, which will enable larger populations to survive.
- Some researchers suggest that it may be possible to meet the food demands of 9 billion people by closing the yield gap, increasing production limits, reducing waste, changing diets, and expanding agriculture.
- The declining levels of hunger and malnutrition are testament to the ability of agricultural production to surpass Malthusian limits.
- Three things may cause famines: the redirection of food, the destruction of productive capacity to grow food, and total neglect of those who are hungry.
- Despite enormous gains in agricultural productivity, there is still disagreement over the world's ability to sustain its population with enough food, including the arguments that there will not be enough productive land or

water resources for agriculture, or ocean fisheries, in the future.

- Climate change may negatively impact the world's ability to sustain its human population.
- Despite gains in the availability of food worldwide, questions arise about the quality of food and the environmental impacts of new production technologies and processes.
- One problem associated with greater availability of processed food is the increasing rates of obesity found around the world, in both affluent and middle-income countries.
- A food movement is emerging that shifts the debate from the basic supply of food to a wider debate concerned with nutrition that includes discussions about the quality of food, the environmental impact resulting from the means for producing food, sustainability of food sources, and access to different communities.
- The idea of overpopulation is a politically charged debate that often ignores the greater environmental impact of more affluent populations. So it is not just a case that there are too many people; rather, too many of them have a heavy ecological footprint, which can be solved only when understanding the influence of politics as well as the opportunity that humans are a potential resource for talent, creativity, and creation of solutions.
- Although the basic food-supply limits first proposed by Malthus may no longer apply, out of 7.5 billion people, 1 billion remain hungry, 800 million malnourished, and another billion are obese, confirming that there are still food-related concerns far beyond what Malthus or other early researchers may have considered.

5

Population and Resources

LEARNING OBJECTIVES

Explain how humans have come to value nature and commodified resources.

Summarize the history of global coal consumption and production, along with its contemporary uses and their consequences.

Recall the five laws of resource use and relate them to specific resource examples.

Outline the twentieth-century history of oil production and consumption followed by present-day changes affecting the industry.

Consider the environmental impacts and growing costs associated with the continued use of oil and gas.

Economic development is dependent on resource use. How nature is turned into a resource depends on many factors, including changing technologies and modes of activity. Coal and oil provide examples of the changing relationship between population and resources.

In this chapter, we will consider some of the general relationships between people and resources, using the **nonrenewable resources** of coal and oil as primary examples. Let us begin with some definitions. In the *Oxford English Dictionary*, among the many definitions of "resource," the primary one is "a means of supplying a deficiency or need." This sense emerges first in the seventeenth century and is linked with the development of words and ideas associated with markets, money, buying, and selling. From its origin, then, the term "resource" is embedded in social considerations of money and capital. The definition also hints at the relational nature of resources; things are resources because they are linked to scarcity and desire.

Resources are socially constructed, which does not necessarily mean that humans make resources; oil and coal, for example, were made by eons of geologic processes. "Socially constructed" in this case means that a resource is a function of social, technical, and economic considerations. For centuries, oil was just a messy black substance that oozed from the surface of the Earth. It became a valuable resource only with the creation of a **carbon economy**. The history and geography of resources are the unfolding tale of how "nature" becomes a commodity, with all the social and political consequences involved in this valuation and subsequent revaluation.

The translation from inert material to resource occurs through technical change and commodification. "**Commodification**" means that the material is valued and traded, it has a price equivalent, and it enters the arena of things bought and sold. For things to become commodities, technical change is often required. The movement of oil from sticky liquid to valuable commodity requires the design and construction of a vast technical apparatus to transform the raw material of oil into fuel and to produce its valuable byproducts, such as plastics. Technical change and commodification are not external to each other but are inexorably interconnected. Technical developments occur through the increased valuation of the resource. Let us further consider the case of two primary resources, coal and oil.

The Case of Coal

When the Romans first invaded what they called Britannia around two thousand years ago, they found coal being used as raw material for jewelry. When carved and polished, it adorned the bodies of affluent Britano-Romans. Some local tribes in south Wales also used the material to cremate their dead leaders, but it was only in the twelfth century that people in Britain began using it as fuel. Before then, and for five centuries more, wood was the principal fuel for cooking and heating. As the population expanded, the forest reserves became depleted and, as so often happens in the history of resources, an alternative resource had to be exploited. Coal that was close to the surface was readily available, and soon there was a vigorous trade in coal mined in northeast England, around Newcastle, then shipped to the large urban centers such as London. The burning of coal became such a common domestic practice by 1600 that there was rising concern with air quality. In 1661 John Evelyn described London as similar to an active volcano, as smoke and noxious fumes belched out from the city's chimneys to create unhealthy air. The smoke blocked out the Sun, and sooty particles covered buildings. Despite its side effects, coal was preferred because wood was two to five times more expensive. And here we come to a second characteristic of resource use: the difference between private costs and benefits and social costs and benefits. For individual London households, it was cheaper to use coal than wood. As more households adopted this strategy to cope with private costs, they imposed **social costs** on their neighbors by harming London's air quality.

As the demand for coal increased, it became profitable to make larger outlays to dig deeper into the Earth. There was a problem, however. The deeper the mines were sunk, the more vulnerable they were to flooding. Scientists and inventors tackled the problem with attempts to build a pumping machine using piston action fueled by the steam from burning coal. It worked, but it used tremendous quantities of coal; mines could easily access this fuel, but otherwise it was not suitable for wider usage. Then James Watt devised an improvement to the pumping machine to make it more energy efficient. He added a condenser that enabled the piston's cylinder to remain hot and thus able to pump more efficiently. His machine did not need such large quantities of coal; the innovation freed the machine from the coal mine. In 1776, Watt created two coal-driven, steam-powered engines, one to pump water from a coal mine, the other to power an iron foundry. It was a momentous year: the Industrial Revolution was inaugurated, Adam Smith published *The Wealth of Nations,* and the fledgling American Republic declared independence.

Coal and the Industrial Revolution

Coal powered the **Industrial Revolution** and would prove an important ingredient in the future manufacture of iron and steel. In 1830 Britain produced four-fifths of the world's coal, and by the middle of the nineteenth century, it was the world's leading industrial power and later the world's unrivaled superpower. Coal was not the only factor in the stunning national success, but it was an important ingredient. Coal mining was also at the center of a nexus of capital–labor relations. Coal-mining areas, with their large pools of workers living side by side, produced not only coal but also an organized force of working-class politics. Political radicalism as well as coal came from the coalfields of south Wales, northern England, and central Scotland.

By 1900, the United States had become the world's largest coal producer, and, as in Britain, coal was a vital resource in the country's rise to industrial prominence. Coal powered the growing industrial nation and was an essential part of the subsequent creation of a mighty iron and steel industry. Emissions from coal-burning power plants also figured in the nation's history of **environmental legislation**. There is no simple path between growing environmental health hazards and resultant environmental legislation. Smoke-belching factories also implied economic growth to such an extent that political scientist Matthew Crenson writes of the "un-politics of air pollution" in many industrial cities in the United States. Factory chimneys were seen as a sign of progress and full employment, so in many industrial cities air quality did not become a political issue; attempts to restrict emissions were often resisted and disputed, and the problems of polluted air downplayed. In 1970, however, the landmark Clean Air Act established new standards of air quality; it was just the opening salvo in the ongoing struggle to control and restrict air pollution. Coal producers and owners of coal-fired power stations continue to lobby to restrict further legislation. And here we come to another element in resource use: the producers always trumpet the benefits and try to ignore or downplay the costs.

Coal Production

The United States remains a major coal producer; it ranks second in the world after China (Figure 5.1). In 2015 the United States produced over 900 million tons of coal, almost

5.1 The Hubbert Curve

The Hubbert Curve: ideal

The **Hubbert Curve** is named after a geophysicist who first employed it in 1956 to predict the decline of US oil production. Essentially a symmetrical bell-shaped curve, it suggests that nonrenewable resource use rises quickly from zero to a peak and then falls off; at the top are peak production levels. Hubbert used it to predict that oil production in the continental United States would peak in 1965. The curve is perfectly bell-shaped; in reality, resource use is rarely so symmetrical. And there is always the possibility that new technologies will increase the efficiency of extraction. However, it raises the issue of immediate and rapid falloff from peak production. The curve has been used to model historical resource exploitation, such as coal production in the United Kingdom. It is also part of the considerable debate about the existence and timing of a global oil peak. The figure below shows the gradual tailing off of most sources of global oil production.

REFERENCE

Bardi, U. (2009) Peak Oil: The Four Stages of a New Idea. *Energy* 34: 322–326.

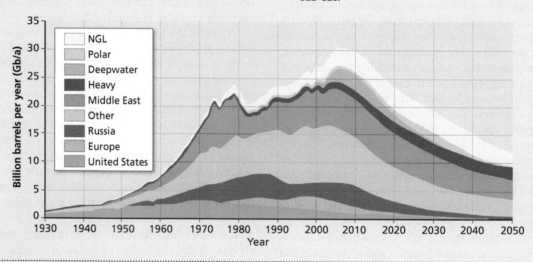

The Hubbert Curve: empirical (NGL = natural gas condensates)

40 percent from the state of Wyoming and 12 percent from West Virginia. In the past thirty years, there has been a rapid growth in shallow mining (close to the surface), as opposed to deep mining, and more of this mining in the states west of the Mississippi, a rapid expansion caused by more efficient and powerful mechanical digging equipment and the need for lower-sulfur coal. In 1970 almost 70 percent of US coal came from Appalachia; now it is just 14 percent and falling. Coalfields in the Illinois Basin in the central United States, and the Powder River Basin in the western states, are now the principal sources of US coal production. Coal is dug out close to the surface by giant machines and then

transported by train. Day and night the coal trains move the fuel from the coal mines to the power plants. Almost 90 percent of coal produced in the United States is destined for power plants that generate electricity. The country's coal reserves are difficult to estimate with accuracy. While there may be 480 billion tons, largely in western states, probably only 10 percent of that is recoverable. The rest may be too expensive to mine in current economic circumstances. Resources only become recoverable resources when their exploitation is economically feasible.

In 2015 the top five coal producers in the world were, in order, China, the United States, India, the European Union,

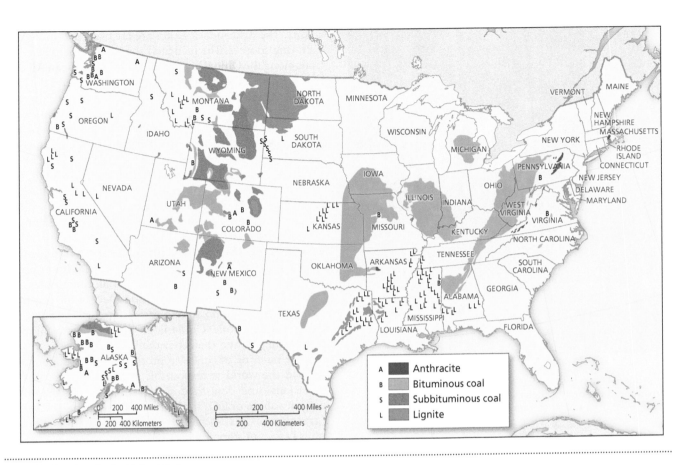

5.1 Coal regions in the United States

and Australia. Coal production in China, as in Britain and the United States before it, quite literally has fueled the country's very rapid economic growth (Figure 5.2). In 2010 China produced over 3 billion tons of coal, three times more than the United States. Even this is not enough: China is also the second largest importer of coal, after Japan. China's substantial and accelerating industrial growth draws in imports from all over the world. Coal from the United States' Powder River Basin is now being sold to China, and the large US coal companies are seeking to build and expand more coal terminals in Washington and Oregon. Australia has experienced a commodity boom as it exports vast quantities of coal and iron ore to China.

However, since 2008 and especially since 2012 coal production has leveled off and in some cases declined. Two main reasons: The first is that cheaper power is now available from underground sources of gas and oil. As power costs decline, more expensive coalfields, such as in Appalachia, become less profitable. The second is due to China's economic growth slowing after three decades. The headlong growth was quite literally fuelled with coal from domestic and overseas coalfields. The age of coal is not coming to an end, but it is being eclipsed as other sources of fuel provide cheaper and sometimes cleaner sources of energy.

Problems with Coal

Coal is still responsible for 42 percent of the world's electricity. Every time we recharge our cell phone, plug a device into a socket, or turn on a light, a coal-fired generator probably produced the power. Coal production continues to increase. While the basic supply of coal is not yet an issue, the wider implications of its burning and exploitation remain problematic. There are proponents of clean coal, but that remains more a fantasy than an empirical fact. Coal has two major problems.

First, burning coal produces noxious fumes that pollute the skies and cause global climate change. Two particular gases, sulfur dioxide (SO_2) and carbon dioxide (CO_2), cause these problems. SO_2 is a major byproduct of burning high-sulfur coal, the kind of coal that is found in the eastern United States and throughout much of China. It is the main ingredient in acid rain, as the SO_2 combines with elements in the atmosphere to become sulfuric acid. The term "acid rain" was first used in 1872 by an English inspector of chemical works, Robert Smith, who noticed how the poisonous fumes from British factories turned into acidic rain that eroded buildings and polluted the water. The term and the underlying process were little used or discussed apart from a paper published in 1955

5.2 Air pollution in Shanghai. Burning coal produces airborne pollution. Chinese cities, despite massive cleanups, still have high levels of air pollution from coal-fired plants.

by Eville Gorham that identified its significance in the Lake District of England. Then, as SO$_2$ levels mounted, a flood of papers began to note its significance. A Swedish researcher, Svante Odén, noted its effects in Scandinavia in destroying forests and making lakes and rivers more acidic. Much of the acid smoke that fell as acid rain in Scandinavia came from coal burned in Britain and Germany. Airborne pollutants do not respect international boundaries. And here we come to another element in resource use: while the benefits may accrue to an individual company, region, or country, the costs are often borne by other companies, regions, and countries. There are social costs to private enterprise. While the benefits may be privatized, the costs are often socialized.

SO$_2$ became a major problem in the United States especially because of the burning of high-sulfur coals. In the upland areas of the Adirondacks, it was discovered that by the 1980s, 80 percent of lakes had suffered from acidification that killed fish and microorganisms. Because of the region's geology, mainly igneous and metamorphic rocks with limited soil cover, there were few carbonate rocks or soils to buffer the effects. The small lakes at higher elevations were particularly vulnerable because they had less buffering soil; the higher elevation also created more rain and snowmelt, which led to more acid rain. The very small lakes that received great runoff were being doused in acid baths that killed the fish stock. Things got so bad that lime was dumped into the lakes to counter the rising

acidity. The Adirondack Lakes are far from the immediate polluting sources. The principal source was the power stations along the Ohio River. In the wake of the Clean Air Act, they were forced to build taller chimneys to reduce the local impact of emissions. Increasing the height of the chimneys did take the SO$_2$ from the immediate area but pushed it higher into the atmosphere, where it drifted eastward and eventually fell as acid rain on the mountains of upstate New York. Resource use often has unforeseen consequences.

While acidity levels in some lakes remain high, SO$_2$ levels across the United States have decreased in the past twenty years. An Acid Rain Program was adopted in 1990 that required power plants to halve their SO$_2$ emissions by 2010. The process was aided by the greater use of low-sulfur coal from the western states. The net effect: SO$_2$ was much reduced. In China, by contrast, because it has lower pollution standards and uses sulfur-rich coal, the problems of acid rain continue to be a major side effect of the country's economic growth, poisoning the air, polluting water, and contaminating soils. China is on the same path followed first by the UK and then later the USA.

The burning of coal also produces CO$_2$. Coal burning around the world is responsible for one-third of all CO$_2$ emissions, and coal-fired power stations are the single biggest source of CO$_2$ in the Earth's atmosphere. CO$_2$ is a "greenhouse" gas, so named because it produces a greenhouse effect of trapping heat in the atmosphere.

The link between CO$_2$ and global climate change is now well established. The steady rise of CO$_2$ in the atmosphere is recorded in the **Keeling Curve**, which plots the concentration of the gas in the atmosphere from 1958 to the present day (Figure 5.3). The curve is named after Charles David Keeling, who initiated the measurements from the Mauna Loa Observatory in Hawaii. The data from this site, free from many local sources of pollution, are a good indicator of global patterns. The yearly variation occurs because more coal is burned in the Northern Hemisphere during winter. The curve is on a seemingly inexorable upward swing. When James Watt invented the steam engine, CO$_2$ levels were roughly 280 parts per million (ppm). At the start of Keeling's measurements in 1959, the figure had risen to 315 ppm. At the time of this chapter's writing it was 404 ppm. (The latest figure is available at http://co2now.org.) The level is accelerating. The consensus among most climate scientists is that climate change becomes significant at 350 ppm.

A second problem is that coal mining often involves environmental destruction as the land is poked and prodded, vegetation is destroyed, and waste material pollutes the soil and the air. One of the most environmentally damaging forms of coal mining in operation in the Appalachia region of the United States is **mountaintop removal** (MTR). The process involves removing the entire peak to gain access to coal seams and dumping the material into the neighboring valleys. The environmental

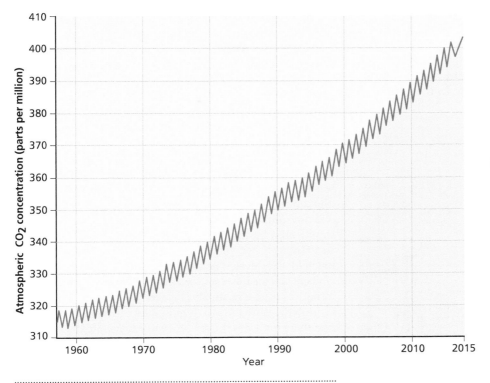

5.3 The Keeling Curve

damage includes disruption of local **ecosystems**, loss of **biodiversity**, and polluted air, soil, and water. By 2015 over 1.2 million acres of MTR mining had occurred. A study by geographers Bard Woods and Jason Gordon showed that while the negative environmental consequences were obvious, the positive economic benefits to local communities, often touted by supporters of MTR, were less apparent.

In some cases reclamation is possible. Across the old industrial landscapes of Europe and North America, land reclamation projects are turning mining wasteland into recreational opportunities. In Leipzig, Germany, for example, a former coal mining area has been turned into a theme park; it opened in 2005, and by 2017 a recreational area of seventeen interlinked lakes was created. The old mining area has been reimagined and remade as a place of recreation and pleasure rather than pollution and work.

Laws of Resource Use

We can summarize some of the themes we have considered so far as five laws of resource use:

1. Resources become resources because previous resources become exhausted, unavailable, depleted, or too expensive. Coal was used in Britain because wood was becoming more expensive. Nuclear power was developed in countries such as Japan and France because they had few coal reserves or oil supplies. Wind and wave are being exploited as energy sources because of the rising cost of nonrenewable sources and as their costs decline.

2. Resource use creates new geographies, whether it be in the eighteenth-century creation of plantation economies in the West Indies to feed the sugar habit of Europeans, the nineteenth-century creation of coalfield landscapes across Europe and North America, or the late twentieth and early twenty-first century creation of vast mining operations in Australia to feed China's demand for energy and raw materials.

3. All resource usage tends to have wider social and political consequences. Nineteenth-century coal production, for example, was also the setting for organized labor movements and political radicalism. The national competition for resources is an important element in the interactions between states. The United States' insatiable demand for oil makes it curry favor with oil-rich countries such as Saudi Arabia even though it has vastly different cultural and political values.

4. All resource use has implications for public spaces and the wider community as well as for private enterprises. The use of fossil fuels, for example, has many positive consequences but also pollutes the air we breathe and is a cause of climate change that affects us all. The greater use of fracking has the very real possibility of polluting public water supplies.

5. Initiatives to regulate and diminish the negative public consequences of resource use and exploitation always come up against powerful interests that firstly deny the problem and then minimize it. In 1980 the head of the US National Coal Association stated that the negative effects of acid rain were only in the imaginations of those who wanted to promote regulation. When major oil companies fund those who deny climate change, they are taking part in a long and constant story of resistance to regulation.

The Limits to Growth?

The relationship between population and resources, like the relationship between population and food, is dominated by the notion of **limits to growth**. In the middle of

5.4 Oil derricks in Los Angeles. This oil field, like many in the United States and around the world, is coming to the end of its life.

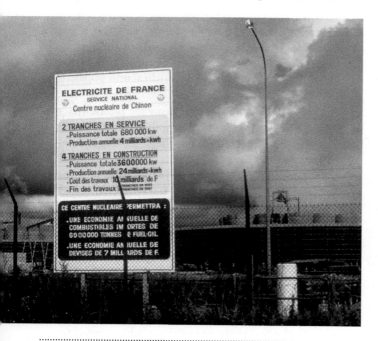

5.5 Nuclear power plant construction near Chinon, France. Without oil reserves and with coal reserves long since depleted, France relies heavily on nuclear power as a source of electricity. Fifty nuclear power plants produce almost four-fifths of its power.

the twentieth century, biological concepts such as **carrying capacity** were employed to indicate that perhaps we were coming up against the limits of growth. The basic idea was that the world could carry only so much population and sustain so much growth. There was a limit to the carrying capacity of the Earth. For neo-Malthusians such as Garret Hardin and Paul Ehrlich, the limits had been reached by the 1970s; the only response was to severely limit population growth.

In 1972 an influential book entitled *The Limits to Growth* was published. Its central argument, summarized in its title, was that there were limits to continuous economic growth. The supply of nonrenewable resources, such as iron and oil, was not infinite, and on current trends, the limits to supply would soon be reached (Figure 5.4). The book's argument was strengthened the next year when oil prices increased fourfold. However, the 1973 **oil shock** led to much more efficient use of oil resources and a renewed search for alternative sources of supply (Figures 5.5 and 5.6). The limits-to-growth argument also highlights another law of resource use: there is always the tendency to exaggerate the risks and preach a form of resource Armageddon.

The book was dismissed by critics for its neo-Malthusian pessimism compared to the reality of human ingenuity. Resources did not run out. Critics such as the economist J. L. Simon argued that as we come close to "limits," price changes stimulate new forms of resource use, alternative supplies, and new ways of doing things. Rather than simply walking over the edge of fixed supply, we negotiate an increasing price curve.

However, the book did stimulate a debate about longer-term prospects and strengthened the notion of sustainability. The arguments have shifted and morphed since the early 1970s. Rather than mechanistic limits to growth, there is an emerging notion of the desirability of more sustainable resource use.

While the idea of limits to growth has its weaknesses and major resource scarcities have yet to materialize, there is a more general point to be made. If there is a heavy reliance on one particular resource, then supply issues can become problematic. We will illuminate this with reference to the case of oil.

The Case of Oil

The Never-ending Demand for Oil

There is an insatiable demand for oil. In 1970, around 45 million barrels of oil were consumed worldwide every day. By 2015 it was closer to 91 million. Even with energy efficiencies, the raw demand keeps on growing. Rapid industrialization is based on oil. In 1970 China consumed only half a million barrels of oil a day. By 2014 it was close to 10.4 million. The US has the world's biggest thirst for oil; it consumed 19.1 million barrels a day in 2014. Our modern world runs on oil.

Despite the increase in demand, the price of oil remained stable for decades as cheap oil was exploited around the world. From 1881 to 1973 the price of oil, in real terms, remained flat. Oil prices were low and, more importantly, stable. Economic growth in the first three-quarters of the twentieth century was quite literally lubricated by the steady supply of cheap, easily available oil. Adjusting for inflation, between 1881 and 1973 the price of a barrel of oil averaged under $20 (in 2008 prices). On October 16, 1973, the **Organization of Petroleum Exporting Countries (OPEC)**, an **oil cartel** dominated by Middle East oil producers such as Saudi Arabia, increased the price of oil by 70 percent to $5.11 a barrel (over $31 in 2016 terms). Oil was central to the economies of richer countries, and, since supply was limited and very largely controlled by the OPEC members, the oil-producing nations effectively used their power to increase the price of oil and hence increase their revenues. The result was dramatic. Between 1972 and 1977, as the current account balances of the world's rich countries increased only from $8 billion to $9 billion, those of OPEC countries increased from $1.5 billion to $7 billion. Non-oil-producing poor countries, meanwhile, increased their deficit from $5 billion to $36 billion. The oil price increase was one of the largest and quickest transfers of global wealth in the history of the world.

The 1973 oil shock had a number of rippling consequences. The era of cheap fuel was over, and energy had to be used more efficiently. Car companies, for example, had to change to more fuel-efficient models. Car companies in Japan, where oil prices were always much higher, already had experience in fuel efficiency. In the United States, a generation of cheap oil meant that car companies had few fuel-efficient models and were slow to realize the seismic shift in oil prices and associated consumer demand. This opened the door to the much more fuel-efficient Japanese cars. The long decline of the Detroit car industry began with the 1973 oil price shock.

As the demand for oil increased, even after the OPEC price hike, oil companies sought out new reserves. Higher prices also meant that oil patches previously considered only marginally feasible now became potentially profitable. The oil companies went long and they went deep in their search for oil.

Consider the case of Brazil. With its rapid economic growth, Brazil has a huge demand for oil. Reserves were discovered off the coast, and daily production increased from less than 1 million barrels of oil in 1990 to 2.7 million barrels by 2010. It is tough to extract, since the largest oil reserves lie 200 miles offshore at depths of three miles below the seabed. There are consequences to such high risk endeavors: in November 2011, 3,000 barrels of oil leaked from a site in the Frade oil field.

The Changing Supply of Oil

The supply of oil on the world market is still dominated by the Middle East. Saudi Arabia remains the largest single exporter of oil at 11.9 million barrels a day, followed by Russia (4.5 million), Canada (2.7 million), the United Arab Emirates (2.5 million), Nigeria (2.4 million), and Iraq (2.3 million).

While demand seems infinite, the supply is fixed. Estimating oil reserves is a difficult business. There is always the possibility of discovering reserves as yet unknown. And even the knowledge of existing reserves is more an estimate than a fine calibration. Yet the long-term prospect does not look good. We may be at **peak oil**, the high point in the Hubbert Curve. In 2015 the total world supply was estimated at 1.65 trillion barrels of oil. It sounds like a lot—it is a lot—but with oil consumption rising and most of the world now surveyed for oil reserves, there is the possibility that oil supply may soon be peaking. The rate of extraction is now double that of new discoveries. Estimates of oil's peak years vary from 2010–2015, a view shared by US military intelligence, to a more optimistic 2050.

Even if we take the most optimistic estimates and put the peaking of oil into the distant future, we are still in the era of tough oil. Most of the **easy oil** reserves close to the surface have been tapped. Many of the big fields are in steep decline. Mexico and Indonesia used to be major oil exporters but are now importers. More than one-third of the world's oil comes from large fields producing more than 5 million barrels a day. Discoveries of these fields declined from 130 in the 1960s to only 34 in the 2000s.

Even if the exact timing of global peak oil is difficult to assess accurately, it is easy to see that we are in an era of tough oil, not only in terms of the cost and risks of deep drilling into ever more dangerous places, but also in the search for immediate alternatives. The tar sands of Alberta are being exploited for their oil at the cost of large amounts of water and heat; elevated pollution rates, as the tar sand oil produces between 5 to 15 percent more CO_2 emissions than other oil; and the need for a 1,700-mile pipeline through much of North America.

We have moved from an era of easy oil, one of large, shallow reserves that were easily exploitable, to an age of **tough oil** in which reserves are deeper, further offshore, and more expensive to exploit. The new fields are smaller and not as long-lasting; they peak fast, then decline sharply, so there is a continuous search for new fields. Companies are persistently at the furthest extent of their experience and technical capabilities as they drill deeper in ever more inaccessible places.

There is considerable debate about peak oil. There are those who focus on the narrow debate of estimating the precise date of the oil peak. This is a discussion that sheds more heat than light, since the oil supply business is one

5.2 The Geopolitics of Oil

Oil is a vital resource. The need for energy security, for ensuring long-term and reliable sources of oil, is an increasingly vital national security consideration. Oil demand creates a strong nexus of mutual interest between major oil importers and large oil companies. It also dictates foreign policy. The USA, for example, has to take into account relations with major oil-producing nations, sometimes creating alliances with dictatorial and authoritarian regimes, other times being actively involved in regime change and direct military interventions. The crucial test is not that foreign oil suppliers treat their own people well, but that the oil keeps coming at a reasonable price. Securing oil supplies was long a vital part of the operation of US global power.

China, as a leading oil importer, also tailors its foreign policy to its energy needs, but in the process may get dragged into regional disputes such as between Sudan and one of the newest countries in the world, South Sudan. The new pipeline that takes oil from Russia to China represents a recent reversal of old enmities. Linking up the supply and demand for oil can turn centuries-old enemies into today's trading partners.

Large and powerful oil importers need to ensure the supply of oil. Treaties, interventions, wars, in some cases supporting regimes and in others undermining them, all become part of the changing tactics devised to meet the need for energy security. The geopolitics of oil links importers and exporters in complex, ever-changing relationships. The suppliers band together in cartels to regulate the supply and price, while the importers draw upon a wide variety of tactics and use their political, economic, and sometimes military power.

REFERENCE

Yergin, D. (2011) *The Quest: Energy, Security, and the Remaking of the Modern World*. New York: Penguin.

---

of estimates rather than solid empirical facts. Then there is the debate that widens out to look at the consequences of oil supply peaking sometime in the future. We can consider three of them.

The first is the search for alternate sources of supply, from alternative sources of oil and gas to **renewable energy** sources such as biofuels, wind, waves, and solar. A list of **biofuels** is given in Table 5.1. Brazil, for example, even though it has huge oil supplies, also uses half of its entire sugar cane crop to produce fuel ethanol. Skylines across the world are sprouting metal windmills that look like modernist mobiles. Solar panels are being installed on roofs in numerous cities. In many cases, the costs of alternative energies are still high, but as more efficient renewable energy technologies are introduced, we may be at the cusp of a major shift in energy supply. Solar energy costs in the US are fast approaching par with power from the electric grid—so-called grid parity.

A second consequence is the long-term future of countries that derive most of their revenue from oil. In oil-based regimes such as Venezuela, whose peak production was in 1996, the decline of oil export revenue, as global demand declines and prices fall, has been the source of political conflict and social unrest. Saudi Arabia derives all of its wealth from oil exports, yet oil production peaked in 2005. When your major and almost only source of revenue begins to decline, national economic stability is put at risk. The possibility of moving past peak production into a world of possibly declining revenue raises issues of social stability and regime longevity. In places such as Dubai and Abu Dhabi, the frenetic pace of building and development are in part an attempt to build an alternative economic future before the oil runs out.

The third consequence is the implication for economies and socio-spatial assemblages that are predicated on relatively cheap oil (Figure 5.7). Much of the United States' suburbanization relies on low and stable oil prices. All those low-density family homes, out-of-town shopping centers, and long-distance commutes arose at a time of dependably cheap gas prices. But as prices rise and become unstable, the suburbs, with their heavy reliance on private automobiles, now look like a landscape predicated on oil prices that may never return to their former stability. The reliance of a built form precariously balanced on one fossil fuel with high and

---

5.6 Windmills off the coast of Denmark, a leader in wind power generation. Wind power from turbines such as these provide almost one-fifth of all electricity production in the country, which now has a technological lead in the provision of this alternative energy source. About half of all wind turbines used around the world are made in Denmark.

TABLE 5.1	Biofuels	
CROP	**FUEL SOURCE**	**MAIN PRODUCING COUNTRIES**
Sugar cane	Ethanol (from sugar)	Brazil, India, China, Thailand
Sugar beet	Ethanol (from sugar)	France, United States, Germany, Russia
Cassava	Ethanol (from starch)	Nigeria, Brazil, Thailand, Indonesia
Maize	Ethanol (from starch)	United States, China
Oil palm	Biodiesel	Malaysia, Indonesia, Nigeria, Thailand
Rapeseed	Biodiesel	China, Canada, India, Germany
Soybean	Biodiesel	United States, Brazil, Argentina, China

fluctuating costs raises issues of long-term sustainability. The suburbs were built on gasoline from oil priced on the order of $27 a barrel (at 2007 prices). In one month in the summer of 2008, they reached over $140. Prices will remain deflated during recessions but will then tend to rise. There are few large oil reserves left, and the price will inevitably rise when the global economy ticks upward. Where does that leave low-density suburban sprawl, which is so reliant

5.7 This tiny car in Florence, Italy, is ideal for managing the narrow streets of the old city. It also has the advantage of very high gas mileage, a necessity in a country with high oil prices. Small cars, hybrid vehicles, and electric cars become more popular as gas prices increase.

on large-scale private car usage? The general answer: in a very precarious position. This same question has been asked before. In 1984 I noted, in a section entitled "Will Suburbs Become a Thing of the Past?":

> *The growth of the suburbs was based upon a number of conditions which are now disappearing into recent history . . . If these changes are long term and large scale this may lead to lack of demand for suburban housing and a rapid fall off in the house prices of suburban districts. Those who live in the suburbs may find it difficult to either sell their houses or recoup their investment.*
>
> (Short, 1984, 23)

The long-term sustainability of low-density, energy-profligate suburbs with a heavy ecological footprint is now a matter of serious consideration.

Fracking in the USA

For decades the USA was a classic case study of a country with a very heavy reliance on oil but a limited supply. Oil production of the traditional sort in the United States peaked at 10.5 million barrels in 1985. By 2008 it was just over 7.8 million barrels. Consumption increased in the same period from 15.7 million to 19.6 million barrels. Because oil is quite literally an essential lubricant of economic growth, the United States had to import an increasing proportion of the oil it consumed. The oil predicament also made the country susceptible to geopolitical as well as economic considerations. US foreign policy had to take into account the sensibilities of such oil-rich countries as Saudi Arabia while also positioning itself to secure access to oil. Its desperate need for oil meant sidling up to some dictatorships while undermining others. The one constant in US foreign policy in the Middle East was the need to secure long-term access to oil. This need for a steady supply was an important shaper and determining factor in US foreign policy.

The gap between demand and domestic supply entailed not only importing from overseas but also exploring tougher sources of oil, such as the expensive-to-extract oil in the Arctic and tar sands in Alberta. Such extraction is costly and environmentally damaging to produce. The oil is extracted either through strip mining, which tears away the surface vegetation, releases heavy metals, and creates air pollution; or by large injections of water and solvents, which leads to water pollution.

The **oil crunch**, the large and growing gap between the national supply and demand, led to a vigorous search for domestic oil reserves. All the easy shallow reserves in places like Pennsylvania and California had long dried up.

5.3 Commodity Cartels

OPEC is just one of a number of **commodity cartels**, organizations of primary producers who band together to ensure high prices for their resource. They often arise from national dissatisfaction with the private companies that control the trade. The oil trade used to be controlled by the large oil companies, the Seven Sisters as they were called. In 1959 they sought to decrease the price by 10 percent to stimulate demand. A rational move on their part, but for countries that relied on oil exports it was a major cut in revenue. Venezuela, a country where oil exports made up 90 percent of foreign earnings, held talks with other oil-producing countries to resist the price increase. In 1960 OPEC was formed; the initial members were Iraq, Iran, Kuwait, Saudi Arabia, and Venezuela. In 1973, with more members and control of over 70 percent of the global oil supply, the cartel raised the price of a barrel of oil from $2.48 a barrel to $11.65.

The success of OPEC in raising and controlling prices stimulated other commodity cartels, which were soon established: coffee (1973); bauxite, the raw material of aluminum, (1974); bananas (1974); and iron ore (1975). Commodity cartels exercise greatest power when there are a small number of producers with shared value systems, demand for their commodities is high, substitution costs are prohibitive, and they control most of the world supply. Their strength also depends on the nature of the commodity. Oil can be kept in the ground and still retain its value. Bananas, in contrast, rot if kept, so the suppliers have limited power to keep their products off the market. OPEC is by far the most successful cartel. It has remained a very powerful organization in controlling world oil prices, as it controls such a large proportion of such a valuable commodity. OPEC members now hold 81 percent of the world reserve of crude oil. The other cartels either proved short-lived, the coffee cartel ending in 1989 and the bauxite cartel in 1994, or, if they did survive, like the natural rubber cartel, they have proved less influential in shaping world prices.

The problem with any cartel is the differing bargaining positions of its members. In the case of OPEC Saudi Arabia can make money when oil is higher than $10 a barrel. Venezuela needs at least $80 a barrel to break even as its oil needs expensive treatment. So when the price of barrel of oil is greater than $10 a barrel but less than $80 a barrel, Saudi Arabia can make money but Venezuela cannot. Therefore, when Saudi Arabia keeps oil prices low, by maintaining supply and thus market share in the face of competition from other sources of energy, Venezuela loses out.

REFERENCES

Khan, R. (2011) In the Footsteps of the OPEC: Trends in Collective Bargaining over Natural Resources. *Strategic Analysis* 35: 162–168.

The OPEC website is a useful source of data: http://www.opec.org/opec_web/en/

The search then looked to places that only decades before were considered too difficult or too expensive to extract oil from.

First there was offshore drilling. The history of oil extraction in the Gulf of Mexico highlights the shift. Drilling first started in 1937, just off the Louisiana coast. The first well was located in 14 feet of water less than a mile from the shoreline. By 1993 approximately 12 percent of all oil produced was from **deepwater wells**, defined as 5,000 feet below the water's surface or deeper. Only 2.5 percent of new wells, forty in total, were deepwater. By 2009, 80 percent of oil came from deepwater wells and over a third of all new wells sunk were deepwater. Between 2006 and April 2010, the number of deepwater rigs grew by 43 percent. As they ventured further offshore and drilled deeper and deeper, oil companies were always at the ever-extending limits of their technical capabilities. Massive and potentially dangerous machinery has to be placed on small platforms far out to sea. At the deeper levels, the oil is subject to up to 10,000 pounds of pressure per square inch, making the safe and steady release of the oil a difficult proposition. It is easy to fracture the surrounding geologic formations. Oil is a hazardous, highly flammable liquid. Pumping millions of gallons of it at high pressure though a small pipe more than three miles down is a high-risk activity. On the ocean floor, where the wellheads are located, temperatures are low and visibility is hazy. Adjustments have to be made using remotely guided machinery. Blowout preventers on the ocean floor are connected to the surface platforms by long pipelines exposed to strong undersea currents. In summary, it is a difficult and dangerous operation to extract deep-sea oil from a small, floating platform so far above the wellhead. The very recent shift to deep offshore drilling, made urgent by the need for oil, means that the technology is continually evolving, which generates new risks and accidents. On April 20, 2010, an explosion aboard the Deepwater Horizon drilling rig led to oil gushing uncontrollably from the underground reservoir. For almost two months, 53,000 barrels a day, 5 million barrels in all, gushed from the well into the Gulf of Mexico, poisoning

5.8 (A) Shale deposits in the USA. **(B)** Flex drilling rig in North Dakota.

gas. This injection technique was developed more than fifty years ago but another innovation made it more powerful: the development of horizontal drilling in the 1990s that allowed drillers to go deep and then sideways to exploit widely dispersed reserves. By 2010 fracking had transformed the US energy sector.

The basic fracking technique is to drill down up to two miles and then drill horizontally as far as a mile. Fracking fluid, a mix of water and chemicals, is then pumped down at very high pressures to shatter rock formations to release the fossil fuels. The oil and natural gas are then pumped back to the surface along with the millions of gallons of flow-back. In 2000 there were 276,000 gas wells in the USA. By 2010 the number had doubled. Approximately 13,000 new wells are now drilled each year.

The impact of the fracking revolution was dramatic. In 2008 the USA produced 7.8 million barrels of oil a day and ranked third in the world. By 2015 it was producing 13.7 million barrels of oil a day and had become the largest oil producer in the world. Because of fracking the peak oil fear was (temporarily at least) squashed, a larger though not total energy security was achieved, and the price of energy to US industry and consumers

the water, contaminating the beaches, and killing sea creatures and animal life.

Second was the **fracking** revolution that involved the exploitation of oil and natural gas locked up in shale deposits. The distribution of major shale deposits in the USA is shown in Figure 5.8. Two of the biggest sites are the Marcellus field in the Northeast and the Bakken field in North Dakota. The oil and gas locked up in these shale deposits are released through hydraulic fracturing—high-pressure application of water, sand, and chemicals that essentially crack the rocks and release the oil and

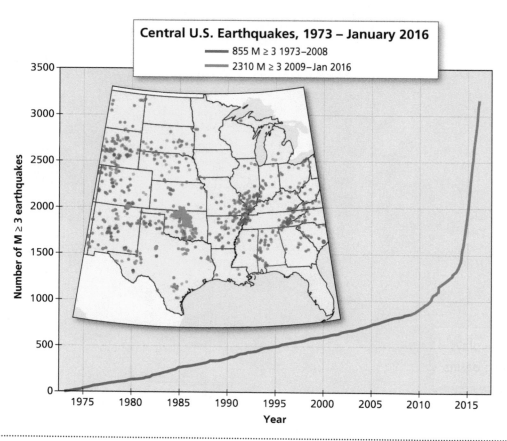

5.9 Central US Earthquakes, 1973–January 2016

fell. The cheap energy put pressure on other oil producers to reduce their prices. The price of a barrel of oil on the world market fell from $143 in 2008 to $30 in January 2016. There were many other factors at work, including the cooling of China's red-hot growth and consequent decline in global oil demand as well as oversupply, but the fracking revolution that provided cheaper oil and gas to the world's single biggest consumer was a significant factor.

There are downsides to the fracking revolution. Fracking pollutes the local water supply. The fluid that is used to fracture the rock contains acids, detergents, salts, lubricants, and disinfectant as well as other chemical additives to ensure that the oil and gas are released. A number of studies have identified benzene in local water supplies where fracking occurs. Fracking causes air pollution as some of the natural gas leaks into the atmosphere. There is also the strange case of fracking-induced earthquakes. In Ohio, Oklahoma, and Texas the number of earthquakes has increased in direct association with the increase in fracking (Figure 5.9). The culprit is the injection deep underground of the millions of gallons of wastewater byproducts. Before 2009 and large-scale

fracking in the state, Oklahoma experienced two earthquakes a year. Since then, it now experiences two a day. Earthquakes of magnitude 3 or higher have increased from 20 to 700 a year.

The Limits to Growth Revisited

The initial limits-to-growth argument was misplaced. The real issue is not that supplies of resources will run out but that their continued supply may create other costs and negative consequences. But what we have to ask is not "Is it going to run out?" but "Is continuing its supply worth the rising costs?" The ultimate law of resource use is that we always pay a price for what we get. The cost can be shifted from private to public, from local to national, or from one country to another, but the world is a closed system. The cost and benefits are always exacted, although they may be shifted across space. There is no free lunch.

Cited References

Almedia, P. D. and Silva, P. D. (2011) Timing and Future Consequences of the Peak of Oil Production. *Futures* 43: 1044–1055.

Bridge, G. and Wood, A. (eds) (2010) Geographies of Peak Oil: Special Issue. *Geoforum* 41: 523–605.

Crenson, M. A. (1971) *The Un-Politics of Air Pollution: A Study of Non-Decision Making*. Baltimore: Johns Hopkins University Press.

Gorham, E. (1955) On the Acidity and Salinity of Rain. *Geochimica et Cosmochimica Acta* 7: 231–239.

Jenkins, J., Roy, K., Driscoll, C., and Buerkett, C. (2007) *Acid Rain in the Adirondacks: An Environmental History*. Ithaca: Cornell University Press.

Llewellyn, G. T., Dorman, F., Westland, J. L., Yoxtheimer, D., Grieve, P., Sowers, T., and Brantley, S. L. (2015) Evaluating a Groundwater Supply Contamination Incident Attributed to Marcellus Shale Gas Development. *Proceedings of the National Academy of Sciences* 112: 6325–6330.

Meadows, D. H. (1972) *The Limits to Growth*. New York: Universe.

Odén, S. (1976) The Acidity Problem: An Outline of Concepts. *Water, Air and Soil Pollution* 6: 137–166.

Short, J. R. (1984) *An Introduction to Urban Geography*. Henley: Routledge.

Witze, A. and Stillwater, O. (2015) Artificial Quakes Shake Oklahoma. *Nature* 520: 418–419.

Woods, B. R. and Gordon, J. S. (2011) Mountaintop Removal and Job Creation: Exploring the Relationship Using Spatial Regression. *Annals of the Association of American Geographers* 101: 806–815.

Select Guide to Further Reading

Burns, S. S. (2007) *Bringing Down the Mountains: The Impact of Mountaintop Removal on Southern West Virginia*. Morgantown: West Virginia University Press.

Freese, B. (2003) *Coal: A Human History*. Cambridge: Perseus.

Holzman, D. C. (2011) Mountaintop Removal Mining: Digging into Community Health Concerns. *Environmental Health Perspectives* 119: a476–a483. http://dx.doi.org/10.1289/ehp.119-a476

Klare, M. (2010) *Rising Powers, Shrinking Planet*. New York: Henry Holt.

Kolbert, E. (2006) *Field Notes from a Catastrophe: Man, Nature and Climate Change*. New York: Bloomsbury.

Mehany, M. S. H. M. and Guggemos, A. (2015) A Literature Survey of the Fracking Economic and Environmental Implications in the United States. *Procedia Engineering* 118: 169–176.

Stokstad, E. (2014) Will Fracking Put Too Much Fizz in Your Water? *Science* 344: 1468–1471.

Willow, A. and Wylie, S. (2014) Politics, Ecology, and the New Anthropology of Energy: Exploring the Emerging Frontiers of Hydraulic Fracking. *Journal of Political Ecology* 21: 222–236.

Zimmerer, K. S. (ed) (2011) Geographies of Energy: Special Issue. *Annals of Association of American Geographers* 101: 705–980.

Websites

Alternative views on fracking
http://www.what-is-fracking.com
http://ecowatch.com/news/energy-news/fracking-2/
Data on coal production is available at:
http://www.worldcoal.org/resources/coal-statistics/
http://www.nma.org/index.php/coal-statistics
Data on global oil production and prices is available at:
http://www.iea.org/statistics/topics/oil/
http://www.api.org/statistics/USA
https://www.iea.org/aboutus/faqs/oil/

US induced earthquakes
http://earthquake.usgs.gov/research/induced/
US oil figures can be obtained at:
http://205.254.135.7/petroleum/
Check out the current CO_2 level at:
http://co2.earth

Interactive Map Websites

https://www.washingtonpost.com/graphics/national/power-plants/

http://www.marcellus.psu.edu/images/PA%20Animated%20Permit%20Map.gif

Learning Outcomes

- Understand how "nature" becomes commodity, with all the social and political consequences involved in this valuation and subsequent revaluations.
- The translation from inert material to valuable resource occurs through the interconnected processes of technical change and commodification.
- Coal was first used as a resource when timber reserves were depleted, and eventually came into common enough use that it became a significant air pollutant.
- The use of resources creates private costs and benefits as well as **social costs** and benefits. While the benefits may be privatized, the costs are often socialized.
- Coal powered the Industrial Revolution and would prove an important ingredient in the future manufacture of iron and steel, as well as an economic driver for many of the world's most successful economies, today producing approximately 42 percent of the world's electricity.
- The top coal producers in the world are, in order, China, the United States, India, the European Union, and Australia.
- Two major environmental concerns result from the use of coal as a major supplier of the world's energy: (1) burning coal produces noxious fumes that pollute the skies and cause global climate change; and (2) coal mining often involves environmental destruction as the land is poked and prodded, vegetation is destroyed, and waste material pollutes the soil and the air.
- Coal burning worldwide is responsible for one-third of all CO_2 emissions, and coal-fired power stations are the single biggest source of CO_2, a greenhouse gas, in the Earth's atmosphere. The increase in CO_2 emissions now has a well-established link to global climate change.
- The five laws of resource use are: (1) Resources become resources because previous resources become exhausted, unavailable, depleted, or too expensive; (2) resource use creates new geographies; (3) resource usage tends to have wider social and political consequences; (4) resource use has implications beyond private enterprises and into the wider community; and (5) attempts to regulate and diminish the negative public consequences of resource use always come up against powerful interests.
- Most of the twentieth century's economic growth was spawned by steady, cheap, and easily available oil.
- As oil prices have risen, alternative supplies of oil and gas have become economically feasible.
- The consequences of a peak in oil supply include the cost and difficulty of implementing alternative fuel sources, the long-term economic outlook of countries that depend on oil as a major source of economic output, and concerns surrounding existing socio-spatial assemblages that favor oil-dependent transportation sources.
- In terms of traditional supplies of oil, we have moved from an era of easy oil, one of large, shallow reserves that were easily exploitable, to an age of tough oil in which reserves are deeper, further offshore, and more expensive to exploit. The new fields are smaller and not as long lasting; they peak fast and then decline sharply.
- Oil is quite literally an essential lubricant of economic growth in the United States. The oil crunch, the large and growing gap between the national supply and demand, has led to a vigorous search for domestic oil reserves.
- Fracking in the US has provided a new and cheap source of oil and gas. However, there are environmental costs and greater risk of earthquakes associated with fracking.
- The initial limits-of-growth argument was misplaced. The real issue is not that supplies of resources will run out, but that their continued supply may create other costs and negative consequences.

6

People and the Environment

The environment is a source of economic opportunity, a repository of meaning, a sometime hazardous context, and always the background to our individual and collective lives. Examining the relationship between people and environment, which I will define as the world around us, is at the heart of human geography. In this chapter I will look at three broad themes: the environment as a source of cultural meaning, the environment's **affect** on human feelings and activity, and the impacts of human activity on the environment.

Environment and Cultural Meaning

We can make a distinction between three basic forms of social-spatial-economic relations: hunting-gathering, traditional agriculture, and urban-industrial systems. Each has overlapping but also distinct conceptions of environmental space. Respectively, they tend to inflate the importance of sacred space, productive space, and managed space.

Sacred Space

In hunting-gathering societies, people derive their individual and collective sustenance by hunting animals and gathering food. There was and often still is a gendered and age division of labor, with men responsible for hunting and women and younger children responsible for gathering other food. In hunting-gathering societies people need to be in tune with the seasonal rhythms of flora and fauna. The close connection with the environment provides both the material basis of life and spiritual sustenance.

Identify the three basic forms of socio-spatial-economic relations and their spatial significance.

Describe ways that humans have managed the environment and how economic dynamics have impacted perceptions of the environment.

Discuss societies' growing environmental awareness and the changing social construction of the environment.

Explain political ecology and relate it to the impacts of human activities on the global environment.

Outline some examples of major natural hazards and disasters by citing both their human and physical features.

The environment is not an inert container of resources but a living organism filled with beings, gods, and spirits that provide comfort and support as well as irritation and trials. Nature is the ground of being, powerful and present, and human–nature relations are not simply functional and economic but also filled with spiritual meaning and cosmological significance.

Cultural Encounters

Hunting-gathering societies came and continue to come into contact with people and societies with different environmental views and cosmologies. Here are just two examples.

Canadian explorer and geographer David Thompson (1770–1857) was traveling in western North America when he recorded the following encounter:

> Both Canadians [French Native Americans] and Indians often inquired of me why I observed the sun, and sometimes the moon, in the day-time, and passed whole nights with my instruments looking at the moon and stars. I told them it was to determine the distance and direction from the place I observed to other places; neither the Canadians nor the Indians believed me; for both argued that if what I said was the truth, I ought to look to the ground.

This encounter encapsulates the difference between the search for the coordinates of the grid and a reliance on a more intuitive sense of place. We now look with more sympathy upon the "Canadians and Indians" than the Enlightenment figure of David Thompson struggling to measure and map the world.

We now have accurate measurements of the Earth. Latitude and longitude can now be easily and precisely calibrated. Handheld GPS (global positioning systems) allow observers to get a very accurate and almost immediate reckoning of their coordinates in the grid. But perhaps because this avenue has been so fully explored and developed, there is a renewed interest in the areas outside the arc of traditional geography. We also need to develop the more empathetic understanding that made the Canadians and Indians tell Thompson to "look to the ground."

Europeans "explored" and mapped large parts of Africa, America, and Asia. Despite the traditional view that Europeans created such maps on their own, local and indigenous peoples helped considerably. It is more accurate to consider the notion of cartographic encounters involving Europeans and local peoples, rather than a simple cartographic appropriation by Europeans. The mapping of the continents was underpinned by native knowledge. There is a hidden stratum of indigenous geographical knowledge that is only now being uncovered.

Take the case of North America. The Europeans landed in a populated place. To find their way and move around the unfamiliar landscape, they relied upon the indigenous people to provide information, advice, and guidance. In return, the indigenous people obtained trade goods, spatial information, and the possibility of new alliances against longstanding enemies. This exchange was invaluable to the Europeans over both the short and long term. For the Native Americans, the exchange over the short and medium term could prove useful, but over the long term it sealed their fate. The cartographic encounters resulted in Native Americans losing control over their territory. No longer needed, they were moved and marginalized, and often killed. In the case of North America, the mapping of the land embodied a *symbiotically destructive* relationship. It allowed Native Americans to parlay their deeper and wider knowledge of the land into a strong bargaining position, but in this very exchange, where they had some leverage, lay the roots of their loss of the land. They used what resources they had to gain a short-term advantage, but over the long term the cartographic encounter gave the newcomers more power. There were cartographic encounters between Europeans, European Americans, and Native Americans. The maps made by the explorers drew heavily upon an indigenous cartographic contribution.

Productive Space

Joseph Campbell made a distinction between the environmental cosmology of the hunter-gatherers, concerned more with the "way of animal powers," and that of agricultural peoples, concerned more with the "way of the seeded earth." It is a shift from the notion of sacred space to that of productive space. The distinction is perhaps too clear-cut. In reality, many traditional agricultural societies grew out of hunting-gathering groups and still retained elements of the earlier environmental beliefs. Along the Atlantic seaboard of what is now the United States, people hunted the mammoth, bison, and other megafauna, but as the climate warmed at the ending of the Pleistocene Ice Age, farming became more intensive. Hunting-gathering cosmologies, concerned with the relations with animals, were carried over into agricultural cosmologies concerned with the fertility of the land. Maize, beans, squash, tobacco, potatoes, and corn were grown throughout the central Atlantic region. The Iroquois referred to corn, beans, and squash as the "Three Sisters" or "Our Supporters." Growing the three crops in the same field was an agricultural innovation that increased productivity and crop yields among indigenous peoples. The corn provided structure for the other plants, the beans were valuable source of nitrogen for the soil, and the squash provided protective mulch.

An annual cycle of festivals and rituals, especially at planting and harvesting times, respectively invoked continued agricultural plenty and gave thanks for it. Animal spirits were still respected; for instance, the panther was an animal of great spiritual significance for the traditional Iroquois.

The agriculturalists tended more to stress the sacrifice, symbolic and real, in order to promote and sustain fertility. Ensuring the fertility of the land is vital to agricultural societies and central to their belief systems. In some societies, such as the Aztec, the sacrifice took on a literal meaning as humans were offered up to the gods on altars in public displays. From its agricultural roots the practice grew to become an instrument of political coercion and control. More than 4,000 human sacrifices were made at a reconsecration of the Great Pyramid of Tenochtitlan in 1487. The idea of sacrifice continues to this day in contemporary religious practices. The Christian Eucharist, which signifies the body and blood of Christ, is an updated and particularized version of the older and more widespread sacrifice narrative of agricultural societies.

As we move toward more **commodified agriculture** where we grow things for profit rather than to simply eat, we shift from the idea of land as a primary source of meaning and identity to land as a commodity, a vehicle to accumulate wealth. Today, modern industrial agriculture relies on chemicals and fertilizers rather than ritual sacrifice.

Managed Space

The rise of cities and the development of industry create new environmental belief systems. On the one hand, there is a strengthening of the belief in technological mastery of the world. The success of urban industrialization in transforming our experience of the world suggested that nature could be managed and controlled rather than revered and respected. A technological triumphalism saw few limits and no constraints. Nature was less a living thing and more a resource to be exploited, a backdrop for human desires and social needs. As land was commodified, the particularities of place as spiritual significance were replaced by the generality of market space. Particular places had economic value but not cosmological significance. The environment was merely a calculus of economic costs and benefits, and nature was now a commodity to be bought, sold, and traded.

Yet, on the other hand, the managed spaces also included religious structures. And the environmental degradation caused by the exploitative attitude did generate counterclaims and alternative visions. **Romanticism** was

6.1 The productive space of a rice paddy in Bali, Indonesia

a reaction to the scientific rationalism of nature and the commodification of land. It saw nature as a source of profound aesthetic experiences and a perfect vehicle for a personal connection to the infinite. As factories were being built and cities replacing fields, William Wordsworth was singing the praises of daffodils and Henry David Thoreau was espousing the virtues of the woods and lonely places. All kinds of public health measures and environmental movements grew up to contest the dominant narrative of nature simply as resource. In the United States, for example, because industrial growth was accelerating, John Muir was extolling the virtues of protecting wilderness. Rapid industrialization and urbanization generated resistances and new ideologies that stressed preservation and conservation. Scientific discourses such as **ecology** developed out of the concern to consider the links that bind all living things. Public policies such as wilderness preservation and land conservation were also developed. The promotion of sustained economic growth now competed with a belief in environmental protection.

Today we have competing claims. Powerful forces promote continued economic growth and tend to see nature primarily as a commodity to be developed and exploited. Arguments about jobs or economic growth are often used to trump "merely" environmental concerns. Yet we also have an increasingly powerful awareness of the environmental consequences of unrestrained growth. The proponents of sustainability often draw upon the ideas of hunting-gathering belief systems. The "rediscovered" (invented?) traditions of indigenous peoples are touted often as a more sustainable and sustaining environmental ethic suitable for our present predicament. We live in a world where economic growth and global environmental impacts are unprecedented, but also in a time when a new environmental ethic, one shaped both by past belief systems and present practices, is taking shape. Its main ideas include a greater awareness of our environmental impact,

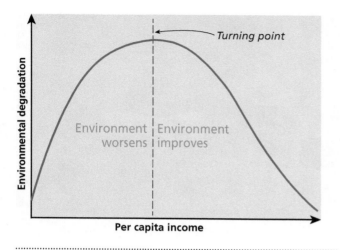

6.2 Environmental Kuznets Curve

a less triumphalist optimism in technology, and a greater concern with long-term sustainability than short-term growth.

The link between economic growth and environmental quality is sometimes represented as a bell-shaped curve termed the **Environmental Kuznets Curve** (Figure 6.2). (The adjective is placed in front of the proper noun because there is another Kuznets Curve that depicts a similar relationship between per capita income and inequality.) At the early stages of economic growth, when per capita income is still relatively low, the environment worsens as aggregate growth is pursued despite the environmental effects. Then, as incomes increase, more people place priority on environmental quality. Evidence can be found that both contests and confirms this dynamic between income and environment. In terms of counterargument, for example, levels of carbon emission continue to increase with rising per capita incomes. Moreover, richer regions and countries may export the negative environmental consequences of their high growth through shipping pollution waste or offshoring the more environmentally damaging economic practices. Poorer countries and regions also may be able to affect environmental quality much earlier than the curve suggests, as social movements and governments are made acutely aware of the damaging consequences and the false nature of the jobs-versus-environment slogan. Eco-communities in regions across China have promoted environmental improvement despite low per capita income.

In terms of environmental awareness and appropriate policy responses, there is now a wide variety of opinion. At one extreme is "**deep ecology.**" The term was first used by Norwegian philosopher Arne Næss to refer to a deep reverence for Nature that implies the radical restructuring of our current lifestyles to be more biocentric. At the other end of the continuum is an environmental

sensitivity that suggests less radical change but a more pronounced commitment to alternative technologies, reducing pollution, and sustainable development. The deep ecologist questions the morality of growth, while the environmentalist/conservationist wants sustainable growth.

There is now a vigorous environmental debate with an increasing number of strands of social ecology that link environmental issues more directly to political issues, and indeed sees the environment as a deeply political issue. For example, **ecofeminism** highlights the gendered nature of many environmental ideologies and practices and tries to engage feminist and ecological issues in a shared dialogue. What they all share is a questioning of the belief in unbridled economic growth and a commitment to a new relationship between humans and nature.

Environmental Impacts on Society

Recent geographical scholarship is developing the idea of "affect," or an emotional effect, to refer to the force that makes us feel, think, or act. At the level of individuals, we also know that certain very local environments have an affect. People recovering in hospital rooms that look out onto attractive scenes of nature get better more quickly than those whose views are only dismal walls. The affect of the micro-environment in influencing moods and behaviors is exploited by restaurants that play discordant music to stop people from lingering too long and by stores that play soothing music to make consumers shop longer. Placing mirrors in public spaces so that people can see themselves reduces vandalism. We act differently in public than when we are in more private spaces. Our mood and behavior are both affected by our environment. A group of Dutch researchers found that there was a very positive relationship between the amount of green space in the local environment and people's self-reported indicators of physical and mental health. The relationship was strongest for reports of anxiety. In other words, people said they felt less anxious if there was more green space in their local neighborhood. Local environments do play a role in people's sense of well-being.

There is also the **bystander effect**. It is a well-known finding that individuals often do not help people in distress if, for example, they scream for help in a public place. The effect is more noticeable in dense urban areas than in rural areas. The difference is not due to rural dwellers being more caring; rather it is the result of more stimuli in urban environment, so that people may not hear or see the distress, and because there tend to be more people, so there is a wider diffusion of responsibility.

Different environments not only exaggerate and encourage certain forms of behavior over others; they also have specific forms of behavior encoded. Our dress and demeanor at a job interview, for instance, are different than when we are in a club with friends.

The context of place is important. People who meet for the first time in stressful situations have greater attraction to each other than if they meet first in less stressful situations: same people, different context, and so different outcomes. The title of Sam Summers's book *Situations Matter* summarizes a very interesting body of work. Place matters in group creativity. Colleagues who work closer together tend to do better work than when they are situated farther apart. Small, flexible spaces are places where creativity between people can flourish, a point not lost on designers of buildings for creative industries, who work to ensure close contact and random encounters in everyday work life.

There is no simple answer to the question of what role the local environment plays in guiding human feelings and behaviors. There is a complex social construction of the environment as people and groups give meanings to space and place, which change over time.

Consider the city of Tunis. While it was under French occupation, a major avenue was constructed to link the city to the port area (Figure 6.3). It was built to represent the French presence, a central boulevard surrounded by roads and bordered by wide pavement that specifically echoed the Champs-Élysées, so that Tunis could be seen as a smaller version of Paris, both linked by their expression of French culture and power. French and local elites paraded along the street to show their embrace and display of modernity. It was built to make visible the connection between France and Tunisia, to show the rationality and technological superiority of France compared to medieval religiosity. After independence, the street was renamed after a revolutionary hero and was known as Avenue Habib Bourguiba, which came to represent a modern independent Tunisia and was the place for celebrations of national identity and solidarity. In late 2010 it became the scene of demonstrations against the regime of Zine El Abidine Ben Ali. As part of the Arab Spring, young people flooded into the street, gaining strength from each other's presence, hardening their resolve of the protestors, and highlighting the lack of support for the status quo. The street has now become a place for young people to show their defiance of the regime and display their resistance. One street with multiple meanings and very different affects.

6.3 Avenue Habib Bourguiba in Tunis, Tunisia

Human Impacts and Environmental Change

The relationship between society and nature can also be considered from the opposite direction by looking at the impacts of humans on the environment. There are two distinct approaches. The first is an accounting of human impacts on the environment. This has a long history. George Perkins Marsh (1801–1882) was one of the first to argue that deforestation could lead to desertification. His book *Man and Nature*, first published in 1864, indicates some of the connections between human actions and environmental impacts. A classic and foundational work by William Leroy Thomas, *Man's Role in Changing the Face of the Earth* (published 1955), deals with the rising dominance of nature by humans and looks at such issues as deforestation, urbanization, and the depletion of resources (Figures 6.4 and 6.5). This work has influenced generations of geographers. An updated edition of *The Earth as Transformed by Human Action* (1990) covers subsequent research. An important aim of this ongoing body of work is to examine and measure the major impacts of human action on the environment, including:

1. destruction of habitats and loss of biodiversity
2. pollution of water and soil
3. land use changes (e.g., desertification, deforestation, loss of productive farmland urbanization)
4. global climate change.

The best geographical research is not simply an accounting and mapping of impacts but also a deeper consideration of the social and political effects and the drivers of these changes. We will consider just one of the very many examples: a study of land use changes in Colombia since 1500.

6.4 Land use change 1: creating sugar cane plantations from tropical rainforest in Queensland, Australia

The researchers identified seven periods of landscape change, listed in Table 6.1. From 1500 to 2000, there was marked long-term decline in forest cover. There was also a drastic reduction in the nomadic hunting-gathering in the savannas and a large increase in the amount of grazing on cleared land and natural grasslands. The periodization used

6.5 Land use change 2: the urbanization of the rural landscape, Maryland

in Table 6.1 references political and economic conditions such as conquest, colonialism, the development of the state, and the growth of international markets. These changes have an impact on land use (Figure 6.6). Land use change, as this example clearly shows, is linked to political and economic changes.

Ground Truthing

The impact of human land use on ecosystems is measured by using traditional methods, such as map comparison, as well as more modern means like **remote sensing**. If we use remotely sensed data, it is important to identify finer-grained processes in order to see the effects and impacts on households and their vulnerabilities and coping strategies. This is sometimes referred to as **ground truthing**. It is an evocative phrase that encapsulates the very nature of human geography. Geographers seek to identify the truth on the ground by utilizing different scales of analysis, from the telescope to the microscope, and by employing aggregate analysis as well as detailed interview techniques.

There is bias in much of ground truthing. Laura Martin and colleagues looked at the location of 2,573 terrestrial study sites that were the basis of studies published from 2004 to 2009 in the top ten major ecological journals. They found a distinct locational bias toward protected areas and temperate deciduous woodlands in wealthy countries. Even when studies considered urbanized and settled areas, they focused on fragments of "natural" areas. Large parts of the global south and the densely populated areas of the world—the places where most of us live and work—are underrepresented in ecological studies. Not quite truthing of all the ground.

Political Ecology

There is now a large body of work that looks specifically at the relationships between socioeconomic factors and environmental changes. One line of work is termed **political ecology**, since it highlights political as much as ecological issues. The term was first used in the 1930s but came to higher prominence in the 1980s. Piers Blaikie looked at soil erosion in developing countries. In his work on Nepal, he showed that rather than being the result of overpopulation or mismanagement, soil erosion resulted from marginalization and poverty. Soil erosion was not a function of peasant farmers' mismanagement but a reflection of their precarious position in commodity markets. In his 1983 book *Silent Violence: Food, Famine, and Peasantry in Northern Nigeria*, Michael Watts looked at the political context of household vulnerability to hunger. He built on Sen's work on the social construction of hunger and famine that we discussed in chapter 4, and in particular examined the role of

TABLE 6.1 Land Use Changes in Colombia 1500–2000

ERA OF LANDSCAPE CHANGE	DATE	MAIN LAND USE CHANGES
Pre-Spanish	Before 1500	80 percent nomadic hunting-gathering in savanna
Conquest	1500–1600	Reversion to forest due to demographic collapse
Colonial period	1600–1800	Forest clearance
Independence and state formation (1800–1850)	1800–1850	Clearance of land for cattle grazing
International markets	1850–1920	Marked forest clearance for grazing
Early urbanization	1920–1970	Deforestation
Industrialization	1970–2000	Deforestation

Source: After Etter et al., 2008.

colonialism and agrarian capitalism in subjecting the Hausa-speaking peasantry to food insecurity. Hunger was not the result of drought or resource scarcity but of political relations.

Political ecology looks at the struggle over resources, who controls them and how the costs and benefits are apportioned. Political ecology situates land use changes in their socioeconomic context and links strategies of corporate profitability to how households are made more or less vulnerable to environmental hazards such as deforestation, social erosion, or irregular food supply. Rather than a neutral backdrop, the environment is the very field in which power is revealed, contested, and enforced.

While many of the early political ecology studies focused on developing societies and rural land use changes, more recent work widens the approach to look at the city. Urban political ecology is now a vibrant part of geographical research.

Climate Change

One of the most significant impacts of human activities on the environment is global climate change. There is now a large body of work that measures and seeks to explain climate change. Global warming, for example, is caused by many factors, but one important factor is heavy use of fossil fuels that increase atmospheric carbon dioxide, which traps sunlight that warms the atmosphere and the Earth's surface. In broad outline, the effect includes an increase in global temperatures by about 1.5°F since 1880, a decrease in the amount of sea and land ice, and rising sea levels on an average of 3.19 mm per annum. Much of the basic data and some striking images are available at the NASA website (http://climate.nasa.gov/). Climate change produces a variety of different effects. The official federal site that collates data on the United States has a regional breakdown. In the American Southwest, key issues include decreasing water supplies, increasing forest wildfires, and flooding. In the Southeast, major elements include significant sea level rise, higher seawater temperatures, and ocean acidification. In the Northeast, changes include increased warming and poorer air quality, severe flooding due to sea level rise, and reductions in fish stocks, especially cod. Some of the complicated suite of impacts on the United States is illustrated in Figure 6.7.

A volume on climate change, edited by Richard Aspinall, gives a sense of the range of global dynamics. It contains articles examining glacier melt in the Andes and the effects on household vulnerability; increasing drought in Jamaica and the impact on local farmers; the "**double exposure**" of climate change and economic restructuring in

6.6 Land use change in Colombia

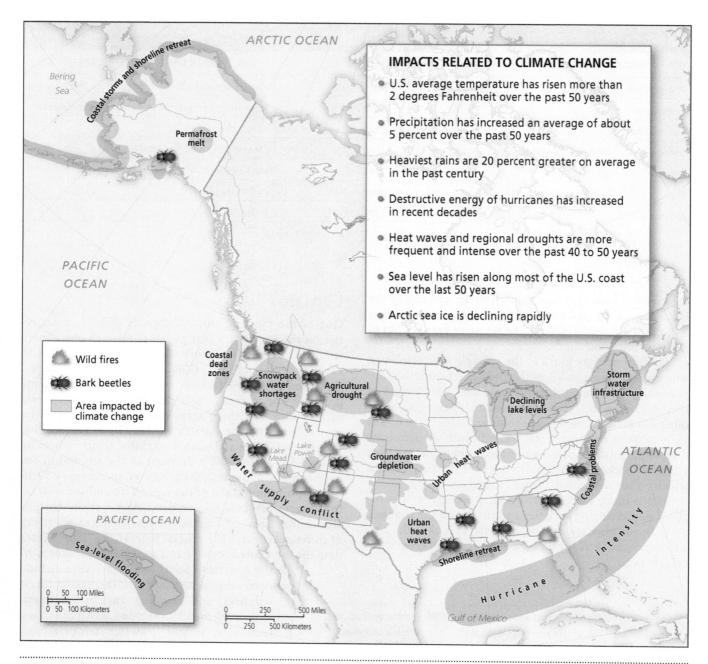

IMPACTS RELATED TO CLIMATE CHANGE

- U.S. average temperature has risen more than 2 degrees Fahrenheit over the past 50 years

- Precipitation has increased an average of about 5 percent over the past 50 years

- Heaviest rains are 20 percent greater on average in the past century

- Destructive energy of hurricanes has increased in recent decades

- Heat waves and regional droughts are more frequent and intense over the past 40 to 50 years

- Sea level has risen along most of the U.S. coast over the last 50 years

- Arctic sea ice is declining rapidly

6.7 Impact of climate changes in the United States

California's Central Valley; and the renewed geopolitical rivalries in the Arctic because warming is leading to new shipping routes, easier access to natural resources, and the ability to project a military presence. The general conclusion to this exciting body of work is that climate change is enmeshed in social structures and economic processes at all levels from the households to the firm and the state. Climate change generates new household vulnerabilities and stimulates new coping strategies as well as changes in geopolitical relations.

Landscape

One line of research in the society–nature theme looks more specifically at landscapes. In his classic 1955 study *The Making of the English Landscape*, W. G. Hoskins explored the English landscape much as an archeologist explores an excavation. The different levels in the landscape, beginning with the pre-Roman and continuing all the way through the industrial, were brought to the surface for inspection, cleaned of dust, and examined. In direct reference to Hoskins's work, the geographer Michael Conzen's 2010 book is entitled *The Making of the American Landscape.*

Landscapes can be viewed as archive, nature, habitat, artifact, ideology, history, place, and aesthetic. Landscape is a complicated text, telling us who has power and how and why that power is wielded. Landscape making has connection with deeper economic and political relations. We can make a distinction between dominant and alternative cultures. Landscapes reflect the dominant culture, but have traces of alternative cultures. Don Mitchell connects the making of the California landscape to agricultural workers and their labor processes. This labor history of landscape highlights the conflicts between workers and their employers.

Landscapes can be interrogated for their connection with capitalism, national identity, social justice, race, and class. Landscape is important in the creation of personal, regional, and national identity and historical memory.

Hazards and Disasters

One of the most visible and dramatic examples of people–nature relations is the experience of hazards and disasters. The tsunami wave that sweeps away the coastal settlements, the storm surges that flood cities, and the flash flood that washes away entire neighborhoods are vivid and searing images that remind us of our environmental vulnerability. We live in a hazardous world of earthquakes, floods, volcanic eruptions, and violent storms. We have a better sense of the underlying dynamics and hence predictability of storms, floods, and eruptions. Yet our knowledge is not absolute. The earthquakes that struck Haiti in 2010 and Italy in 2016 or the tsunamis that overwhelmed towns across the Indian Ocean coastline in 2004 and the Fukushima power plant in Japan in 2011 were unforeseen events. We do know the location of fault zones, but precise plate shifts cannot be predicted within actionable certainty.

We need to be very careful in using the term "natural disasters," because they tend to be very social events in terms of who is affected most. Vulnerability to hazards magnifies as the population increases, settlements become more dense, and more people live in marginal places. I use "marginal" in the double sense implied by "on the edge of things." First, places like the Caribbean and southern Atlantic seaboard of the United States are zones of hurricane activity (Figure 6.8). If we build houses and cities in hurricane zones, it should come as no surprise that someday storms will wash away beachfront property. Hazards become disasters when we build on fault lines and develop along hurricane coasts—especially if we build cheap and quickly without due deference to possible hazards. Marginality also occurs when people are forced to inhabit places such as the squatter settlements on steep slopes or slums on floodable plains and thus are more vulnerable to environmental hazards. The poor are made more vulnerable because they are often forced to live in the more environmentally hazardous areas. "Natural" disasters are less acts of nature and more the result of human actions. It is important to consider the social context of vulnerability to hazards.

How hazards impact households or become disasters is mediated through the prism of social and economic power. Disasters do not affect everyone equally. Based on a sample of 141 countries, Neumayer and Plumper show that disasters killed more women than men and that the higher the women's socioeconomic status, the smaller the gender gap in mortality. The most impacted were low-income women. "Natural" disasters, on closer inspection, are very social in their outcomes and effects.

Bear in mind the fault lines of power, class, and race revealed in these three brief examples:

1. In the Grand Cayman Islands, situated firmly in a zone of hurricane activity (Figure 6.9), there are two responses to the threat of hurricane. The very wealthy, with the warning provided by accurate forecasting, hop on a plane and simply fly away from the island to safety. The poorer inhabitants of the island, unable to flee, have to wait out the storm.

2. Rising temperatures brought about by **global warming** have different effects. Increasingly hot days have less impact on mortality in the United States because it is a relatively rich country with access to air conditioning, either individually or collectively. In rural India, in contrast, an increase from 70 to 90 degrees increases mortality by more than 1 percentage point above baseline rate. The same hazard of global warming, but with very different outcomes, depending on who you are and where you are.

3. The drowning of New Orleans in the wake of Hurricane Katrina in 2005 was anything but a natural disaster. The levees were overwhelmed because they were inadequately built and poorly maintained. The poor environmental management that allowed marshland to erode, and the construction of a navigation canal, heightened the damaging impact of the storm surge. The disaster was not a natural disaster caused by unforeseen, unpredictable events. It was the result of building a city in an area of environmental hazard, pursuing a destructive environmental management policy that enhanced risk, failing to build and maintain the necessary infrastructure to protect the city, and doing little to help the most vulnerable citizens of that city in the event of flooding.

The experience of environmental hazards and the way that hazards become disasters tell us as much about the human geography of class and power as they do about the physical geography of weather and landforms.

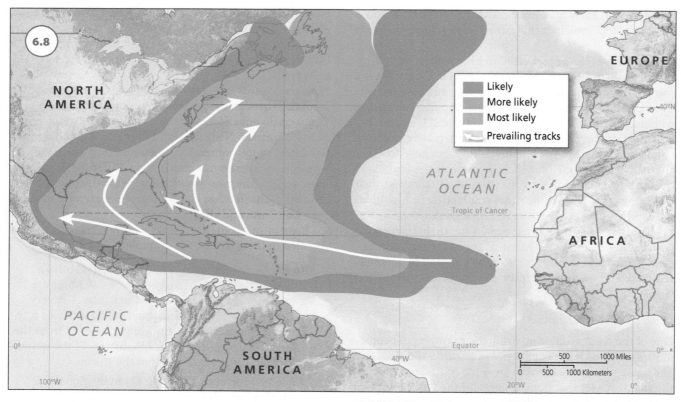

6.8

Likely
More likely
Most likely
Prevailing tracks

NORTH
AMERICA

EUROPE

~40°N

ATLANTIC
OCEAN

Tropic of Cancer

AFRICA

PACIFIC
OCEAN

0°

SOUTH
AMERICA

40°W

Equator

0°

100°W

20°W

0°

0   500   1000 Miles
0   500   1000 Kilometers

6.8 Hurricanes in the Caribbean and Eastern United States
6.9 Hurricane effects in Grand Cayman

6.9

The Anthropocene: Living in a Modified Earth and Socially Constructed Nature

A popular text that summarizes much of the research on people–nature themes is simply titled *The Human Impact on the Natural Environment*. It provides a good survey of material and case studies. But look again at the title. It assumes that people are impacting a "natural" environment. The increasing amount of human impact raises the question: is there such a thing as the natural environment, or have we as humans so transformed the Earth that most of us now live in human-modified environments? Bill McKibben, a popular writer on ecology, wrote a book with the suggestive title *The End of Nature*. More recently, in his book *Eaarth*, he argues that we have wrought such tremendous change that we need a new term—the title of his book. Think of it as Earth 2.0, an Earth so modified that it counts as a new form. He focuses on climate change to

show that changes are creating permanent drought, rapid melting of the polar ice caps, an expansion of the tropical region by as much as two degrees of latitude north and south, and increased ocean acidity. This is not, he argues, some short-term change that will soon revert back to a more "natural" state. We live more and more in a world quite literally of our own making. It is also a world of our own writing. A number of scholars point to the social construction of nature, by which they mean that Nature is not just something out there; it is not only appropriated and used but also represented and understood by humans. The title of an edited book sums up the basic idea: *Social Nature*.

6.1 The Tragedy of the Commons?

In a paper published in 1968, Garret Hardin described the tragedy of the commons. The setting for the tragedy occurs, according to Hardin, when individuals use a collective resource. He gives the example of a common pasture used by individual herders. The tragedy happens because each individual, seeking to maximize their personal interest, tends to overuse the commons, and the result is deterioration of the common space. The solution lies either in some form of government regulation or a privatization of the common property.

The image invoked by Hardin's paper is compelling, yet problems remain with the concept. Hardin describes the case of open-access resources, whereas in many traditional societies the commons were, and sometimes still are, either a common pool resource or, more often, common property. The tragedy of the commons occurs in traditional societies less frequently than Hardin suggests because they were regulated through collective property agreements.

In the most sustained examination of the issue, Nobel Prize winner Elinor Ostrom argues against the simple solution of either privatization or government regulations. She proposes a number of principles for common-pool resource management, including cheap and easy access to conflict resolution, community rights linked with higher levels of regulation, encouragement of collective decision making, and effective exclusion of unentitled parties.

The environment can be considered as a common-pool resource that is susceptible to overuse and deterioration. The depletion of deep-sea ocean fisheries, for example, is a case where users, acting rationally on the individual level to harvest as many fish as possible, collectively undermine the resource base. Global climate change, to take another example, results from the action of carbon polluters who do not pay the full cost of their actions. If we see the environment as a common-pool resource, a very valuable and fragile one, then we can imagine a range of policy responses along the line suggested by Ostrom.

REFERENCES

Hardin, G. (1968) The Tragedy of the Commons. *Science* 162: 1243–1248.

Lant, C. L., Ruhl, J. B., and Kraft, S. E. (2008) The Tragedy of Ecosystem Services. *BioScience* 10: 969–974.

McWhinnie, S. F. (2009) The Tragedy of the Commons in International Fisheries: An Empirical Example. *Journal of Environmental Economics and Management* 57: 321–333.

Ostrom, E. (1990) *Governing the Commons: The Evolution of Institutions for Collective Action*. Cambridge: Cambridge University Press.

Ostrom, E. (2009) A Polycentric Approach for Coping with Climate Change. *Social Science Research Network*. http://papers.ssrn.com/sol3/papers.cfm?abstract_id=1934353

We are now in a new geological age, the **Anthropocene**, a name used to indicate the profound and fundamental restructuring caused by human agency. The term was first coined by Nobel Prize winner Paul Crutzen. Erle Ellis and colleagues have mapped the world's anthromes, biological areas that are shaped by human impacts.

This profound transformation comes with costs and benefits. The loss of more natural ecosystems, while providing benefit in the short term, may have longer-term costs. A major report by the United Nations, *The Global Biodiversity Outlook*, suggests that some ecosystems are close to the tipping point at which they will become less useful to humans. One indicator of the changes is the decline in biodiversity. More than one in five of all mammal species and close to one in three of all amphibian species are close to extinction. And yet it is only recently that we have started to put a specific value on nature and develop an economic metric for measuring biodiversity.

There is also an emerging movement known as **rewilding** that aims to create landscapes and ecosystems similar to those that existed in Paleolithic times. This ecological strategy consists of the reintroduction of predators such as wolves and cougars into core reserves, with migratory corridors linking the cores. It is often summarized up as "cores, corridors, and carnivores." A 15,000-acre reserve in the center of the Netherlands known as Oostvaardersplassen was created in the 1980s and stocked with a variety of animals in an attempt to recreate a European wilderness at the end of the last Ice Age.

We inhabit a modified Earth and a socially constructed nature. We now live in a world of our own making.

Cited References

Aspinall, R. (ed) (2010) Special Issue: Climate Change. *Annals of Association of American Geographers* 100: 715–1045.

Blaikie, P. M. (1985) *The Political Economy of Soil Erosion in Developing Countries.* London: Pearson.

Burgess, R., Deschenes, O., Donaldson, D., and Greenstone, M. (2009) Weather and Death in India. http://www.iig.ox.ac.uk/output/presentations/pdfs/E12-weather-and-death-in-india.pdf. Accessed January 24, 2012.

Campbell, J. (1983) *Historical Atlas of World Mythology.* San Francisco: Harper & Row.

Castree, N. and Braun, B. (eds) (2001) *Social Nature: Theory, Practice, and Politics.* Malden and Oxford: Blackwell.

Conzen, M. (ed) (2010) *The Making of the American Landscape.* New York: Routledge.

Crutzen, P. (2002) Geology of Mankind. *Nature* 415: 23.

Ellis, E. E., Goldewijk, K. K., Siebert, S., Lightman, D., and Ramankutty, N. (2010) Anthropogenic Transformation of the Biomes, 1700 to 2000. *Global Ecology and Biogeography,* 19: 589–606.

Etter, A., McAlpine, C., and Possingham, H. (2008) Historical Patterns and Drivers of Landscape Change in Colombia since 1500: A Regionalized Spatial Approach. *Annals of Association of American Geographers* 98: 2–23.

Global Diversity Outlook https://www.cbd.int/doc/publications/gbo/gbo3-final-en.pdf. Accessed January 20, 2012.

Goudie, A. S. (2013) *The Human Impact on the Natural Environment: Past, Present and Future.* 7th ed. Oxford: Blackwell.

Hoskins, W. G. (1955) *The Making of the English Landscape.* London: Hodder and Stoughton.

Lui, L. (2008) Sustainability Efforts in China: Reflections on the Environmental Kuznets Curve through a Locational Evaluation of "Eco-communities." *Annals of the Association of American Geographers* 98: 604–629.

Maas, J., Verheij, J. R. A., de Vries, S., Spreeuwenberg, P., Schellevis, F. G., and Groenewegen, P. P. (2009) Morbidity Is Related to a Green Living Environment. *Journal of Epidemiological Community Health* 63: 967–973.

Martin, L. J., Blossy, B., and Ellis, E. (2012) Mapping Where Ecologists Work: Biases in the Global Distribution of Terrestrial Ecological Observations. *Frontiers in Ecology and the Environment* 10: 195–201.

McKibben, B. (2010) *Eaarth: Making a Life on a Tough New Planet.* New York: Time Books.

Meinig, D. (ed) (1979) *The Interpretation of Ordinary Landscapes.* New York: Oxford University Press.

Mitchell, D. (1996) *The Lie of the Land: Migrant Workers and the California Landscape.* Minneapolis: University of Minnesota Press.

Neumayer, E. and Plümper, T. (2008) The Gendered Nature of Natural Disasters: The Impact of Catastrophic Events on the Gender Gap in Life Expectancy, 1981–2002. *Annals of Association of American Geographers* 97: 551–566.

Short, J. R. (2000) *Alternative Geographies.* New York: Prentice Hall.

Short, J. R. (2009) *Cartographic Encounters: Indigenous Peoples and the Exploration of the New World.* London: Reaction.

Summers, S. (2011) *Situations Matter: Understanding How Context Transforms Your World.* New York: Riverhead.

Thomas, W. L. (ed) (1955) *Man's Role in Changing the Face of the Earth.* Chicago: University of Chicago Press.

Turner, B. L., Clark, W. C., Kates, R., Richards, J. F., Mathews, J. T., and Meyer, W. B. (eds) (1993) *The Earth as Transformed by Human Action.* Cambridge: Cambridge University Press.

Watts, M. J. (1983) *Silent Violence: Food, Famine and Peasantry in Northern Nigeria.* Berkeley: University of California Press.

Select Guide to Further Reading

Benton, L. M. and Short, J. R. (1999) *Environmental Discourse and Practice.* Oxford: Blackwell.

Benton, L. M. and Short, J. R. (2000) *Environmental Discourse and Practice: A Reader.* Oxford: Blackwell.

Cosgrove, D. (1984) *Social Formation and Symbolic Landscape.* London: Croom Helm.

Cosgrove, D. (2008) *Geography and Vision: Seeing, Imagining and Representing the World.* London: I. B. Tauris.

Foley, S. F. et al. (2013) The Palaeoanthropocene—The Beginnings of Anthropogenic Environmental Change. *Anthropocene* http://dx.doi.org/10.1016/j.ancene.2013.11.002.

Hartman, C. and Squires, G. D. (eds) (2006) *There Is No Such Thing as a Natural Disaster: Race, Class, and Hurricane Katrina.* New York: Routledge.

Hecht, S. B. and Cockburn, A. (2011) *The Fate of the Forest: Developers, Destroyers, and Defenders of the Amazon.* Chicago: University of Chicago Press.

Howard, P., Thompson, I. H., and Waterton, E. (eds) (2012) *The Routledge Companion to Landscape Studies.* New York and London: Routledge.

Mitchell, D. (2012) *They Saved the Crops: Labor, Landscape, and the Struggle over Industrial Farming in Bracero-Era California.* Athens: University of Georgia Press.

Montz, B. E. and Tobin, G. A. (2011) Natural Hazards: An Evolving Tradition in Applied Geography. *Applied Geography* 3: 1–4.

Olwig, K. and Mitchell, D. (eds) (2008) *Justice, Power and the Political Landscape.* New York: Routledge.

Peet, R., Robbins, P., and Watts, R. (eds.) (2010) *Global Political Ecology.* New York: Routledge.

Wisner, B., Blaikie, P., Cannon, T., and Davis, I. (2004) *At Risk: Natural Hazards, People's Vulnerability and Disasters.* London: Routledge.

White, L., Jr. (1967) The Historical Roots of Our Ecological Crisis. *Science* 155: 1203–1207.

Websites

The Intergovernmental Panel on Climate Change
   http://www.ipcc.ch/
NASA website on global climate change
   http://climate.nasa.gov/
US climate change and impacts
   http://www.globalchange.gov/
Measuring global diversity
   http://www.teebweb.org/

NASA Earth observatory
   http://earthobservatory.nasa.gov/
NASA land use changes
   http://climate.nasa.gov/images-of-change?id=591#591-
   older-thicker-arctic-sea-ice-declines
United Nations Environment Program
   http://www.unep.org/

Interactive Map Websites

Interactive maps on climate change
   http://environment.nationalgeographic.com/environment/
   global-warming/gw-impacts-interactive/

http://www.climatehotmap.org
http://www.cgdev.org/page/mapping-impacts-climate-change
http://www.npr.org/news/specials/climate/interactive/

Learning Outcomes

- The relationship between people and environment (also referred to as society–nature) is at the heart of human geography.
- The three basic forms of social-spatial-economic relations include: hunting-gathering, traditional agriculture, and urban-industrial systems.
- For many hunting-gathering societies, human–nature relations are not simply functional and economic but also filled with spiritual meaning and cosmological significance.
- The shift from hunting-gathering societies to agriculture production also leads to a shift from the notion of sacred space to a view of spaces as being productive.
- There are competing claims on the importance of the environment. On the one hand there are powerful forces that promote continued economic growth and tend to see nature primarily as a commodity to be developed and exploited; on the other, we also have an increasingly powerful awareness of unrestrained growth's environmental consequences.
- The environmental Kuznets Curve suggests a relationship between a growing environmental awareness and increasing income.

- There are now a variety of opinions on the appropriate policy responses to promote environmental awareness, ranging from "deep ecology" to an environmental sensitivity that emphasizes alternative technologies, pollution reduction, and sustainable development.
- Research has shown that our mood and behavior are both affected by our environment, as local environments play a role in our sense of well-being.
- One of the most significant impacts of human activities on the environment is global climate change.
- One line of research in human geography looks specifically at landscapes as an embodiment of the people/nature relationship.
- Of the most visible and dramatic examples of people–nature relations is the experience of hazards and disasters.
- The experience of environmental hazards and the way that hazards become disasters tell us as much about the human geography of class and power as they do about the physical geography of weather and landforms.
- We live in the Anthropocene, a world of our own making.

PART 3

The Economic Organization of Space

The economic organization of space is the subject of Part 3. In chapter 7, we identify the economic geography of the primary, secondary, and tertiary sectors. In chapter 8, we take a different tack, looking at the economic geography of uneven development at the global and regional scales.

OUTLINE

7

The Geography of Three Economic Sectors

Economic geography looks at the changing nature and spatial distribution of different sectors of the economy. The mix of sectors varies over time and space. Examining the changing configurations of economic growth at the regional and global scales provides an intriguing insight into economic change and its consequences.

Three distinct sectors of economic activity can be identified: primary, secondary, and tertiary. These are outlined in Figure 7.1. Some care should be taken with this division, because it suggests strong demarcation lines between the sectors, whereas in reality there is considerable interaction. Consider the case of a financial institution (tertiary sector) that lends money to a farmer to buy machinery to grow potatoes (primary) that are then transformed in a factory into potato chips (secondary). All three sectors are connected in complex arrangements, networks, and flows. In the rest of this chapter, we will explore some of the socio-spatial dimensions of each of these sectors separately but will be mindful of the flows and connections that link them together.

As economies grow and mature, the secondary and tertiary sectors tend to become more important. Table 7.1 highlights the changes in the sector mix in China. As the economy developed, the **tertiary sector** became an increasingly important element. The relative decline of the primary sector and rise of the secondary and tertiary sectors in a national economy is linked to profound socio-spatial changes, including the commercialization of agriculture, high levels of rural-to-urban migration, and increasing levels of industrialization and urbanization.

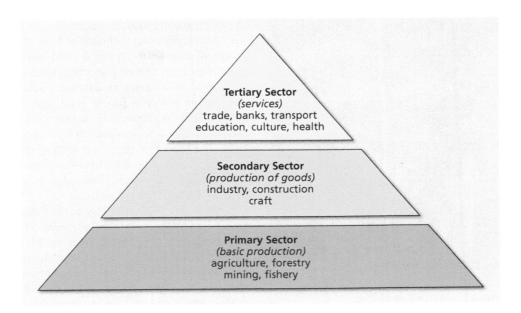

7.1 Three sectors of the economy

TABLE 7.1 Changing Sector Composition in China, 1978–2015			
YEAR	1978	1996	2015
GDP	100	100	100
Primary	28.1	20.2	5.8
Secondary	48.2	49.0	42.6
Tertiary	23.7	30.8	51.6

Source: China Statistical Yearbooks.

Agriculture

The **primary sector** includes agriculture, forestry, mining, and fishing. Here we will focus on agriculture.

Agriculture is very much influenced by location. The type of agriculture possible depends largely on climate and weather, soils, and water availability. Natural rubber production, for example, takes place in tropical areas, especially in Thailand, Indonesia, Malaysia, India, and Vietnam. Wheat growing, in contrast, is concentrated in the temperate grasslands of the world. The **wheat belt** countries include Canada, Russia, and the United States in the Northern Hemisphere and Argentina and Australia in the Southern Hemisphere.

Geography obviously plays an important role in the distribution of primary activities. But it would be incorrect to see the primary sector only as the outcome of locational opportunities and constraints. Take the case of sugar. Traditionally, it was a tropical plant that could be grown only in the hot, humid areas of the world. By the seventeenth century, a market for sugar had developed in Europe to such an extent that the search for sugar-growing areas constituted an important element in the European colonization and annexation of Caribbean islands. The work was brutally hard, and diseases from the Old World soon killed off the indigenous people. The need for labor fueled the slave trade. European countries without tropical colonies were at a huge disadvantage, having to pay others for an expensive good. It is not incidental that sugar beet was developed in Germany, a country with no tropical possessions. By the early 1800s, the first beet sugar factory opened in Germany. Sugar beet is a plant with a tuber that contains a high concentration of sucrose. The plant can be grown in colder climates, and so the tropical source became less important. As beet sugar production increased in Europe and North America, the sugar plantation economies of the Caribbean dramatically declined.

The von Thünen Model

Markets play an important part in affecting agricultural production. In 1826, wealthy German landowner Johann Heinrich von Thünen (1783–1850) proposed a model of land use around a city (Figure 7.3). His model was highly idealized in order to identify the independent role of location. The "isolated" city of his model was situated on a flat plain with homogenous fertility and unvarying transportation costs. The city was the main market for farmers. Since farmers closer to the city paid less for transport, they

7.2 Primary sector: quarry in Pennsylvania

could bid more for land, which resulted in higher land costs closer to the city. Only farmers growing the more intensive crops, with high returns, could thus afford the land closer to the city. The net result, according to von Thünen, was a concentric ring pattern with more expensive land and thus more intensive agriculture closer to the city. At the time he was writing, market gardening and firewood production were intensive land uses, while livestock farming was more extensive.

While the **von Thünen model** operates on the assumption of a flat plain with unvarying fertility, conditions that are rarely met with in the real world, it has been useful as a platform for subsequent studies. The von Thünen model is useful to explain historical developments. Environmental historian William Cronon, in his 1991 book *Nature's Metropolis*, examines the relationship between Chicago and its hinterland from 1850 to 1890. He shows how the physical world was turned into a commodified human landscape as grain, lumber, and meat production transformed prairies and woodlands into the physical basis for

the city's growth and development. More detailed analyses of contemporary conditions look at the rural–urban fringe of cities. In capitalist land markets, the ring of land closest to the city is often underutilized, in contrast to von Thünen's predictions, because of the heavy shadow of possible future land use conversion. With rural-to-urban conversions a very real possibility, land-owners make few investments in agricultural production. The land use planning system that directs which areas are available for rural-to-urban conversion will clearly affect the simple linear-distance model proposed by von Thünen.

In the case of Megalopolis, as the urbanized seaboard of the United States is often called, there is empirical evidence of the validity of the von Thünen model. Land closer to the cities is much more expensive than land on the periphery of Megalopolis. If farmers do locate in areas closer to the city, they need to engage in intensive high-yield farming, such as market gardening of fruit and vegetables. In some cases, it is not so much distance as time of transportation that is important. Large cities require quick and immediate supplies of fruit and vegetables and other specialized agricultural products. Restaurants, for example, require daily supplies for their diners. Proximity to the city provides swift access, but at the price of high land values, which in turn means that only intensive farming with high yields makes economic sense. Organic fruits and vegetables is another emerging market that yields a high dollar return per acre. Affluent consumers are willing to bear the greater costs associated with these products. High-yield counties are

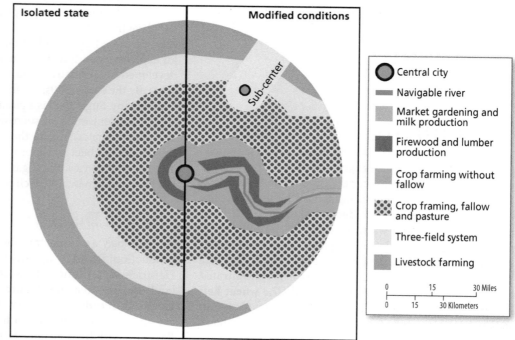

Isolated state   **Modified conditions**

Sub-center

Central city

Navigable river

Market gardening and milk production

Firewood and lumber production

Crop farming without fallow

Crop framing, fallow and pasture

Three-field system

Livestock farming

| 0 | 15 | 30 Miles |
| 0 | 15 | 30 Kilometers |

7.3 The von Thünen model

located around New York City, Philadelphia, and Boston. There are other areas of relatively high yield such as New Jersey (giving credence to the nickname Garden State), Rhode Island, and southern Connecticut. Agricultural land is being lost as the suburban spread flows across the landscape, but the remaining farmland that is safe from urban development, even temporarily, and close to the city has become so expensive that very intensive forms of farming are required, including nurseries, greenhouse crops, and higher-value specialty products.

There are now many case studies of land use change that draw upon the von Thünen model, including tropical forest degradation close to Dar es Salaam in Tanzania and crop production outside of Yaoundé in Cameroon. In their study of land use changes in forest areas of Indonesia, Miet Maertens and her colleagues used satellite imagery, GIS data, and survey techniques. The von Thünen model, while useful, did not explain all the variation in this Indonesian forest frontier area. Not just distance to markets was important; so were population levels, access to technology, and individual household characteristics.

The Agrarian Transition

The **agrarian transition** occurs as agriculture shifts from subsistence to commercial and moves from meeting local market demands to provisioning national and global **food supply chains**. It involves changes to agriculture, including intensification and expansion, market integration, and environmental change.

The extension and intensification of agriculture are pronounced (see Figure 7.4). There is a global expansion of arable land, an expansion of irrigated surfaces, and the development of intensive mixed farming systems in dense population areas. In developed countries, there is large-scale mechanization, biological selection, and increasing use of chemicals to boost yields. In some developing countries, there is a marked commercialization of agriculture involving the marketization of peasant economies, land grabs for commercial farming, and overall, a greater integration into global food supply chains.

Market integration takes two main forms. On the one hand, there is the commercialization of the peasantry, as subsistence smallholders become commercial farmers. Geographer Sarah Turner looked at the contemporary process among Hmong farmers in the upland areas of northern Vietnam. She conducted more than 200 interviews and measured the expansion of rice paddies as a replacement for opium cultivation, which was banned by the government.

There are enormous barriers to this process, as smallholders and subsistence farmers often lack access to capital and technology. There are many cases, however, where traditional peasant societies respond creatively to the new commercial realities despite the barriers. Small-scale

7.4 San Joaquin Valley, California

farmers can, for example, market their ecological sensitivity to sophisticated consumers, as in the case of smallholding organic coffee growers in Costa Rica. The growth of fair-trade initiatives, for example, often links ethical consumers with small-scale producers.

On the other hand, there is also evidence of a **land grab** as peasant farmers are displaced in the wake of the forced commodification of land with the resultant commercialization of agriculture. This is the second and very brutal form of market integration. This has a long history. In England, for example, the **Enclosure Movement** privatized the common land of fields, marshes, heaths, and woodlands. Between 1760 and 1820, almost 20 percent of England's total acreage was enclosed, and the English peasantry was destroyed. A system of traditional rights and obligations was replaced with the cash nexus of agrarian capitalism.

Similar processes are occurring now throughout the world, reinforced by a growing commercialism of agriculture, demand for goods such as sugar and soya, and turbocharged by the spike in food prices in 2008 and demand for biofuels. Land grabbing involves the buying, leasing, or simple appropriation of large parts of land illegally, unfairly, or in an underhanded manner. It is especially evident in the developing world where indigenous peoples and peasant societies often do not have legal documentation of their land claims, and if they do, they are easily pushed aside by powerful foreign investors in alliance with corrupt government officials. Estimates suggest at least 81 million acres in the past decade were purchased by foreign investors that led to the subsequent eviction of local people. In just one example among many, 769 families were forced off the land in Guatemala's Polochic Valley in 2011 to make way for sugar cane plantations.

Land grabs worldwide are fueled by the commercialization of agriculture as food production in many regions

7.1 Food Supply Chains

Supply and demand for food are linked through food supply chains, which have become longer and more complex for three interrelated reasons. First, with urbanization, more people live in towns and cities. This growing urban population needs to be supplied with food produced elsewhere. Long and complex transportation links connect urban consumers with rural providers. Second, there is also dietary transition. As household incomes grow, so does the consumption of dairy, meat, and fish. **Bennett's Law** states that the consumption of starch staples declines as household income increases. In China, for example, economic growth and the development of a middle class has led to increased consumption of meat and dairy products. Third, there is the increased globalization of trade in foodstuffs. Food supply chains are increasingly multinational and global. The longest chains are often associated with very large retailers, such as supermarkets that are large enough to have substantial economies of scale. One study that looked at exports of horticultural products from sub-Saharan Africa to Europe found important positive welfare effects as rural providers were able to make more money and have a more dependable source of income.

There are least four types of food supply chains. In **buyer-driven chains**, a small number of retailers consolidate their supply around a few suppliers, although there are opportunities for smallholders where crops need intensive care, as in Peru's asparagus sector. In **producer-driven chains**, such as coffee, cocoa, and tomatoes, producers can establish cartels. In **bilateral oligopolies**, such as bananas and pineapples, large-scale producers combine with large-scale retailers. In **traditional markets**, smallholders cater to domestic and local markets.

Consumers, especially more affluent food consumers, can access different types of food chains, shopping at supermarkets as well as at local organic markets and buying food from distant international food supply chains as well as from local producers.

REFERENCES

Lee, J., Gereffi, G., and Beauvais, J. (2012) Global Value Chains and Agrifood Standards: Challenges and Possibilities for Small Holders in Developing Countries. *Proceedings of the National Academy of Sciences* 109: 12326–12331.

Maertens, M., Minten, B., and Swinnen, J. (2012) Modern Food Supply Chains and Development: Evidence from Horticulture Export Sectors in Sub-Saharan Africa. *Development Policy Review* 30: 473–497.

becomes more linked to global markets and the growing demand for biofuels. When agricultural land becomes more valuable as a commodity, rich and powerful interests seek to possess the land. This often involves the dispossession of local communities with long-established links to particular parcels of land.

The land grab is particularly pronounced when the peasantry is of a minority ethnicity compared to the ruling elites. Often marginalized by political structures and denied full access to political power, peasant societies struggle to maintain their land rights against well-connected corporations and powerful state bureaucracies. For example, the Mapuche tribal group was restricted to land reserves by the Chilean government in 1883. Even this "reserved" land is now subject to state and private appropriation as hydroelectric dams are built, logging companies move in, and plantations are created to grow sugar and soya.

In many countries, legal land rights are uncertain. The traditional claims to land often do not hold up in modern legal frameworks more concerned with adjudicating private property rights than communal rights. In the Philippines, for example, only 3.5 million out of a total of 12 million hectares are private property. The rest of the land has uncertain formal legal status, making it more susceptible to a land grab.

There are also instances where peasant communities resist the land grab. The Bajo Aguán region in Honduras has been a scene of intense conflict over the last two decades. In the 1990s, large landowners purchased land from farmer cooperatives to harvest palm oil for export. Local activists complained that the deals were unfair and people were not made aware that they were signing away land rights. One company owned by a member of the very wealthy Honduran elite amassed one-fifth of all the land in the region. In 2009 peasants invaded and occupied company land. Troops were sent in, people were evicted, and more than forty people were killed. The peasants resisted, and the bad publicity led one German bank to withdraw loans to some of the large companies. In 2011 the Honduran government passed a law that allowed peasant farmers in Bajo Aguán to purchase 4,000 hectares at favorable interest rates. This particular story has all the elements of the global land grab: marginalized peasants, rich landowners, and a growing market for food exports and biofuels. In this case, however, the peasants resisted and forced a change in the government attitude.

The Commercialization of Agriculture

The commercialization of agriculture involves greater capital investment in machinery and biotechnology, which leads to greater productivity. To take just one example: the average American farm produced 40 bushels of grain per acre in 1900. A hundred years later, the same acre was producing 100 bushels (see Table 7.2). This enormous increase in productivity is the result of increased use of pesticides, genetically engineered crops, and heavy doses of fertilizers. All of these come at a price.

Pesticides, for example, increase productivity by reducing the harm of insects on crop yields. In her 1962 book *Silent Spring*, Rachel Carson highlighted the negative environmental consequences of heavy pesticide use. The title of the book describes a spring devoid of birdsong after pesticides have decimated songbirds. Almost 3 billion kilograms of pesticides are used around the world each year, 500 million kilograms in the United States alone. One study looked at the impact on public health, the development of pesticide resistance in pests, honeybee and bird losses, and groundwater contamination, estimating the total costs in the United States in 2002 at $10 billion. The costs are probably much higher in countries with less stringent environmental protection legislation and enforcement.

Increased use of fertilizer, especially those that restore soil nitrogen, also has deleterious effects on groundwater contamination and air quality through emission of nitrous oxides. Runoff from fertilized fields finds its way into streams, rivers, and seas, creating environments superenriched with nutrients, where algae and plant overgrowth can reduce the oxygen so much that dead zones occur. **Eutrophication** is the name given to the process of enriching ecosystems with chemical nutrients, which creates algal blooms that result in dead zones. Nutrient enrichment also causes new patterns of disease for humans and wildlife. In Caribbean wetlands, for example, increased nutrient runoff creates lusher vegetation that provides a home for malarial mosquito larvae.

The issue of using **genetically modified organisms (GMOs)** to grow crops is controversial. Plants are genetically altered to make them more resistant to disease and predators and to produce higher yields. While the proponents of GMO highlight the advantages of cheaper food, critics point to the risks of unintended health effects on consumers and of contamination between GMO and non-GMO crop varieties with unknown consequences. The risks and full costs of GMO agriculture have yet to be fully assessed.

The most sustained commercialization of agriculture in the developing world that involved large inputs of biotechnology was the **Green Revolution**, the name given to a series of innovations and transfer, especially marked from the late 1940s to the 1970s, that increased agriculture production through new high-yielding varieties of crops, irrigation, and liberal helpings of pesticides and fertilizers. It began in Mexico and was then implemented in India, Philippines, and selected African countries. The Green Revolution did improve yields, but at the cost of environmental damage and harming the health of local communities. It is now seen as a deeply flawed experiment with some positive consequences, such as increased crop production and cheaper food for consumers. We need a real green revolution, one that is sensitive to environmental impacts, community concerns, and translating sustainable and safe practices into mainstream agriculture.

Manufacturing

The manufacturing of goods constitutes the bulk of the **secondary sector**. While craft industries have existed for centuries—glass making in Venice, for example, dates from the fourteenth century—the large-scale manufacturing of

TABLE 7.2	**Corn Production in the United States**		
YEAR	**YIELD (BUSHELS PER ACRE)**	**HOURS REQUIRED FOR 100 BUSHELS**	**EQUIPMENT USED**
1850	40	75–90	Plow
1900	40	35–40	Gangplow, disk, harrow, 2-row planter
1950	50	10–14	Tractor, 3-bottom plow, disk, harrow, 4-row planter, 2-row picker
2000	100	2.5	Tractor, 5-bottom plow, 25 ft. tandem disk, planter, 25 ft. herbicide applicator, 15 ft. self-propelled combine, trucks

Source: American Farm Bureau.

7.2 The Industrial Revolution

The Industrial Revolution occurred around 1800 and was centered in Britain. At its core, the Industrial Revolution was a new way of making things. Production capacities were increased by mechanization. Steam power released capacities beyond the limits of even the most strenuous human exertion. Textile production was increased tenfold by mechanization. Once started, the Industrial Revolution took off in an upward cycle of increased growth and expansion as inventions were turned into innovations that improved, increased, and streamlined the making of things. The Industrial Revolution was also an institutional revolution, as economic transactions shifted from one based on patronage and personal contact to a greater reliance on anonymous markets, performance measurements, and impersonal exchange.

Britain was the first industrial nation. A number of reasons lay behind its transformation. Britain was a relatively small, politically stable, densely populated country. Colonial expansion ensured cheap imports and secured export markets. Because of high wages, British companies were forced to introduce labor-saving techniques. It is also important to note the role of London. In 1800 the city had a population of almost one million, almost double the size of any other European city. The city's population growth provided a huge, growing, and secure market for food producers and manufacturers, and this effective demand was the basis for further investment in agricultural and industrial production.

The very early Industrial Revolution was centered in Manchester. One compelling reason is the importance of the culture of innovation: between 1600 and 1800, a proto-industrialization system of economic organization was established, there was a capacity for continuous innovation, and there was a large middle class of small capitalists able to employ a range of entrepreneurial talent, social networks, and local cultures that allowed a constant improvement in products and processes. The city's location was important: it was close to the coal that provided the raw power to animate the incessant spinning and weaving machines. But other places were also close to coalfields, so this single locational factor has limited explanatory power. Manchester became the first industrial city because it was the first truly innovative city.

REFERENCES

Allen, D. W. (2011) *The Institutional Revolution: Measurement and the Economic Emergence of the Modern World.* Chicago: University of Chicago Press.

Allen, R. C. (2009) *The British Industrial Revolution in Global Perspective.* Cambridge: Cambridge University Press.

Hall, P. G. (1998) *Cities in Civilization.* New York: Pantheon Books.

goods as a major component of an economy is a comparatively recent phenomenon associated with the Industrial Revolution.

The Industrial Revolution is one of several long waves of production based on the clustering of innovations. The Soviet economist Nikolai Kondratieff first identified these long, fifty-year cycles in the 1920s. Kondratieff made a distinction between inventions and innovations. Inventions are new ways of doing things; they occur almost randomly, but tend to be adopted into production techniques in waves of innovation. The basic nature of these **Kondratieff cycles** is noted in Table 7.3. Four cycles are commonly noted: textiles, iron and steel, mass production, and high-tech. Each Kondratieff cycle is associated with key innovations that structure society and space. The first two, from 1785 to 1895, are associated with the development of factories, the

TABLE 7.3 The Four Kondratieff Cycles

	TEXTILES	IRON AND STEEL	MASS PRODUCTION	HIGH TECH
	1787–1845	1846–1895	1896–1947	1948–2000 (?)
Key innovation	Power loom	Steel making	Electric light, automobile	Transistor, computer
Key industry	Cotton, iron	Steel	Cars, chemicals	Electronics
Industrial organization	Small factories	Large factories	Giant factories	Large and small factories
Labor	Machine minders	Craft labor	Deskilled	Segmented
Geography	Towns	Towns	Conurbation	New industrial regions

emergence of industrial districts, and the growth of towns and cities.

The First Cycle

The first Kondratieff cycle, from 1787 to 1845, is associated with textile manufacturing and saw its full flowering in Manchester, England. In 1750, Manchester was just one of hundreds of textile towns located all over Europe. What made it the first industrial city were developments in technology, new sources of supply and demand, and social networks that fostered innovation and risk taking. In the 1760s, new steam-powered machines increased production levels. In 1774 the population of Manchester was only 41,032; by 1831 it was 270,901. Mills were working day and night. Other ancillary industries also developed. Railway links were built to transport its goods, and by 1840 the city was served by six railway lines and was a center of locomotive construction.

The Second Cycle

The second and third cycles were also experienced in other parts of the developed world, especially Germany and the United States (see Figures 7.5, 7.6, and 7.7). New industrial regions were established in North America and Europe. The city of Paterson in New Jersey was a center of textile production and then the making of locomotive engines. Pittsburgh in Pennsylvania typifies the second and prefigures the third cycle. It was surrounded by bituminous coal; the state produced almost a quarter of the nation's coal, over 10 billion tons in 200 years. By the mid-nineteenth century, over a thousand factories in Pittsburgh were

using coal as their power source, including at least sixty-five glass factories. During the Civil War, the city became a center for armaments manufacturing. Scottish-born industrialist Andrew Carnegie founded the first steelworks there in 1873, and by 1910 the city was producing a third of the nation's steel. Aluminum was first made in Pittsburgh in 1888. The city grew from a population of only around 46,000 with a metropolitan population of 100,000 to a city population in 1920 of almost 600,000 with a metro population of almost 1.5 million. Pittsburgh was the epicenter of US industrialization, and even today major employers include Alcoa, U.S. Steel, and Wheeling-Pittsburgh Steel. As the economy matured, the city of Pittsburgh was only the central point in a wider suburbanization of people and jobs throughout a wider urban region. With the creation of large integrated industrial establishments such as large-scale steel works, industry, jobs, and people were widely distributed throughout a forty-mile radius from downtown Pittsburgh. As firms increased in size, the plants grew larger. The result was a complex suburbanized metropolitan economy. The rapid and heavy industrialization also created unhealthy pollution.

The Third Cycle

The third Kondratieff cycle was based on automobile manufacturing, giant factories, and standardized deskilled manufacturing processes. The exemplar city was Detroit as it became home to car manufacturing and new forms of industrial production. At the beginning of the twentieth century, cars were luxury items, handcrafted and designed for the wealthy. Detroit was only one of many car-making cities. Even before producing cars, Detroit had been a manufacturing center; by 1900 it had almost a thousand machine shops making ships, stoves, engines, and mining machinery. There was a pool of skilled labor and a network of local financiers. Detroit was soon producing two-fifths of the nation's car output, concentrating on the cheaper end of the market.

7.5 Paterson, New Jersey, was a site of textile production in the US industrial revolution.

7.6 Secondary sector: steel-making plant in Baltimore

Henry Ford transformed the manufacture of cars. He was born in 1861 and grew up close to Detroit. In 1879 he went to work for the Michigan Car Company, which was building ten cars a day. In 1899 he formed the Detroit Automobile Company, but it was not a success. Ford was too much of a perfectionist and had yet to hone his market sensibilities. Fine-tune them he did, however, when he founded the Ford Motor Company in 1903 with twelve shareholders, all from Detroit. In 1908 the first Model T appeared on the market. It was a simple yet robust model that went through numerous design improvements to become one of the first mass-produced cars. Assembly lines and mass production had been developed over the previous century in a range of manufacturing industries, including bicycles and watches as well as the Chicago meat processing and packing industry. Ford did not invent mass production, but he refined and improved it. Car production became standardized, precise, and continuous. Mass production allowed reduction in the final price: the cost of a Model T in 1908 was $805, but by 1924 it was down to $290. With each price decrease, new markets were created, and so mass production not only met demand but created new demand. Cars became less a luxury item and more a regular purchase, especially in the rural hinterland, where farmers

used them for a variety of purposes. By 1920 every second car in the world was a Model T Ford. In order to reduce labor turnover, Ford also paid high wages for the time. A well-paid industrial workforce was essential to maintaining worker allegiance and limiting labor turnover. Ford was instrumental in creating the highly paid blue-collar sector of the new industrial city.

Detroit was the forerunner of a form of production named after Henry Ford. Fordist production is controlled by a small number of very large companies operating under oligopolistic conditions. Fordist production is mechanized and repetitive, with long production lines. Under **Fordism** there were outbreaks of labor unrest when the business cycle softened demand and management implemented layoffs and wage cuts. But, by and large, a stable system of capital–labor relations was established as a relatively affluent working class arose. The high point of this wave lasted from 1945 to the mid-1970s.

Global Shift

There is a changing global geography to the third cycle of manufacturing mass production. Since the 1970s there has been a **global shift** as manufacturing moved from the high-wage economies of the developed world to the lower-wage

7.7 The Ruhr; one of the major industrial regions of Europe

economies of the developing world. As manufacturing became more mechanized and routinized, employers shifted operations to areas of cheap, semi-, and unskilled labor. Transport improvements, such as container shipping and integrated transport connected at ports, made it cheaper and easier to ship goods. As transport costs plummeted, attention shifted to moving manufacturing plants to cheaper labor regions of the world. First Japan, then South Korea, Singapore, and Thailand became centers for textile and shoe production. These newly industrializing countries were centers of rapid economic growth and were sometimes known as **Asian Tigers**. The regional economic complexes then became the basis for a move up the **value-added chain** to more complex manufacturing such as cars and electronics. As labor costs increased, low-value manufacturing, such as textile sand shoes, shifted to even cheaper areas such as Bangladesh, China, Indonesia, and Vietnam.

As manufacturing becomes even more mechanized and transport improvements flattens the world, these industries move to even cheaper accessible areas. This global shift has had three major consequences. The first was the decline of manufacturing employment in Europe and North America. They experienced **deindustrialization** as manufacturing jobs were lost through a combination of efficiency improvements as well as global shift. For cities and regions based almost entirely on manufacturing such as Detroit, the net effect was massive job loss and economic decline. Across the developed world, from Birmingham, Alabama, to Birmingham, UK, and from Newcastle, UK, to Newcastle, Australia, there was a massive loss of manufacturing employment and a decline in industrial regions and cities. There was also a decline in organized labor and an effective hollowing out of the blue-collar middle class.

The second was that the destination regions of this global shift now became places of sustained economic growth. Japan, then South Korea, and more recently China have witnessed rapid economic growth based on the increase in manufacturing.

Third, there was also the creation of new class and gender identities. In the developed world there was a reduction in male, blue-collar employment opportunities and a decline in organized labor. Meanwhile in regions and countries that saw a manufacturing upswing because of global shift, a female working class was created, and economic growth enabled an enlarged and more prosperous middle class. Much of the labor in China's new manufacturing, for example, is young female workers. The global shift resulted in new forms of class and gender identities.

In summary, then, global shift meant, on the one hand a decline of older industrial regions and cities and of the organized labor class in the Europe and North America,

and on the other hand, in the developing world, the creation of new industrial cities and regions, the emergence of a new female working class, and the great enlargement of the middle class.

The Fourth Cycle

The fourth Kondratieff cycle is associated with the development of high-tech information technology (IT) industries. The Kondratieff distinction between inventions and innovations is apparent in this fourth cycle. The inventions associated with the IT sector occurred in a steady stream throughout the twentieth century but became a wave of high-tech innovation that transformed the economy and society only by the very end of the twentieth century (Table 7.4). The IT wave took over a century to turn inventions into the wave of innovations that occurred from 1980 onward. By 1994 the World Wide Web had become a powerful information portal for business and personal computer users.

Each of the previous waves has an exemplar city or region: Manchester, Pittsburgh, and Detroit. For the fourth wave it is Silicon Valley, a high-tech cluster centered on Palo Alto in California (see Figure 7.8A), where personal computing was developed for mass markets. Nearby Stanford University played a major role as a source of smart computer scientists and as the force behind the establishment of Stanford Industrial Park, where two Stanford graduates established Hewlett-Packard in 1953. Technological innovation was funded by a steady stream of government defense contracts and facilitated by corporate initiatives, such as Xerox's Palo Alto Research Center. Over time, the area became home to law firms and venture capital firms that facilitated and funded the startup and establishment of innovative companies. By 2013,

TABLE 7.4	Inventions in Computer Technology
APPROXIMATE DATE	**INVENTIONS**
1890	Punch cards and mechanical calculating machines
1939	Vacuum tubes and electromechanical calculating
1952	Transistors and electronics
1964	Mainframe computing
1971	Microprocessors
1981	Personal computers
1994	The Internet

7.8 (A) Silicon Valley. **(B)** Google headquarters.
(C) Apple headquarters.

major computing companies such as Adobe, Apple, Cisco, Google, Hewlett-Packard, Intel, and Yahoo! were all head-quartered in the region. Over time, the initial early advantages solidified into a major competitive edge for this region as a center for innovative high-tech computing.

Services

In mature economies there is a shift from manufacturing to services. In the United States, for example, service employment now accounts for almost 78 percent of GDP. Services are defined as selling, assistance, and expertise rather than making a tangible product. In some cases, the distinction is clear; when you buy a car you are buying a good, whereas when you hire someone to clean your house you are buying a service. At other times, the distinction is fuzzy; the classic example is a restaurant, where you buy food but also a service. The term "services" is best considered a loose and hazy one, with slippage at the edges. It also covers a wide range of activities, from health care and financial consultancy to computer information companies. The sector includes a range of wildly differing job experiences. At one end are Wall Street brokers working in high finance and international currency dealing, all making high wages and lucrative annual bonuses that fuel the local housing markets (Figure 7.9). At the other end are the nighttime contract cleaners of the offices that house these executives.

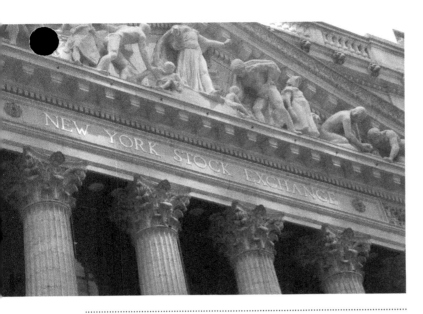

7.9 Tertiary sector: Wall Street is a center of financial services

Evolution of the Service Sector

Part of the services sector broke away from the secondary sector. In the nineteenth and early twentieth centuries, the factory contained both the assembly line and the business service sector that administered the whole process of buying raw material, hiring workers, and selling finished products. These sectors have been hived off into separate divisions and companies. We can think of a simple model of contemporary multinational manufacturing business. At the base is the routine assembly plant that needs all those things noted in the standard location models of industry, such as cheap labor and low taxes. As economic globalization has created a flatter world, these plants can be located in a wide range of countries. Another level of the company that has separated out from the production plants is the research and development sector that tests new products. This sector needs to be close to pools of highly skilled labor, knowledge centers, and the amenity-rich locations that attract such workers. Then there is the company headquarters, which needs a metropolitan location to maximize face-to-face business contacts and be close to business services such as advertising, financing, legal services, and other important services and decision makers. Consider the case of Boeing, a company with almost 156,000 workers around the world, almost 66,000 of them in the state of Washington. In 2001 the company moved its headquarters from Seattle to Chicago and separated its corporate headquarters from its production base. According to the company, the move was made in order to secure ready access to global buyers and financial markets. It was also lubricated by subsidies of over $60 million from the state of Illinois and the city of Chicago, which was in competition with Denver and Dallas–Fort Worth.

Characteristics of a Service Sector Economy

A service-dominated economy tends to have a number of characteristics. It produces a more polarized job market, with high-paying jobs with full benefits on the one hand and minimum-wage employment with few benefits on the other. At the core of the new service economy are highly paid knowledge-based professionals, analysts of data, trends, and ideas, such as business consultants and investment bankers. Paid generously, they have employment security, good working conditions, and generous benefits. They often work long hours with brutal deadlines, but they are firmly located in the middle to upper-middle class, and they can ensure their children's similar economic success by being able to buy good private education or make the right residential choices to ensure the best public education. At the very upper levels of this sector is a world of affluence. It is the world of the corporate jet and the million-dollar-plus stock options that provide a deep financial cushion against the vagaries of life. Outside of this core group, there are two peripheries. A semi-periphery consists of full-time workers with less income, status, and prestige. Their jobs are less secure, and too often they are able to hang on to middle-class status only if more than one person in the household is working. Beyond this, a periphery comprises people on short-term or part-time contracts. They may work from home on their computer and relish the flexibility of such work, or be picked up each day at the street corner by landscapers to manicure the lawns of the wealthy.

The shift from a manufacturing-based economy to a service-based economy also has profound effects on income distribution. A strong manufacturing base allows low-skilled workers to obtain relatively good wages. The service economy, in contrast, provides high-paying jobs for those with marketable skills, but more limited opportunities for low-skilled workers. For those lacking high educational attainment, the opportunities shrink to the low-wage service sector.

This shift also goes hand in hand with an increase in female employment. In the United States in 1950, the female participation rate in the formal economy was only 30 percent, but by 2012 it had increased to almost 58 percent. By 2012, out of a total employed labor force of 155 million, 72.6 million were women. The workforce has been feminized in the shift to a more service-based economy.

As service employment increases, there is the danger of what is called **Baumol's disease**, after the economist William Baumol, who first noted that while innovation and efficiency occur rapidly in the manufacturing sector, leading to price reduction, they are less pronounced in the **service sector**. You can make a car quicker and more efficiently, but it still takes a string quartet the same time to

7.3 The Cultural-Creative Economy

An important and growing sector of the economy is the **cultural industry** sector, which includes the artistic industries of music, dance, theater, literature, the visual arts, crafts, and many newer forms of practice such as video art, performance art, and computer and multimedia art. There are also the "cultural-products sectors," which include high fashion, furniture, news media, jewelry, advertising, and architecture.

The work of creative artists enhances the design, production, and marketing of products and services in other sectors. Cultural industries concentrate in world cities like Los Angeles and Paris and other large cities, contributing immensely to their economies. The concept of "path dependence" can be used to analyze cultural production as well as technological innovation. Path-dependent theories claim that small historic events or locational advantages can affect macroeconomic consequences that privilege certain paths to development and limit others. This idea sheds light on the prestigious success that Paris enjoys in high fashion, New York in advertising, and Los Angeles in motion picture entertainment. Their leading roles in these industries have long enjoyed wide recognition, and their advantages over potential competitors relate not only to the quality of their products, but also to the "symbolic images"—such as authenticity and reputation—that those products carry.

Creative talent drives economic development. Richard Florida argues for a new social class, the creative class, which leads the new creative economy. The creative class divides into two groups of people. The first makes up the "super-creative core," a broad and diverse group including scientists, engineers, university professors, poets, novelists, artists, entertainers, actors, designers, and architects, as well as nonfiction writers, editors, cultural figures, think-tank researchers, analysts, and other opinion makers. The other group consists of "creative professionals" who work in a wide range of knowledge-intensive industries such as high-tech sectors, financial services, the legal and health care professions, and business management. The former group produces new forms or designs that are readily transferable and widely useful, while the latter group engages in creative problem solving.

There is a vigorous debate about the rise of the so-called creative class, its contingency and flexibility, its scalability from a few well-known case studies, and its ideological underpinnings as a form of urban economic policy. A convincing argument could be made that the poor are the real creative class, because of their ingenuity and resourcefulness in the face of limited income and restricted employment opportunities.

REFERENCES

Currid, E. (2007) *The Warhol Economy: How Fashion, Art, and Music Drive New York.* Princeton, NJ: Princeton University Press.

Florida, R. (2002) *The Rise of the Creative Class: And How It's Transforming Work, Leisure, Community, and Everyday Life.* New York: Basic Books.

Hutton, T. (2014) *Cities and the Cultural Economy.* New York: Routledge.

Peck, J. (2005) Struggling with the Creative Class. *International Journal of Urban Regional Research* 29: 740–770.

Scott, A. J. (2008). *Social Economy of the Metropolis: Cognitive-Cultural Capitalism and the Global Resurgence of Cities.* Oxford: Oxford University Press.

Scott, A. J. (2012). *A World in Emergence: Cities and Regions in the 21st Century.* Cheltenham, UK: Edward Elgar Publishing.

Kim, Y. H. and Short, J. R. (2008) *Cities and Economies.* New York: Routledge.

play a Beethoven quartet. As the service sector comes to dominate an economy, it is subject to rising costs. While prices for goods decline due to manufacturing efficiencies, the costs of services keep rising in absolute and relative terms. In his book *The Cost Disease*, Baumol suggests that while we can work to reduce service costs, they are the price of a successful economy.

Quaternary Sector

The service sector has grown so much in size and sophistication that different subsections can be identified. Some argue for a distinct new sector, the **quaternary sector**, that involves use of knowledge and information. It includes communications, consultancy, and research and development. This sector is concerned with the increased and strategic use of information and knowledge.

Producer services is another term used to describe a range of knowledge-based activities including accounting, advertising, business consultancies, and financial and legal services.

Although an easy demarcation between the tertiary and quaternary sectors is not possible, we can make a broad distinction between the highly skilled workers in financial services and less skilled service workers such as a restaurant server. While both may work equally hard, the levels of compensation and work condition are often very different.

Some even make a further distinction and define a **quinary sector** that is composed of key decision makers, the small group of people perched atop the information- and knowledge-based economy who make the executive decisions. Even in the quaternary sector there is world of difference in wages, working conditions, and social status between someone at a desk in bank inputting data into a spreadsheet and the bank's CEO who is part of the the top 1 percent of who have done so much better than the vast majority.

The terms "quaternary," "quinary," and "**advanced producer services**" are attempts to designate a particularly specialized and often highly paid sector of the economy. As economies grow and become more sophisticated, shifting from primary through secondary to tertiary, there is a greater reliance on knowledge and information. It is not enough just to make things; it is also important to identify markets, assess consumer preferences, keep records of transactions, mobilize bank loans, influence government policy, and undertake advertising campaigns. The more an economy develops, the greater the reliance on the quaternary sector of knowledge- and information-based services.

Global Shift in Service

From the early 1970s, global shift affected mostly the manufacturing sectors. In more recent years, there has been a significant increase in the global shift in service employment. A new round of economic globalization, made possible by changes in technology, is sending a range of service employment overseas from the developed world to the developing world. Back offices in Bangalore, India, now process home loans for US mortgage companies, while many insurance claims made in the United States are routinely processed in offices situated in New Delhi. The economics are simple. Software designers in the United States cost $7,000 a month, while experienced designers in India cost only $1,000 a month. US companies now routinely outsource work previously done at home. In the 1970s and 1980s, engineers would come to the United States and Europe; now the jobs come to them.

Selected cities in developing countries are developing as centers of service employment. In part they develop as hubs of corporate national headquartering, which generate service employment. In China, for example, Beijing and Shanghai are the favored sites for headquarters of foreign companies. Corporations are attracted to Beijing because of the closer proximity to key decision makers, an important criterion for working effectively in a centrally planned economy. However, even in the secondary city of Shanghai there has been a shift from manufacturing to service employment. In 1990 manufacturing accounted for 60 percent of the city's GDP while services only accounted for 31 percent. By 2014 the respective figures were 31 and 65 percent. While still an important manufacturing center, Shanghai has emerged as a predominantly service-based urban economy.

Summary

A global trend is evident for each of the three sectors. There is an increasing intensification and commercialization of agriculture. Manufacturing is becoming more automated and also relocating toward a select group of developing economies. In much of the developed world, there is a decline of manufacturing employment and a rise in service employment. Around the world, the more dynamic economies are shifting from agriculture to manufacturing and from manufacturing to services.

Cited References

Ahrends, A., Burgess, N. D., Milledge, S. A., Bulling, M. T., Fisher, B., Smart, J. C., and Lewis, S. L. (2010) Predictable Waves of Sequential Forest Degradation and Biodiversity Loss Spreading from an African City. *Proceedings of the National Academy of Sciences* 107: 14556–14561.

Baumol, W. (2012) *The Cost Disease: Why Computers Get Cheaper and Health Care Doesn't*. New Haven and London: Yale University Press.

Folefack, A. J. J. and Adamowski, J. F. (2012) Application of the von Thünen Model in Determining Optimal Locations to Transport Compost for Crop Production Outside of Yaoundé, Cameroon. *Journal of Human Ecology* 39: 125–143.

Maertens, M., Zeller, M., and Briner, R. (2011) Agriculture Land-Use in Forest Frontier Areas: Theory and Evidence from Indonesia. *Environmental Research Journal* 5: 505–522.

Turner, S. (2012) "Forever Hmong": Ethnic Minority Livelihoods and Agrarian Transition in Upland Northern Vietnam. *The Professional Geographer* 64: 540–553.

Select Guide to Further Reading

Borras, J. and Franco, J. (2010) Towards a Broader View of the Politics of Global Land Grab: Rethinking Land Issues, Reframing Resistance. *Initiatives in Critical Agrarian Studies Working Paper 001.* http://www.tni.org/paper/towards-broader-view-politics-global-land-grabbing. Accessed January 19, 2012.

Evenson, R. E. and Gollin, D. (2003) Assessing the Impact of the Green Revolution, 1960–2000. *Science* 300: 758–762.

Hall, T. (2010) *Earth into Property: Colonization, Decolonization, and Capitalism.* Montreal: McGill-Queens University Press.

Harris, C. (2003) *Making Native Space: Colonialism, Resistance and Reserves in British Columbia.* Vancouver: University of British Columbia Press.

Horlings, L. G. and Marsden, T. K. (2011) Towards the Real Green Revolution? Exploring the Conceptual Dimensions of a New Ecological Modernization of Agriculture That Could "Feed the World." *Global Environmental Change* 21: 441–452.

Knudsen, M. T., Halberg, N., Olesen, J. E., Byrne, J., Iyer, V., and Toly, N. (2006) Global Trends in Agriculture and Food Systems. In Niels Halberg (ed), *Global Development of Organic Agriculture: Challenges and Promises.* Wallingford, UK: CABI Publishing, 1–48.

Lyon, S., Aranda Bezaury, J., and Mutersbaugh, T. (2010) Gender Equity in Fairtrade–Organic Coffee Producer Organizations: Cases from Mesoamerica. *Geoforum* 41(1): 93–103.

Morgan, K., Marsden, T., and Murdoch, J. (2006) *Worlds of Food: Place, Power and Provenance in the Food Chain.* New York: Oxford University Press.

Pearce, F. (2012) *The Land Grabbers: The New Fight Over Who Owns the Earth.* Boston: Beacon.

Pimentel, D. (2005) Economic and Environmental Costs of the Application of Pesticides Primarily in the United States. *Environment, Development and Sustainability* 7: 229–252.

Scholtz, C. S. (2006) *Negotiating Claims: The Emergence of Indigenous Land Claim Negotiation Policies in Australia, Canada, New Zealand, and the United States.* New York: Routledge.

Sinclair, R. (1967) Von Thünen and Urban Sprawl. *Annals of the American Association of Geographers* 57: 72–87.

Sen, D., & Majumder, S. (2011) Fair Trade and Fair Trade Certification of Food and Agricultural Commodities: Promises, Pitfalls, and Possibilities. *Environment and Society: Advances in Research* 2(1), 29–47.

Selected Websites

For the US economy, see:

Federal Reserve Bank of Richmond—economic indicators
http://www.richmondfed.org/research/national_economy/national_economic_indicators

US Department of Commerce, Bureau of Economic Analysis
http://www.bea.gov

US Department of Labor, Bureau of Labor Statistics
http://www.bls.gov/eag/eag.us.htm

US Department of Commerce, Center for Economic Studies
http://www.census.gov/ces/dataproducts/economicdata.html

US Department of Commerce, Economics & Statistics Administration
http://www.esa.doc.gov/

Learning Outcomes

- The three distinct sectors of economic activity are primary, secondary, and tertiary.
- The primary sector of the economy includes the agriculture, forestry, mining, and fishing industries.
- The von Thünen model predicts land use around a city. The more expensive land closer to the city is used for more intensive market-oriented agriculture.
- The agrarian transition occurs as agriculture shifts from subsistence to commercial and moves from meeting local market demands to provisioning national and global food supply chains. The shift involves changes to agriculture, including intensification and expansion, market integration, and environmental change.

- Food supply chains have become longer and more complex due to urbanization, changing diets, and increased globalization, and can be categorized as buyer-driven, producer-driven, bilateral oligopolies, or traditional markets.
- Market integration of agricultural products happens when subsistence smallholders become commercial farmers or via land grabs and forced commercialization.
- The commercialization of agriculture involves greater capital investment in machinery and biotechnology, which leads to greater productivity, but this may have specific, unintended, and deleterious environmental effects.

- The most sustained commercialization of agriculture in the developing world that involved large inputs of biotechnology was the Green Revolution, which increased agriculture prediction through new high-yielding varieties of crops, irrigation, and liberal helpings of pesticides and fertilizers.
- The four Kondratieff cycles of manufacturing depict fifty-year waves of increased innovation and invention.
- The tertiary sector of the economy is the service industry, which comprises selling, assistance, and providing expertise as opposed to making a tangible product.

- The knowledge- and information-based economies are sometimes referred to as advanced producer services or designated as a distinct sector, the quaternary sector, with the executive decision making levels defined as the quinary sector.
- A region's success rests less on manufacturing employment and more on the extent to which it can generate, retain, and attract knowledge-based employment.
- The shift from a manufacturing-based economy to a service-based economy has profound effects on income distribution, gender relations, and urban growth.

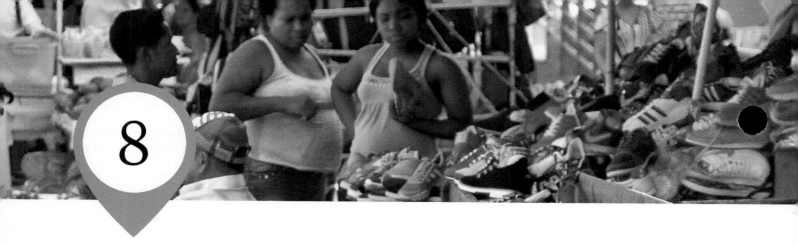

The Economic Geography of Uneven Development

In this chapter, we will look at the economic organization of space through a discussion of uneven development at different scales.

Global Differences

Across the globe, uneven development of economic activity has caused marked disparities in living standards and quality of life. Table 8.1 highlights data from three countries, Brazil, Cameroon, and the United States. In Cameroon, many people do not live past their mid-fifties, while in the United States people regularly live into their seventies and beyond.

To explain these differences, some, such as Jared Diamond, take a very long-term view—13,000 years to be precise—and focus on how initial geographical advantages were turned into economic advantages. Geography plays a part in shaping the global economic order of marked disparities, but it is rarely pivotal. Japan, with few natural resources, has reached a high level of economic development and good living standards, while much of tropical Africa, despite a bounty of resources and minerals, remains poor.

The patterns not only vary over space, they change over time. In the fifteenth century, China was far ahead of Europe, and had enough scientific and technical knowledge to inaugurate its own industrial revolution.

TABLE 8.1 Global Differences			
	CAMEROON	**BRAZIL**	**USA**
GDP per capita (USD)	3,200	15,800	56,300
% Labor force in agriculture	70	15	0.7
% Labor force in manufacturing	13	13	20
% Labor force in services	17	71	79
Life expectancy at birth (years)	57	75	79

Source: CIA World Factbook.

This did not happen, and China focused inward. In contrast, technologically backward Europe became the center of massive economic change as Britain's Industrial Revolution transformed the old patterns into a new global order in which manufacturing and economic vitality became the monopoly of the West. In recent years, even this order has changed with the global shift of manufacturing and the rise of new industrial powers such as Brazil, China, and India. The global economy is dynamic and ever-changing.

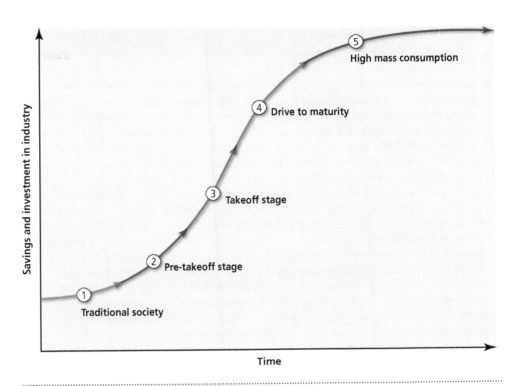

8.1 The Rostow model

The Rostow Model

American economist Walt Rostow identified different stages of economic development (see Figure 8.1). The **Rostow model** has different stages. In traditional society, stage 1, the majority of people are employed in the primary sector, with very few in the secondary and tertiary sectors, much like Cameroon. In stage 2, the transitional or pre-takeoff stage, there is some specialization and a more developed infrastructure. Stage 3 is the takeoff stage, which refers to rapid industrialization. The takeoff stage is different for different countries: 1783 to 1802 for the United Kingdom, 1843 to 1850 for the United States, and since 1990 for China. Stage 4 is the drive to maturity, involving diversification and innovation with less reliance on imports. In stage 5, high mass consumption, the domestic demand of consumers, is a major economic engine and the service sector is dominant.

The stages are less an explanatory model than a historical sequence with, at its heart, the notion that through time and encouragement of private investment, countries can achieve economic growth. The assumption is that every economy can be like the United States if it follows the same path of encouraging private investment and ensuring returns to investors. Global inequalities and differences are thus a function of time, with some countries more advanced along the trajectory than others. That raises the questions: why are some countries where they are on this historical trajectory, and why are some countries rich and some so poor? The answer, according to some, lies in the differential ability to adopt economic opportunity and harness technological possibilities.

Cultures that promote innovation and openness tend to do better than others. Small differences over time became sources of major divergences. For instance, when the Ottoman sultan forbade the establishment of a printing press in 1485, it reduced the opportunities for a spread of learning and an increase in technical proficiency. Institutions, especially institutions that secure property rights and allow broad economic and political participation, play a significant role in economic growth. The Industrial Revolution in Britain was due to many factors, but there were important institutional shifts, including the creation of an efficient banking system to connect savings to investment and the establishment of a mass education system to train people for work and industrial employment.

The Core-Periphery Model

The **core-periphery model** is an alternative to the Rostow model (Figure 8.2). The global economy has winners and losers not just as a historical accident but as part of the very process of economic development. Some countries are rich not because they are further along the Rostow trajectory or have a more innovative culture, but because other countries are poor. The initial advantage of some countries was reinforced by colonialism and economic imperialism, which led to the differential distribution of wealth and poverty. Instead of being located at different stages in time, countries are part of a spatially connected order, a

8.1 Different Economies

We can distinguish between different types of economies. Employment in the **formal economy** involves the sale of labor in the marketplace, formally recorded in government and official statistics. The **informal economy** is the unrecorded sector where few, if any, taxes are paid. The informal sector is a response to the lack of formal employment opportunities. This sector is just as complex as the formal sector. Based on his work in Latin American cities, Ray Bromley identifies different sectors, including retail distribution, small-scale transport, personal services, security services, gambling services, and recycling enterprises. Work conditions vary from short-term wage work to precarious self-employment. The workers in this sector may be disguised wage workers (they are paid a wage but it is not recorded), workers on short-term wage work, dependent workers who may lease a personal transport vehicle from an employer, and the self-employed. In his research on the city of Cali, Colombia, Bromley found that 40 to 45 percent of people who worked in the street were in precarious self-employment, 39 to 43 percent were disguised wage workers, 12 to 15 percent were dependent workers, and only 3 percent were short-term wage workers.

The divisions between the formal and informal sectors fluctuate according to circumstance. When and where the formal economy provides many employment opportunities, the informal sector becomes less significant. However, if the formal sector provides few jobs or only jobs with low wages, then the informal sector is a sophisticated coping strategy as people use a variety of tactics to make a living.

The **illegal economy** is the shadowy world of prostitution, illicit drug sales, racketeering, and the like. Bromley noted the importance of prostitution, begging, and property crimes involving illegal appropriation through stealth (theft), the threat or use of violence (robbery), or deception (conning). It is difficult to estimate the sector's full extent. Much of the global illegal sector consists of the production and distribution of goods and services for which there is a demand even though they are illegal. The demarcation of legal and illegal varies: alcohol is not illegal in the United States but cocaine is; prostitution is legal and regulated in both Germany and the Netherlands but is illegal in most of the United States; one can buy marijuana in Amsterdam without breaking the law, but the same transaction in New York would be considered criminal. The line between legal and illegal is constantly being crossed: police who take bribes; chemical companies that illegally dump waste; corporations that form illegal cartels to charge high prices to the federal government. The distinction between the three sectors, formal, informal, and illegal, is at times fuzzy. Rather than look at the differences, a more rewarding strategy may be to look at how transactions transgress the lines.

A **social economy** consists of not-for-profit activities; it is also sometimes referred to as the "third sector." It includes cooperatives, charities, voluntary groups, trusts, and religious organizations. This sphere of economic activity, somewhere between the market and the state, plays an important role in certain parts of the economy as both funding source and employer.

The **communal economy** involves the cashless exchange of goods and services. It is common in neighborhoods and extended family systems. If someone wants a babysitter for an evening, they have a number of possibilities: they can hire someone and record the transaction and pay taxes (the formal economy); they can hire a teenager and pay them an agreed sum without informing the authorities (the informal sector); or they can ask a neighbor, with the often unspoken understanding that they will return the favor. The last, and often the most common, response is an example of the communal economy in action. It can range from reciprocal favors, such as babysitting, grass cutting, and garbage removal, to a host of household chores, from building maintenance to carpooling.

The **domestic economy** is the amount, type, and division of labor within the home. A hundred years ago, there was much greater use of paid domestic labor for middle and upper-income groups. The increasing cost of labor has meant the decline of mass domestic labor, though it still persists among the wealthy; in rich countries, foreign, vulnerable, and illegal workers do a high proportion of such work. The division of labor is a source of change, conflict, and negotiation within the households. As more women have joined the formal labor force, they feel the double demand of work and home because they carry much of the domestic load. The division of domestic labor is not simply an apportionment of necessary work; it involves questions and representations of femininity, masculinity, and the family.

Recent years have also witnessed the rise of the **sharing economy** in which owners of assets such cars or rooms rent them out in peer-to-peer exchanges. Examples include Uber and Airbnb. The sharing economy has grown with the widespread adoption and use of social media and specific apps that link providers with consumers. The sharing economy disrupts traditional markets: the introduction of Uber, for example, is seen as a threat to many cities' traditional taxi systems. The consequences of the widespread diffusion of the sharing economy on traditional economies are only just unfolding.

REFERENCE

Bromley, R. (1997) Working in the Streets of Cali, Colombia: Survival Strategy, Necessity or Unavoidable Evil? *Cities in the Developing World: Issues, Theory and Policy*, 124–138.

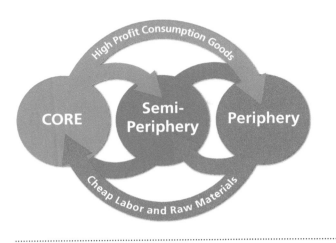

8.2 The core-periphery model

world system comprising core, periphery, and semi-periphery. For two hundred years the core was in Western Europe and later in North America. Economic transactions are marked by an unequal exchange of cheap raw materials from the periphery to the core and the export of manufactured goods from the core to the periphery. This essential dynamic explains the colonial expansion of core countries as they sought to establish monopoly control over regions that were then transformed into colonial peripheries used to feed and sustain the core economies by supplying cheap raw materials and purchasing manufactured goods. Britain was a core country that incorporated much of India into commercial domination and political control. Raw cotton grown in India was shipped to Britain, processed into cloth in Lancashire cotton mills, and then exported abroad. The indigenous cotton manufacturing industry in India was destroyed, while the domestic cotton industry in Britain was a leading force in the Industrial Revolution, which propelled national economic growth.

This core-periphery model allows us to globalize such issues as the Industrial Revolution, presenting them less as unique national experiences and more as part of a global drama. The model is dynamic. The category of semi-periphery is something of a transitional condition situated somewhere between the dominance of the core and the relative weakness of the periphery. Some countries can move from the periphery toward the core. In the past fifty years, China, South Korea, and others have undergone marked industrialization to place them closer to the core. And even some strategic primary producers of raw materials, such as the oil-producing countries of OPEC or mineral-rich Australia, can use their collective interests and combined power to increase prices and force redistribution in global wealth. The core-periphery model has the enormous advantage of globalizing historical events and adding a much-needed dose of political economy to standard historical narratives while being flexible enough

to explain the moves of some national economies from periphery to semi-periphery, with the possibility of core membership changing and shuffling. From their initial advantages, the core countries became richer.

Regional Differences

Economic disparities are also evident at the regional level. Consider the case of China. China's economy, like most national economies, is a patchwork quilt of very different regional economies, some growing faster than others, some more industrialized than others. Figure 8.3 shows the marked regional variation in income levels. The fastest-growing coastal regions, where the bulk of export-oriented manufacturing takes place, are wealthier than the more distant inland regions with much less manufacturing activity. These differences are the principal reason behind China's large-scale internal migration from inland provinces to the coastal regions.

Models of Regional Economic Growth

Regional disparities are best understood with reference to models of regional economic growth. The process of **cumulative causation** occurs when growth feeds on itself. Growth in a leading region spreads out into other lagging regions through diffusion of innovations and access to markets (Figure 8.4). There are also backwash effects that include the flow of capital and labor from the lagging to the leading regions. In the United Kingdom, the backwash effects are clearly more pronounced than the spread effects, with growth concentrated in London and the southeast part of the country, which attracts more labor and capital that in turn, through cumulative causation, generate even more growth. The flow of capital and skilled labor from the peripheral regions, such as Scotland, limits economic growth and ultimately leads to pronounced regional disparities. As Figure 8.5 shows, Britain has the widest regional disparities among the larger economies.

Uneven regional development is not a temporary condition that will be ironed out through time, but rather is absolutely essential to a capitalist mode of production. Waves of investment are uneven across space, making some regions more developed than others. These investments are spatially fixed and form the basis for subsequent rounds of capital investment. Some regions are then bypassed by subsequent waves of investment as capital searches for areas of greatest profit. The result is the constant production of uneven development.

We can also consider regional economic dynamics from the perspective of stages of product and **profit cycles**. A

8.3 Regional differences in China.

Source: © The Economist Group Limited, London (September 12th, 2015)

product cycle is the trajectory of a brand new product. At early stages of a product cycle, when new products are just being created, it is important to be close to research and development. Sales, shown in red in Figure 8.6, go through a steep rise during the growth and maturity phases and then drop off. Profits, shown in the shallower curve, are negative until the growth period. Innovation occurs in the more specialized regions, but as the products became more standardized the crucial locational component is less closeness to innovation and more reducing labor costs. Consider the case of typewriters, which were introduced in the early twentieth century but then declined by the end of the century as computers replaced them. Typewriter manufacturing shifted from the early specialized centers where labor was expensive, such as Syracuse, New York, to closer but cheaper labor areas such as Cortland, New York, to extremely cheap labor areas in Mexico before finally becoming obsolete. Similarly, today, while new computer technologies are generated in Silicon Valley, the production of actual computer chips can occur in newly industrializing regions with good infrastructure and cheap labor, such as Thailand. In the early stages of

product development, when super profits can be made, new firms are attracted to the location of the initial innovation, but as more firms enter the markets, new spatial forces are at work and firms may relocate closer to markets or away from higher-wage areas in order to maintain and increase profitability.

Regional Economic Clusters

There is a distinct clustering of similar economic activities at the regional scale. Four factors are important:

- locational pull of resources such as proximity to transport links or energy sources;
- the concentration of firms and workers, which provides a pool of specialized labor that benefits both labor and companies;
- subsidiary trades and services that are attracted to the core industries;
- the spillover of innovative ideas from firm to firm.

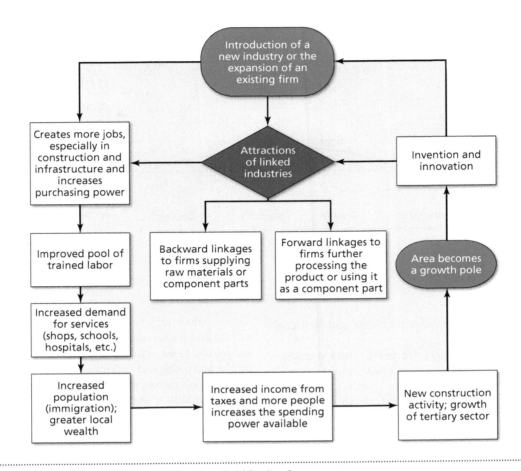

8.4 Cumulative causation. Diagram courtesy of Barcelona Field Studies Centre.

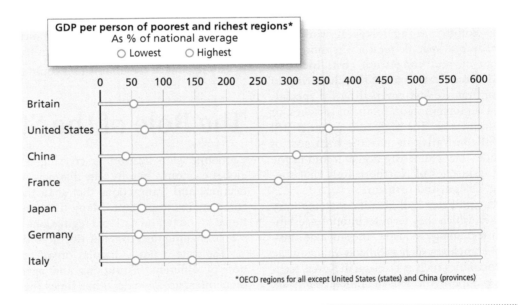

8.5 Regional disparities

Regional economic complexes emerge because of the locational pull of reduced transaction costs. However, these are maintained by the tight social networks between firms that require interpersonal trust, proximity, and regular face-to-face contact. These social networks are particularly important in highly specialized sectors. Even in the information age, the spatial clustering of activities, firms, and workers provides the efficient means

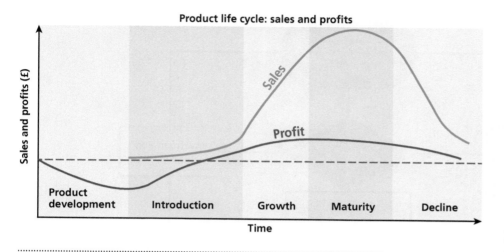

Product life cycle: sales and profits

Sales and profits (£) — vertical axis

Sales

Profit

Product development | Introduction | Growth | Maturity | Decline

Time

8.6 Product and profit cycles

of communication that help solve problems and facilitate mutual cooperation.

Regional economic clusters provide a **"thick market"** (a large pool of labor and specialized firms), market access, and savings in public goods provision. In effect, economic agglomerations, because they concentrate things in the same place, make markets more efficient. There are efficiency gains from the scale effects. And in turn this spatial ensemble of benefits, through cumulative causation, creates path-dependent development. Once the movie industry was established in the Los Angeles area, for example, it continued to attract capital and skilled labor, which made it even more attractive to make movies in that location.

There are also countervailing forces at work. The more firms cluster in one specific region, the more land and labor costs go up and congestion costs increase. These **centrifugal forces** can disrupt the clustering of routine production but are less compelling for specialized sectors such as entertainment, financial services, or cultural-creative industries. The high-tech sector is still concentrated in Silicon Valley despite the high cost of land and labor. Industrial clustering is also found in the small-craft-based industries of Northern Italy that specialize in clothing, design, and furniture.

Manuel Castells and Peter Hall identify contemporary industrial complexes, which they term **technopoles**, established around high-technology industry. They include semi-planned territorial complexes such as Silicon Valley in the United States, science cities such as Daejeon in Korea, technology parks such as Sophia Antipolis in France, twenty-six technopoles spread through Japan, and large metro areas around the globe, including more established ones such as London and Tokyo and newer ones such as Seville, Spain, and Adelaide, Australia. They identify four foundational elements: a determined administration; liminal individuals who straddle the business and academic worlds; an environment of innovation; and easy access to venture capital.

Regional Economic Clusters in the United States

As an example of regional differences, we can consider the new geography of employment in the United States. Enrico Moretti identifies a divergence from around the mid-1980s, when selected cities with propulsive economic sectors kept attracting more growth and generating higher-wage jobs. Meanwhile, certain urban regions failed to attract new growth and were unable to replace the jobs lost by deindustrialization. In these declining urban regions, wage levels dropped and the more mobile and higher-skilled workers left. For urban regional economies to be successful, they need to constantly push up the steep slope of innovation adoption, which requires expertise, education, and knowledge. The more the national economy shifts toward a knowledge-based economy, the more marked the differences become between those regions that attract and those that fail to attract these sectors. There are now three types of economic regions in the United States:

- the brain hubs of a well-educated labor force, higher wages, and innovation; examples include San Francisco and Palo Alto;
- dying manufacturing hubs of deindustrializing cities, such as cities in the **Rust Belt** (Figure 8.7);
- city regions in the middle that could go either way.

The Role of the State

Economic globalization is creating a more integrated global economy, but even in this new world of interconnections and transactions that span the globe, the state still remains important. Many firms operate within national markets, and national regulations play an important role in shaping the economic decision-making context.

The state is not a singular organization but a collection of different institutions and agencies. Sometimes their interests coincide; other times they collide. The US Department of Defense, for example, wants to maintain a large and expensive military presence funded by the government. The US Treasury, in contrast, is concerned with controlling government expenditure. The clashes are resolved through the political process.

Through its multiple and varied roles, including employer, investor, policy maker and regulator, the state shapes

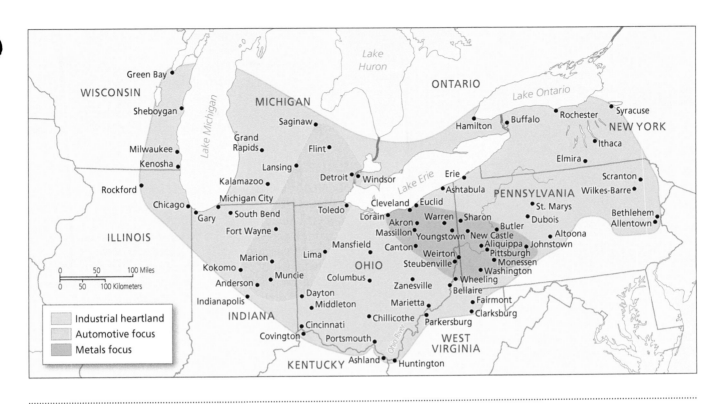

8.7 Rust Belt

economic geographies. Within the nation's borders, the state has fiscal powers of taxation and spending, currency control, fixing exchange rates, and setting interest rates. Increasing interest rates, for example, slows down industries such as house building that rely on long lines of credit. Lowering interest rates in turn can stimulate such industries.

The state sets the regulatory framework, which affects economic activity. The economic geographer Jamie Peck examines the political economy of **workfare**, the welfare-to-work initiatives that have developed since the 1990s. He traces the development of workfare policies in Canada, the United States, and the United Kingdom and shows that workfare is not so much concerned with creating jobs as with deterring welfare claims and facilitating the acceptance of low-paying jobs.

The State and the Global Economy

The state also has a role in shaping the connection between national economies and the global economy. At one extreme are the closed economies that limit trade and economic connection with the outside world; examples include Cambodia under the Khmer Rouge and present-day North Korea. Such instances are rare, because global trade is the lifeblood of most national economies. At the other extreme are more loosely regulated economies that facilitate penetration by foreign capital. Most countries fall somewhere between the two.

The governments of developing countries, seeking to support the industrialization of the national economy,

follow either an **import substitution** strategy in which they protect local industry, or an export-oriented form of industrialization. This was long followed in Latin America up until the 1980s. In contrast to **protectionism**, there is the ideology of free trade. One dominant view is that countries develop through the adoption of free-trade policies. In a provocative argument, Ha-Joon Chang takes issue with this idea and argues for protectionism rather than free trade. He claims that promotion of a free-trade strategy works against the development interests of national economies. Protectionism allows local firms to grow and prosper by borrowing ideas and techniques from more efficient producers. Since the rise of Britain, most countries, even successful ones such as the United States, have competed initially with forms of protectionism for local industries. More recently, the spectacular success of countries such as South Korea has been based on protecting local industries before they became internationally competitive.

The empirical evidence suggests that more open trade leads to rising per capita income. However, per capita income is a crude measure that does not measure the distribution of income, merely the average. The evidence on income distribution is inconclusive. There are examples of countries where income inequality actually increased in the wake of trade liberalization: Argentina, Chile, Colombia, Costa Rica, and Uruguay all saw increasing inequality. In the United States, the declining incomes of the middle class are commonly linked to open trade policies.

The Development State

The **development state** is the name given to countries that have successfully crafted their economic policies to move from the periphery of the global economy closer to the core. The term implies a state that is focused on economic development. Examples include Japan, South Korea, and China. While they adopted a variety of different measures, they all aid specific industries, protect domestic industries in the early years, enhance infrastructure, and spend on education. There is a strong commitment to export-led growth.

The most dramatic and recent example is China. The China model, which consists of directing public spending to foster infrastructure improvement and educational attainment, while lacking democratic accountability, was spectacularly successful in raising 600 million people from chronic poverty in less than two decades.

The development state was primarily an East Asian phenomenon, and that raises the question of whether it is transferable. Francis Fukuyama argues that there are limits to this transferability, as the model is culturally specific to Confucian societies with a meritocracy in public service, a heavy emphasis on education, and a deference to authority. His argument would seem to be undermined by the relative success of countries like Brazil that have different traditions from East Asian societies but are still achieving economic growth.

Does the development state have to be authoritarian? In the early years of industrialization, this tends to be the case, and modernization does not necessarily lead to democratization but may make democratization more possible. The more developed and richer a country becomes, with wealth widely spread, the more it tends to move further to the democratic end of the political spectrum.

The development state can achieve spectacular success as public and private interests are focused on the same goal of achieving rapid growth. However, the reliance on export-led growth makes the economies vulnerable to global shocks.

Capital and Labor

An important element of economic geography is the changing relationship between **capital** and **labor**. The capital–labor relationship is crucial to the modern economy. Capital seeks to keep the costs of labor down and uses its influence over the state to ensure legislation that minimizes the costs and power of labor. Labor in turn seeks to organize in order to have bargaining power to raise wages and improve working conditions (see Figure 8.8). Changes in economic geography can be partly understood in relation to capital–labor relations and their differential mobility. Capital has generally become more mobile than labor, which strengthens the hand of capital in relation to labor.

The differential in mobility has widened. The suburbanization of industry was a response to the growing power of labor within cities. The shift of car manufacturers from Michigan to the southern states was a move by capital to reduce the costs and power of labor. More recently, the global shift in manufacturing is partly a move to cheaper regions of the world with less organized labor, such as China.

Capital

"Capital" has a number of meanings. In the standard literature, it is used to refer to real assets such as factories, property, and machinery, or financial assets such as bonds, securities, and stocks. But this is only the physical expression; capital is also a social relationship.

The mobility of capital has been reinforced by changes in production. In the system known as **Fordism**, plants were huge, fixed capital investments. Bargaining between capital and labor thus took place against a fixed location. Recently, however, a more flexible form of production has been introduced. Nike, for example, has no shoe factories and no Nike workers. Nike's shoes are made under contract by a range of shoe manufactures. Factories compete to obtain Nike orders and are then licensed by Nike if they are capable of making shoes to cost and design specifications. The old model of manufacturers making things and retailers simply selling them has been replaced by the power of retailers and brands. Now retailers tell the manufacturers what to produce. Contracts are for short-run lines rather than long-run batches. This just-in-time production system ensures that goods are made to meet demand. Just-in-time, flexible production allows low prices and high turnover. It also means a marked change in capital–labor relations. Capital is now hypermobile. Workers in one factory cannot bargain in the same effective way that the workers of the old Fordist system could. Capital is no longer fixed in place. Retailers can move their production contracts to another factory in another country. While capital can roam the world, labor is more fixed in place. The result is an uneven bargaining arrangement.

Capitalism is a dynamic, crisis-laden economic system with tremendous vitality. Perhaps the most sustained geographical analysis of capital as a social relationship is the work of David Harvey, who since 1973 has been developing a historical geography of capitalism. In his work he outlines the recurring crises of capitalism. For the system to continue, money needs to be mobilized and invested, technology employed, nature exploited, labor assembled, and credit made available. At any of these stages, problems may occur; there may be a lack of money to make suitable investments, technological innovation may undermine profitability, labor may become too powerful and demand too much in wages, environmental costs may restrict growth, or access to credit may dry up. A profit squeeze or credit crisis can cause blockages in the smooth functioning of the system. In fact, smooth functioning is rare. Capitalism is capable of amazing growth and innovation, but it is subject to recurring crises and continual turmoil as the crisis solved in one area becomes a crisis in another area. The globalization of capitalism has kept the system going but has widened the opportunity for recurring crises worldwide.

Labor

There is a geography of labor that includes the importance of the geography of labor legislation, the political geography of union organizing, and the power of place. Geographer Andy Herod, for example, has shown how organized labor impacts and influences the geography of capitalism with reference to selected case studies of garment workers, steel workers, and dockworkers.

The overall nature of the capital–labor relationship has shifted. From the 1930s to the 1980s, organized labor in the Europe and North America was powerful and played an important role in creating public policies that guaranteed relatively high levels of social welfare. Skilled labor was in short supply and so could use its bargaining power to increase both wages and what is termed the social wage, the range of social benefits provided by the state. There were national variations in the affluent West, with social welfare levels significantly higher in the Scandinavian democracies than in the United States. Since the 1980s, the power of labor has been weakened by global shift and mechanization. The deindustrialization of the economies has led to a weakening of organized labor and a resultant shift in capital–labor relations, with wages stagnating (Figure 8.9).

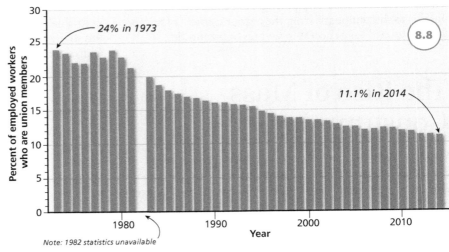

8.8 Declining union numbers 8.9 Union hall in Baltimore

The basic geographic difference that strengthens capital and weakens labor is differential mobility. Capital is freer to roam the globe in the search for profitable investments. Labor, in contrast, is more fixed in place, unable to freely move across national borders. Mobile capital has more power in a globalizing economy than a labor force more fixed in space.

And here we come to a paradox. Even as firms and companies seek to minimize labor costs, the high-mass-consumption economies need consumers to have enough disposable income to buy goods and services. This is the Wal-Mart economy, where companies pay low wages but then have to provide cheaper goods. The goods are cheap because the wages of the producers are low.

Labor is not just an economic entity. It is embodied in people. An important element of the new economic geography considers how social differences are reinforced, undermined, and contested in the workplace. Who works in what jobs and how much they are paid is not a simple labor market process but a wider economic and social differentiator. There is considerable work on ethnicity, for example, that examines ethnic economies (where workers in one sector or trade share common ethnicity), the geography of ethnic networks, and the connection between ethnic networks and accelerating economic globalization. Geraldine Pratt explores the geography of Filipina domestic workers in Vancouver, Canada. She found that women from a variety of occupations were restricted through immigration laws, that did not recognize their national qualifications, to the domestic service sector.

Households get by through use of the different sectors of the economy. The available getting-by strategies will depend on the resources. A household with a large income from the formal economy will not have to sell cigarettes on the street corner. A household with few formal opportunities from the market or state, in contrast, has to look at alternative sources and rely on the informal economy.

At the very bottom are the poor, who have to be creative in their survival strategies (see Figure 8.10). Around 40 percent of the world's population lives on less than $2 a day. One billion people live on less than $1.25 per day. Lacking income, they are trapped in extreme poverty. One proposed solution is the use of microcredit—lending small amounts to people to provide them the financial means to become more self-reliant. Muhammad Yunus pioneered the use of microloans in Bangladesh. One study looked at the detailed diaries of 250 households in Bangladesh, India, and South Africa. It was clear that the problem is not only the meager income but also the unpredictable and irregular nature of that income, which makes saving and planning very difficult. One proposal is to shift emphasis from the encouragement of microfinance toward easier access to savings opportunities.

The Rise of Mass Consumption

We can make a distinction between different types of capitalist economy. In the nineteenth and early twentieth centuries, capitalist economies were based on the production of commodities and most people were relatively poor. Beginning in the United States, in the 1920s, a new economy emerged based on the mass production and mass consumption of goods. In societies of high mass consumption, the consumer emerges as an important element in the making of economic geographies. Consumer spending in the United States is now responsible for two-thirds of all economic activity. The enthronement of the consumer has many implications; here I will discuss only five.

First, as consumer spending shifts from meeting basic requirements to discretionary spending, desire rather than need becomes a more important driver of consumption. Consumers have to be persuaded to buy things, and this basic fact is the reason behind the growth and emergence of advertising and promotion of goods. While we may want clothes, we have to be persuaded to buy particular brands. Consumption is now tied to the creation of identities and the pursuit of lifestyles. To buy a pair of sunglasses

8.10 Informal employment in Panama

is to shield your eyes from the Sun, but to buy a pair of Ray-Bans is to evoke other images and make other claims. Consumption is now tied to aspirations, images, and desires as much as to basic needs and requirements.

Second, forms of consumption are markers of social status and social difference. People are distinguished not only by how much they have but also by what they have and how they use it. Forms of consumption mark different levels of society. While all middle-income groups can have a beach holiday, only the richer can afford a Caribbean vacation, and only the very rich can rent a villa in St. Bart's. Luxury fever can filter down the income scale as the consumption trends of the very rich become embedded in popular consciousness.

Third, the full flowering of a society of high mass consumption is predicated upon access to credit so that consumers can buy big-ticket items. The rise of credit to lubricate mass consumption has its upside, as it enables purchase before the full price is saved. It also has its downside, as it makes the economy vulnerable to credit squeezes. That is what happened after the 2008 meltdown as many housing consumers in the United States had extended credit on homes that were depreciating in value. With consumers unable to gain access to any more credit, the economy sputtered and declined as demand fell. The more consumer spending is reliant on credit, the more the economy is vulnerable to credit crises.

Fourth, the economic geography of places is shaped by the places of consumption. As shopping becomes an increasingly important social and economic activity, retail sites become bigger. The typical mall has increased in size and scope. The mall is the cathedral of

8.2 The Changing Concerns of Economic Geography

The subdiscipline of economic geography has undergone important shifts as its practitioners look at the economy in different ways. Early work, produced at a time of industrial growth, focused on theories of industrial location. One of the earliest studies was Alfred Marshall's *Principle of Economics,* which first appeared in 1891. A strong German contribution is embodied in Albert Weber's *Theory of the Location of Industries* (1909) and August Losch's *The Economics of Location* (1939). By the middle of the twentieth century, geographers sought to explain the location of industry through looking at the various factors of production from the standpoint of the individual enterprises. Estall and Buchanan (1961) discuss the role of various factors, including materials, markets and transfer costs, energy sources, labor and capital, technological change, and the role of government. In their case study of iron and steel, for example, they point to the geographical pull of coal, iron ore, material assembly, market, capital, and labor as well as the importance of inertia.

Critics of standard industrial location theory pointed to the assumption of perfect economic rationality. Decision makers, in the models at least, had all the necessary information and were blessed with perfect rationality. Scholars pointed to the need to incorporate suboptimal decision making and satisfying rather than maximizing behavior. Critics also pointed to the emphasis on single-plant, single-product establishments and raised the need for a fuller understanding of agglomerations, corporate context, and multinational transactions. An institutional approach emerged that stressed the interaction between firms rather than the behaviors of individual firms, and showed that location was the result of complex negotiation and bargaining with a variety of other agents in a wider social and political context.

The concerns of economic geographers have also widened. Some economic geographers have utilized evolutionary economics to refine location and relocation theories. A **Marxist political economy** approach explains how uneven development is an essential part of capitalist economies. A new economic geography now locates economic processes in deeper social, cultural, and political contexts in the exploration of themes such as commodity chains, technology and agglomeration, commodification of the environment, and the role of knowledge-based economies and creativity in regional and city development.

There has also been a fruitful infusion of feminist perspectives into economic geography. Gender relations intertwine with uneven development, globalization, locational strategies, and employment practices. Issues such as the feminization of poverty, the dominance of young women in the factories of coastal China, the feminization of domestic service, and the double bind that affects many women as they work both outside and inside the home all highlight the close connections between gender, work, and the economy. The workplace is structured with distinct sexualized and gendered scripts. Gender relations are created and contested in the workplace.

REFERENCES

Barnes, T. J., Peck, J., and Sheppard, E. (eds) (2012) *The Wiley-Blackwell Companion to Economic Geography.* Oxford: Wiley-Blackwell.

Boschma, R. and Frenken, K. (2011) The Emerging Empirics of Evolutionary Economic Geography. *Journal of Economic Geography* 11: 295–307.

Estall, R. C. and Buchanan, R. O. (1961) *Industrial Activity and Economic Geography.* London: Hutchinson.

Krugman, P. (2011) The New Economic Geography, Now Middle-Aged. *Regional Studies* 45: 1–7.

consumerism, an alternative retail environment where consumers are enticed, entertained, and encased in a totalizing retail space.

Retail establishments are clustered together. We can understand this phenomenon with reference to **Hotelling's Law,** named after economist Harold Hotelling, and his classic paper, first published in 1929. The law states that it is rational for producers to make products similar to their competitors'. If we substitute retailers for producers, then the law explains the clustering of retail distributions. Assume a straight street. If there is only one coffee shop, the owner is indifferent to location anywhere along the street. But if there are two or more coffee shops, it is more rational for them to be together at the center of the street, drawing upon 50 percent of the street's coffee drinkers. Retail establishments cluster together because it makes rational economic sense even for providers of the same product—otherwise they might lose customers (see Figure 8.11). That is why a Burger King and a McDonald's are often close

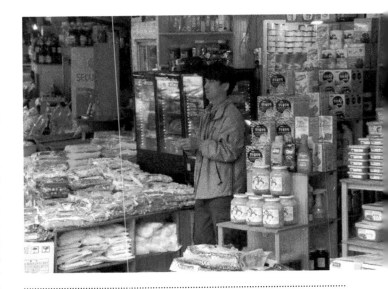

8.11 Retail clustering in market in Seoul

together when you turn off the highway. Even retailers with different merchandise cluster, because they get the extra advantage of possible market capture of consumers of other producers. That is why you often find a Starbucks clustered with the two hamburger sellers.

Finally, there is the growing dominance of retailers over producers. Traditionally, manufacturers made things, and then sold them to retailers who sold them to customers. Retailers were simply in the middle, buying goods from producers and then selling them to customers. But with the rise of the big-box retailer and a more flexible production system, with many small companies producing small runs of goods relatively quickly, power shifts from the large number of small manufacturers to the small number of larger retailers. The very large retailers have such enormous power that they can force producers to compete with each other to meet strict cost limits. Manufacturers are willing to do this because access to a big retailer assures a large customer base. The lower price

per item is offset by the higher turnover of items. Take the case of Wal-Mart, which in 2015 had 11,453 stores in 27 countries. Wal-Mart uses its enormous leverage to drive down the prices of the manufacturers who supply these stores. Wal-Mart reduces the price of goods by relying on producers from low-wage countries, thus pushing North American producers to reduce their costs. A number of North American producers have offshored their production to reduce costs in order to sell to Wal-Mart. Wal-Mart and stores like it are a major cause for the shift in manufacturing to lower-wage regions.

Wal-Mart uses its market power to reduce costs, leading to more offshoring and more imported goods from cheap-labor regions, which in turn leads to fewer jobs and lower incomes in the formerly richer regions and consequently lower wages, forcing more people to seek deals at Wal-Mart. While big-box retailers reduce costs and increase productivity, there is also the associated downward spiral of job loss and wage reduction.

Cited References

Chang, H.-J. (2007) *Bad Samaritans: The Myth of Free Trade and the Secret History of Capitalism.* London: Bloomsbury.

Daryl, C., Morduch, J., Rutherford, S., and Ruthven, O. (2009) *Portfolios of the Poor: How the World's Poor Live on $2 a Day.* Princeton, NJ: Princeton University Press.

Diamond, J. M. (1998) *Guns, Germs and Steel: A Short History of Everybody for the Last 13,000 Years.* New York: Vintage.

Fukuyama, F. (2012, January–February) The Future of History. *Foreign Affairs,* pp. 53–61.

Hall, P. and Castells, M. (1994) *Technopoles of the World: The Making of 21st Century Industrial Complexes.* New York: Routledge.

Harvey, D. (2010) *The Enigma of Capital and the Crises of Capitalism.* London: Profile.

Herod, A. (1998) *Labor Geographies: Workers and the Landscapes of Capitalism.* New York: Guilford Press.

Hotelling, H. (1929) Stability in Competition. *Economic Journal* 39: 41–57.

Moretti, E. (2012) *The New Geography of Jobs.* Boston: Houghton Mifflin.

Pratt, G. (1999) From Registered Nurse to Registered Nanny: Discursive Geographies of Filipina Domestic Workers in Vancouver, BC. *Economic Geography* 75: 215–236.

Peck, J. (2001) *Workfare States.* New York: Guilford.

Rostow, W. W. (1990) *The Stages of Economic Growth: A Non-Communist Manifesto.* 3rd ed. Cambridge, UK, and New York: Cambridge University Press.

Select Guide to Further Reading

Acemoglu, D. and Robinson, J. (2012) *Why Nations Fail: The Origins of Power, Prosperity, and Poverty.* New York: Crown.

Allen, R. C. (2011) *Global Economic History: A Very Short Introduction.* Oxford: Oxford University Press.

Chant, S. (ed) (2010) *The International Handbook of Gender and Poverty: Concepts, Research, Policy.* Cheltenham, UK: Edward Elgar.

Escobar, A. (2011) *Encountering Development: The Making and Unmaking of the Third World.* Princeton, NJ: Princeton University Press.

Frank, A. G. (1966) *The Development of Underdevelopment.* New England Free Press.

Inglehart, R. and Welzel, C. (2005) *Modernization, Cultural Change, and Democracy: The Human Development Sequence.* Cambridge, UK, and New York: Cambridge University Press.

Jessop, B. and Sum, N.-G. (2006) *Beyond the Regulation Approach: Putting Capitalist Economies in Place.* Cheltenham, UK: Edward Elgar.

Landes, D. S. (1999) *The Wealth and Poverty of Nations: Why Some Are So Rich and Some So Poor.* New York: Norton.

Leyshon, A., Lee, R., McDowell, L., and Sunley, P. (eds) (2011) *The SAGE Handbook of Economic Geography.* London: Sage.

Smith, N. (2008) *Uneven Development: Nature, Capital and the Production of Space.* 3rd ed. Athens: University of Georgia Press.

Wood, A. and Roberts, S. M. (2011) *Economic Geography: Places, Networks and Flows.* London: Routledge.

On economic and historical discussions of the core-periphery model, see:

Krugman, P. (1991) Increasing Returns and Economic Geography. *Journal of Political Economy* 99: 483–499.

Wallerstein, I. (1974) *The Modern World System.* Vol. 1. New York: Academic.

Wallerstein, I. (1980) *The Modern World System.* Vol. 2. New York: Academic.

Wallerstein, I. (1989) *The Modern World System.* Vol. 3. San Diego: Academic.

Websites

International Institute for Environment and Development
http://www.iied.org
The International Labor Office
http://www.ilo.org/global/lang–en/index.htm
The International Monetary Fund

http://www.imf.org/external/index.htm
The World Bank
http://www.worldbank.org
The World Trade Organization
http://www.wto.org

Learning Outcomes

- Across the globe, there is an uneven development of economic activity that results in marked disparities in living standards and quality of life.
- In the Rostow model of economic development, five stages are identified: traditional stage, pre-takeoff stage, takeoff stage, drive to maturity, and high mass consumption.
- The global economy has winners and losers, not just as a historical accident but also as part of the very process of economic development.
- The core-periphery model argues that the global economy consists of the historical evolution of a rich core and poor periphery, with some countries making the transition to the semi-periphery.
- Economic disparities are also evident at the regional level, with some regions growing faster and becoming more industrialized than others.
- Uneven regional development is not a temporary condition, but is essential to a capitalist mode of production as uneven waves of investment cause some regions to be more developed than others.
- Four factors that contribute to the regional clustering of economic activity include locational pull of resources, the concentration of firms and workers, subsidiary trades and services attracted to core industries, and the spillover of innovative ideas between firms.
- Regional economic clusters make markets more efficient, but negative side effects include the increased costs of land and labor.
- Regional economic dynamics can also be considered from the perspective of the product and profit cycles. In the early stages of product development, when super profits

can be made, new firms are attracted to the location of the initial innovation. As more firms enter the markets, however, new spatial forces are at work and firms may relocate closer to markets or away from higher-wage areas in order to maintain and increase profitability.
- The governments of developing countries, seeking to support the industrialization of the national economy, follow either an import substitution strategy in which they protect local industry or an export-oriented form of industrialization.
- The development state can achieve spectacular success as public and private interests are focused on the same goal of achieving rapid growth. However, the reliance on export-led growth makes the economies vulnerable to global shocks.
- Capitalist economies have grown from being based on the production of commodities to being based on the mass production and consumption of goods. Implications of mass consumption include: desire trumping need, consumption being equated to social status, different levels of access to credit, geographies shaped by consumption, and a growing dominance of retailers over producers.
- The differential mobility of capital and labor at the national and global level is becoming more pronounced. Capital is more mobile while organized labor is more fixed.
- Capitalism is capable of amazing growth and innovation, but it is subject to recurring crises and continual turmoil, as the crisis solved in one area becomes a crisis in another area.
- Employment in the formal economy involves the sale of labor in the marketplace, formally recorded in government and official statistics, while the informal economy is where unrecorded transactions take place and is a response to the lack of formal employment opportunities.

PART 4

The Cultural Organization of Space

There is an intimate relationship between space, place, and **culture**. Space is given meaning and place is rendered unique through cultural formations. Culture is expressed in space and embodied in place. Part 4 considers the cultural organization of space through an examination of population, religion, and culture. Chapter 9 outlines the basic trends of population distribution and movement. Chapter 10 explores the geography of religion, while chapter 11 examines the geography of language.

9

The Geography of Population

This chapter considers the size, distribution, and movement of population. The population is composed of different categories of gender, age, race, ethnicity and other factors of social difference. The social construction and spatial distribution of these categories reveals how social and cultural geographies are created and maintained.

The world population is close to 7.5 billion people. The basic distribution across the world is depicted in Figure 9.1. Notice how the population pools in certain regions of the world. This is largely a function of the distribution of landmass, but there are also significant geographical macro trends. There is a concentration in the northern mid-latitudes. Nine of out of every ten people live in the Northern Hemisphere. The human population is sandwiched in a relatively narrow terrestrial band. Our inhabited world is centered on the Tropic of Cancer at approximately 24 degrees north. Farther south from this favored location is the historically unhealthy region of the equatorial tropics, while farther north it is too cold to support large populations.

The world's population distribution reflects our historical reliance on agriculture. The centuries-old agricultural regions of East, South, and Southeast Asia continue to have heavy concentrations of people. Population has also concentrated in the urban industrial regions of the world; first in Western Europe, then North America, and more recently in East Asia. The result is a distinct population clustering in long-established agricultural and industrial areas in the Northern Hemisphere and, more recently, in large metropolitan regions across the globe.

9.1 Global distribution of population (courtesy of Bill Rankin, www.radicalcartography.net)

The Distribution of Population

The world's population is distributed and demarcated in cultural, political, and economic space. Human populations are culturally specific; they speak particular languages and have different beliefs, and part of this difference is a function of where they are. A major geographic truth and variant of **Tobler's Law** is that who you are depends on where you are.

International Differences

Figure 9.2 depicts population distribution by countries. More than a third of the world's population lives in just two counties, China and India. Half of the world's total population lives in just six countries. And the ten largest countries combined contain almost 58 percent of the world population.

In the past fifty years the fastest rates of growth were in the global south. High birth rates and declining mortality levels resulted in rapid population growth in countries such as India and China, leading to their present day huge populations. Meanwhile growth slowed in the richer parts of the world. Today, growth has slowed even in countries such as India and China but remains high in sub-Saharan Africa.

Population size is an important component of national power. Very small countries tend to wield less power than larger countries. A large population allows for a bigger economy, a larger military, and a greater ability to project economic, cultural, and political power beyond national boundaries. However, population size is not everything. Switzerland, with a population of just over 8.9 million, has total a GDP of $710 billion, while Uganda with 37.7 million has a GDP of only $27 billion. The difference is due to the stark difference in gross national income per capita: $88,000 in Switzerland but only $670 in Uganda. It is not just the total population that defines national power but also the wealth and productivity of the population.

The most powerful countries are those with a large population and a successful economy. The US wields enormous power and influence not only because of its population size (the world's third largest) but because of its economy (the world's largest). Superpowers have demographic, economic, and military power.

We may be witnessing a change in geopolitics and geo-economics as some countries, such as Brazil, China, and India not only have large populations but also growing economies. They combine a demographic dividend with economic growth. Their growth marks a shift in the relative balance of power in the world.

Population size has advantages and disadvantages. Large countries have greater relative weight in the world. China's global status is reinforced not only by its impressive economic growth of the past twenty-five years but also by the basic fact that this growth occurred in the world's most populous country. But a populous country may also have possible problems maintaining national coherence as very large populations may be diverse. In the case of India, for example, the sheer size of the Muslim minority means there is a significant resistance to Hindu nationalism.

Small countries may have the benefit of greater homogeneity and coherence but are more vulnerable to external shocks. A hurricane landing in the Caribbean island of Dominica, with a population of just over 72,000, has a greater national impact than a hurricane landing in the USA. A single hurricane can paralyze Dominica's entire tourist trade and drag down its whole national economy. By contrast, while a hurricane striking America might harm the tourist trade in Florida or the Carolinas, these are only a small proportion of the US economy.

National Imaginaries and Population Distribution

National populations are often unevenly distributed. Consider the case of the USA. Figure 9.3 highlights the uneven nature of population density. Notice the towers of high density in the major metropolitan areas, the columns of adjacent suburbia and the flat nature of much of the rural USA.

Population distribution often connects with the national imaginaries. **National imaginaries** are a distinct form of geographical imaginary that refer to the spatial nature and representation of the nation state. Figure 9.3 highlights the metro nature of US society, but many of its

Russia, 1.7%
Mexico, 1.7%
Nigeria, 2.2%
Bangladesh, 1.9%
Pakistan, 2.5%
Brazil, 2.6%%
Indonesia, 3.5%
United States, 4.4%
India, 18.1%
China, 18.9%
Other Countries, 42.5%

9.2 Population distribution by country

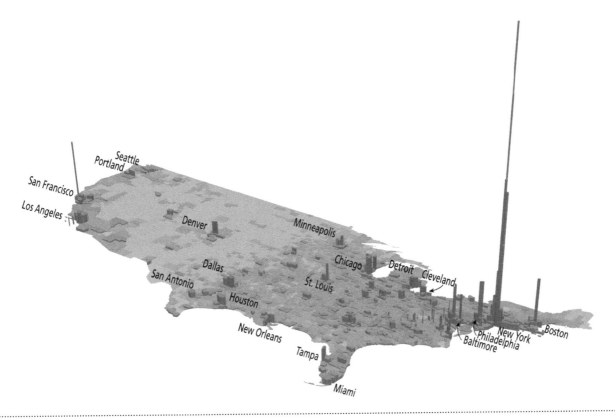

9.3 Population map of USA

cultural representations focus on the American West and its rural regions as the quintessential USA. The notion of a US heartland, the region of the rural interior USA, may be culturally significant, but it is of minor and declining demographic significance. While wilderness is a major theme in how the US sees itself, the demographic reality is of a nation of big metro areas.

And here we come to an interesting phenomenon of the differences between demographic realities and political representation. In the USA, states have equal representation in the Senate. Both California and Wyoming have two senators in Congress, yet they have very different populations. California is the most populous state with just over 39 million people, while Wyoming has closer to 600,000. Thus Wyoming is vastly overrepresented in the US Senate while California is vastly underrepresented.

Because the Senate makes important independent decisions, such as confirming Supreme Court nominations and foreign policy agreements, these different weightings given to the population in different parts of the country have significant political effects. Table 9.1 shows the disparity between the top most populous states. California, Texas, Florida, New York, and Illinois combined hold more than a third of the entire US population but only have 10 percent of seats in the US Senate. The five least

populous states, Wyoming, Vermont, Alaska, South Dakota, and North Dakota have just 1 percent of the nation's population but also hold 10 percent of Senate seats. Rural conservative states are thus given a greater political representation in the Senate than the larger, more metropolitan states.

There are disparities between population distribution and political representation in other countries. In Japan, for example, rural areas are losing population yet retain significant electoral representation. The electoral boundaries and number of representatives was decided decades ago, before the massive rural-to-urban migration. The result is that rural areas are overrepresented in the national parliament while the cities are

TABLE 9.1 Population and Political Representation in the US Senate

	FIVE LARGEST STATES	FIVE SMALLEST STATES
Population	119 million	3.5 million
% National population	37.2	1.1
% of US senators	10	10

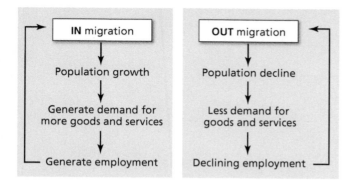

9.4 Demographic cumulative causation

underrepresented. Rural residents have political power way beyond their demographic number or economic importance. Voters in rural areas have on average 250,000 voters per national parliament seat, compared to over a million voters per seat in the big cities. These rural and agricultural areas and their predominantly elderly populations have an outsized role in the national politics of Japan because they are granted disproportionate electoral power.

Demographic Cumulative Causation

Across the world we find regions of population growth and areas of population decline. This uneven distribution is a function of changing demographics; regions with higher birth rates increase population, while regions such as rural Japan witness more deaths than births and so lose population.

There is also a redistribution as people move to areas of greater economic opportunity. We can identify a **demographic cumulative causation**. As people move from a declining region to growing regions, profound changes occur that can reinforce the regional differences (Figure 9.4). It is the most able and active who make the move, adding to the labor power of the receiving region and detracting from the labor stock of the sending region. The receiving regions gain extra sources of labor and effective economic demand, leading to greater regional growth. This may be offset by greater pressure on goods and services in the receiving region, increased overcrowding, and congestion. Meanwhile, the sending region loses labor power and purchasing power. While sending back remittances may offset this, the net effect is often a further exacerbation in economic development disparities between the richer receiving region and the poorer sending region. Continued population movement may exacerbate regional differences. Governments can iron out regional differences through regional polices such as establishing growth poles in or guiding government investment and jobs to declining regions.

Population Differences: Gender, Age, Race, and Ethnicity

"Population" is an aggregate term that covers many different types of people. The sources of difference are many and include age, gender, sexual orientation, race, and ethnicity. We should be careful of imputing biological origins to sources of social difference. While male and female may be biological categories, what it signifies to be man or what it implies to be woman is socially produced and reproduced as well as celebrated and challenged. Our definition of what it means to be old, and when to be old, to take another example, is less a biological fact than a constantly shifting and changing social categorization. It is important to remember that differences are a product of social relations, not unchanging biological verities.

Social differences are codified, reinforced, and policed. They are contested and embraced as they are imposed and adopted, resisted and celebrated, becoming sources of constraint as well as platforms for creativity. They are also subject to change—even the most rigid divisions of gender or sexual orientation. We are in the middle of a period of flux when traditional categories are being redefined.

Here we will consider three major sources of difference: gender, age, and race/ethnicity.

Gender

Space and place are gendered. The **gendering of space and place** is an important topic in human geography. Gender differences have spatial dimensions as men and women occupy space differently, certain places are gendered in that they are associated more with one gender than the other, sexism influences the shape and location of places, and the migration movements of men and women can and do differ. Across much of the world women often have a form of second-class citizenship and are denied equal status with men.

Gender distinctions are often an integral part of cultural and social identity. In Aboriginal Australia, for example, there were separate sacred sites for men and women. Authority and control over these sites was highly gendered. In today's Catholic Church, women are allowed to be members of the Church but are not allowed to become priests, bishops, or popes.

Different cultures assign different roles for men and women in the public and private spheres. Economic activities are often highly gendered. In areas of peasant agriculture, women are responsible for much of the agricultural labor. Traditionally women were not involved in heavy industry such as coal mining and steel making, but the advent of mass assembly work meant more jobs for women.

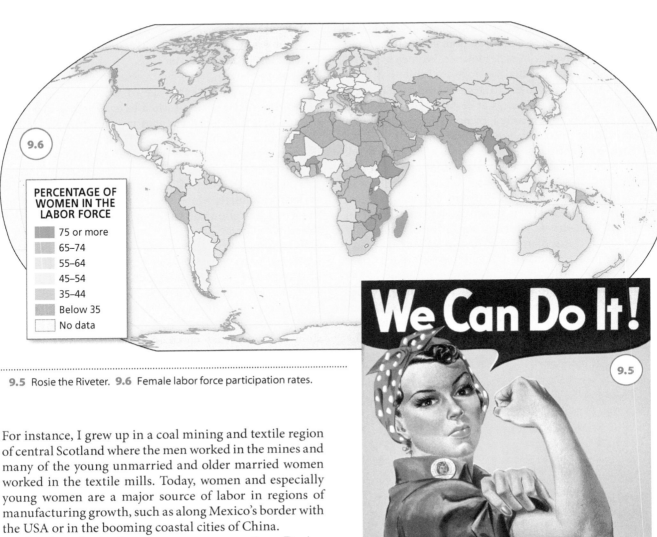

9.5 Rosie the Riveter. **9.6** Female labor force participation rates.

For instance, I grew up in a coal mining and textile region of central Scotland where the men worked in the mines and many of the young unmarried and older married women worked in the textile mills. Today, women and especially young women are a major source of labor in regions of manufacturing growth, such as along Mexico's border with the USA or in the booming coastal cities of China.

Gender roles in the economy vary over time. During the Second World War, for example, women were actively encouraged to enter the manufacturing labor force to help in the war effort. Rosie the Riveter, an image of female muscular power, was use to recruit women (Figure 9.5). After the war ended, women were encouraged to return to the domestic sphere.

In peasant economies and in economies shifting away from manufacturing toward services, women often play a more active role in the public realm. **Female participation rates** measure the amount of women aged 15 to 64 employed in the labor force. The highest rates of around 85 percent are found in countries such as Cambodia, Burundi, and Zambia, where women work in the fields and as traders in the towns. In other regions of the world women are restricted to the domestic sphere. The lowest figures are recorded in traditional Islamic countries such as Afghanistan where it is 16 percent. Figure 9.6 shows that the traditional Islamic countries have the lowest rates of female participation. In the US the figure is 66 percent.

Women also do a great deal of work in the domestic sphere, and writers have referred to the double bind that exists for many women who work hard in the formal economy as well as in the domestic economy.

Space is an arena where gender roles are inscribed and transgressed, codified and undermined. The ability to move across space is also gendered. Many **patriarchal societies** limit women's mobility. An extreme example is the practice of foot binding of young girls in traditional China. This was the ultimate form of limiting women's freedom to move. Even today, women in Saudi Arabia are banned from driving, thus severely limiting their ability to move freely through urban space without a male presence.

There are also the more informal mechanisms. Women's use of urban space, for example, is more constrained than men's because of the fear of male violence. Women are more sensitive to the fear of sexual violence, and this structures their behavior in many cities. Strategies of individual safety include avoiding certain places at certain times, going to others only when accompanied, or not participating in an entire repertoire of activity, especially at night.

Gendered identities are rarely stable and never fixed. Amy Hanser looks at changing conceptions of gender and sexuality in her survey of new service work regimes in China. Under the Maoist regime strong, physically robust, hard-working women were positively portrayed, but in the new, market-driven China other forms of femininity are employed. Hanser surveyed three retail outlets in the Chinese city of Harbin. In the state-owned department store there was little attempt at gendered display of the women workers. Uniforms were bulky and unisex. Older rather than younger women worked in the store. The atmosphere echoed the Maoist era. In a wholesale/retail sector selling cheap clothes to recent rural migrants, the selling was done by young sexily dressed women. There was a hypersexualized, undisciplined atmosphere. In the high-end, privately owned department store, the apex of the retail sector, the emphasis was on a youthful but disciplined femininity. The young women were trained in physical deportment and demeanor. They were presented for the discriminating **male gaze**.

Gender is a site of resistance and contestation as well as a setting for subordination and oppression. Elizabeth Wilson suggests that much early feminist writing on the city underscores the construction of women as passive victims. She argues for a wider debate that recognizes women's more active agency with cities as platforms for liberation where women widen horizons and escape traditional expectations.

There are tensions, connections, and intersections between race, class, ethnicity, age, and gender. Thomas Abowd, for example, examines the gendered politics of residential space in Jerusalem and looks at how unmarried Palestinian women are deemed as "out of place." Gendered relations intersect with other relations of power based on race, class, nationality, and religion.

Age

According to William Shakespeare there are seven stages in the life cycle, from infant through schoolchildren going "unwilling to school" through adulthood to the final stages of a "second childishness."

What it means to be at different stages varies across space and through time. In traditional peasant societies young children were an important form of labor. They were prized for their muscles, not their brains. As society becomes more modern, people tend to have fewer children and devote more resources to each child. Young children are seen less as source of domestic labor force and more as a receptacle of a formal education.

Biology determines when a woman can have a child, but the socially acceptable age of when to have a child varies. The social norm of when to have a child is a result of economic and social changes in what it means to be a woman of a certain age in society. In the US, for example, the average age of a woman giving birth to her first child is now around 26 years, and across much of Europe it is closer to 29. Women are living longer, and thus able to delay childbirth. More women are spending more time in formal education and in the formal economy, although the result of delaying childbirth is a greater risk of birth defects and infertility. By contrast, in Bangladesh, where more traditional roles hold sway, the average age of women having a first child is 18, in Honduras it is 20, and in Hong Kong, where women feel that having children reduces their employment prospects, the figure is closer to 30.

We could examine each of the stages of life cycle and show how what they mean and how they are considered varies across the world, as societies have different religious beliefs and levels of modernity. In richer counties where people tend to live longer, the age of being "old" has shifted upward, expressed in such phrases as "sixty is the new forty." Because people on average live longer than in previous eras when our attitudes were shaped and codified, what it means to be old, and how "old" people should conduct themselves, is now subject to revision and change.

Race and Ethnicity

We need to begin this section with a warning: race and ethnicity are neither natural categories nor are they unchanging. The very definitions of race and ethnicity are social constructions, classifications that themselves are the result of socioeconomic, not biological, processes.

Race is not so much a fact of biology as a social construct, partly adopted, partly imposed, and always constructed and defined in relation to others. Black is defined in relation to white, and Asian to non-Asian. There is no separate biological fact of race other than the relative, socially constructed definitions that we use. Race is the name we give to perceived social differences based on a shifting and unclear combination of half-understood genetics and half-remembered history. We can more accurately consider "race" as a fluid, relational concept rather than a fixed, eternal, biological fact.

Races are identified in relation to other races, and this categorization is not innocent of political and economic considerations. We classify in order to count, to demarcate, and ultimately to control. The term "**racialized groups**" refers to this social construction and use of racial categories; it reminds us that races are artificial constructs that embody political power and exact a social cost on those in the least favored, non-hegemonic categories.

Ethnicity, on closer linguistic attention, is even more vague. Race and ethnicity slide into each other in common discourse. And there are some categories that combine both, such as "Hispanic" in the US Census. Ethnicity is a more general term for different collective identities that are in part imposed and in part adopted. We can distinguish between **imposed ethnicity**, when a group of people are named and treated as ethnically different, and **adopted** or **symbolic ethnicity**, when individuals choose to express an ethnic identity.

People often prefer to live with people like themselves. **Racial/ethnic clustering** occurs around the world as these neighborhoods provide a support system and a platform to the new society. The degree of clustering is a function of the degree of difference between the group and the dominant population. Racial/ethnic clustering may also result from discrimination, as households are forced explicitly or implicitly to live in certain areas, through coercion or social sanctions. The policy of spatial demarcation by race was an integral part of the apartheid system in white supremacist South Africa.

Race and ethnicity is also created and maintained by acts of power. Kay Anderson provides a fascinating case study of the evolution of Chinatown in Vancouver, British Columbia. She shows how the racial category of Chinese was constructed and maintained by white politicians and local elites. Chinatown in Vancouver, for example, was a racial and spatial category created by institutional power. As she notes,

> ..."Chinatown" was a shared characterization—one constructed and distributed by and for Europeans, who, in arbitrarily conferring outsider status on these pioneers to British Columbia, were affirming their own identity and privilege. That they directed that purpose in large part through the medium of Chinatown attests to the importance of place in the making of a system of racial classification.
>
> (ANDERSON 1987, p. 594)

Chinatown in Vancouver, like other Chinatowns in cities around the world, has evolved from a place of compulsion and restriction of ethnic difference to what is now a setting for performance and celebration of ethnic difference.

In 1963 Nathan Glazer and Daniel Patrick Moynihan published their classic work on ethnicity, *Beyond the Melting Pot*. It was primarily a study of New York City and the extent to which various ethnic groups sturdily maintained their distinctiveness. Their detailed review of the city's different immigrant groups showed that urban living did not diminish the ethnic identities, but reinforced them.

The "**melting pot**" notion was always something of a fiction, but so is the idea that identities are unchanging, singular. There is a complex social interaction as newcomers and natives interact with one another, changing each in turn. Rather than a multicultural society, a condition that implies unchanging groups remaining separate, there are different forms of **polyculturalism** as various groups interact with and change one another. Rather than thinking of a simple continuum of assimilation, measured from the initial immigrants being different to successive generations becoming more like the rest of the society, we can think of a complex picture with both the immigrant groups and the host society metamorphosing. Thus immigrants moving from southern Italy in the early nineteenth century were not Italians until they came to the US. The construction of an Italian American identity was not transposed from Italy but created in the US. And the active creation of a new Italian American identity shaped the US.

The Movement of Population

Maps of population distribution are enlightening but can also be deceptive. They freeze the movement of a restless species into a static picture. In reality, people are always on the move. There are the daily rituals of journeying to and from school and work, the weekly movements to places of recreation, the seasonal and annual trips and outings. There are also the more radical ruptures when we change houses, jobs, or even countries. Movement takes place on a daily, weekly, seasonal, and annual basis following circular and noncircular patterns, sometimes involving permanent relocations.

Daily Space-Time Paths

We are a restless species, always moving around. Each day we make traces through space and time. Swedish geographer Torsten Hägerstrand realized that we move through both space and time, and he provided a way to map activity in each. Figure 9.7 traces the path of two people who leave their homes to go to work, then meet up in a cafe around 4 p.m. Such maps can highlight colocations in time and are enormously useful in tracing the paths we weave through space-time.

Our space-time paths are constrained. Much of our activity revolves around fixed pegs in space-time, such as home and work. A prism outlines the space that can be covered in the time available; those with a car will have a much larger prism, as they can cover a larger distance. They can therefore look much further afield for employment. What are the pegs and prisms that constitute your typical day?

9.1 The Age of Distraction

Figure 9.7 depicts the traditional view of space-time paths. However, in the world we live in, people occupy space while also communicating with people and things in other places. We need to amend the figure:

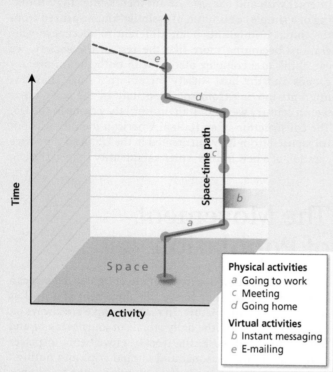

Physical activities
a Going to work
c Meeting
d Going home

Virtual activities
b Instant messaging
e E-mailing

A common experience: you are walking down the street and someone is walking in the opposite direction toward you. You see him but he does not see you. He is texting or looking at his cellphone. He is distracted, trying to do two things at the same time, walking and communicating. There is also the telltale recognition of a car driver on a phone; she's driving either too slowly or too fast for the surrounding conditions, only partly connected to what is going on around her. Connected to someone else in another place, she is not present in the here and now.

These types of occurrences are now common enough that we can label our time as the **age of distraction**.

The age of distraction is dangerous. A recent report by the National Safety Council showed that walking while texting increases the risk of accidents. More than 11,000 people were injured in 2015 in the USA while walking and talking on their phones.

Distracted drivers have more fluctuating speed, change lanes fewer times than is necessary, and in general make driving less safe for everyone. Texting while driving resulted in 16,000 additional road fatalities from 2001 to 2007. More than 21 percent of vehicle accidents are now attributable to drivers talking on cell phones.

Multitasking is not so much an efficient use of our time as a suboptimal use of our skills. Student achievement improves when cell phones are banned, with the greatest improvements accruing to lower-achieving students, who gain the equivalent of an additional hour of learning a week. Students with open laptops learn less and recall less than students with their laptops closed.

We are more efficient users of information when we concentrate on one task at a time. When we try to do more than one thing, we suffer from inattention blindness, which is failing to recognize other things, such as people walking toward us or other road users.

Media theorist Douglas Rushkoff asserts that our sense of time has been warped into a frenzied present tense of what he calls "**digiphrenia**," the social media-created effect of being in multiple places and more than one self all at once.

REFERENCE

Short, J. R. (2015, July 7) The Value of Unplugging in the Age of Distraction. *The Conversation.* http://theconversation.com/the-value-of-unplugging-in-the-age-of-distraction-43572

Seasonal Movements

We move with the seasons. **Transhumance** is the traditional movement of mountain people with their animals from summer to winter pastures. In the Alps and the mountains of Scotland, for example, a traditional form of seasonal movement was to take cattle and sheep up the mountains to feed on the summer pasture. The herders and shepherds lived in modest shelters, known as sheilings in Scotland. In winter people moved back, with their flocks and herds, to the more protected and warmer valleys.

Rural-to-urban migration has swollen city populations across the globe. Sometimes the move is permanent. When it is seasonal it is known as circular migration. In Delhi, for example, seasonal workers are known as *barsati mendaks*, the name of the frogs that appear during the monsoon. In one neighborhood of the city, Bara Tooti:

> *The barsati mendaks, the rain frogs of Bara Tooti, are the seasonal workers from villages in Delhi's neighboring states of Uttar Pradesh, Haryana and Rajasthan. Most of them have land*

9.7 Space-time mappings

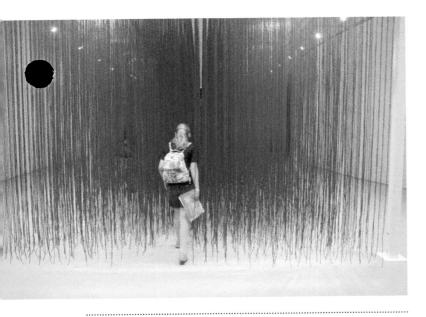

9.8 Space-time traveler. This photograph of a young woman walking through an installation by the artist Jesús Rafael Soto at the Hirshhorn Gallery in Washington, DC, highlights the sense of movement though space.

back home, a few acres that their fathers own, which will soon be divided among brothers. They come first in January after the winter crop has been harvested and the fields lie fallow, and return home for the sowing season in July. Once sowing is complete, they return to Bara Tooti for another few months of work before heading back to the

village around Diwali [Hindu festival of light]. The barsati mendaks work frantically and live frugally to save as much money as they can. In the weeks leading up to Diwali they stop drinking or smoking, and save every last rupee so as to have something to show for the long absence of home. On the day before they leave, the mendaks hurriedly pay off their debts and pile into interstate buses headed homewards.

(Sethi 2012, p. 16)

All of the *barsati mendaks* are men. Their wives and sisters are at home caring for children, farming, and tending to household duties.

There are also the rituals for those who can afford vacations. College students across the USA look forward to spring break. Some concentrate on their studies, or continue working or volunteer, while others take to the beaches and resorts in an annual rite of passage. There are also family vacations over the summer and, for some, ski vacations in the winter. We move in order to recreate and we recreate through moving.

Metropolitan Redistributions

There are various forms of movement between and within metropolitan areas. Urban-to-urban migration occurs as people tend to move from declining urban economies to expanding ones. Thus people in Cleveland or Flint tend to move to better economic opportunities in the bigger cities and Sunbelt cities, such as Chicago or Dallas.

There is also **suburbanization** as people move their permanent residence from the central city to the suburbs. This was the dominant form of metropolitan movement in the USA from the 1950 to 1990s. The move was selective as those with more income were more able to afford the move. The poor were sometimes trapped in the central city neighborhoods.

More recently there has been a marked shift of population back to the city, especially of the more affluent, younger households in search of central city employment and urban lifestyle. Empty nesters whose children have left their suburban homes are now leaving them to return to the city. As the number of single person and non-child households increases, a move to dense, walkable neighborhoods is also likely to continue. And again, the move is selective, with those who have more income better able to move into central city accommodations.

Rural to Urban

There are also relocations that involve a permanent change. The largest single permanent relocation of the past fifty years across the globe is rural-to-urban migration. The move to the cities is particularly pronounced in counties experiencing rapid population growth, industrialization, and modernization. Across the world cities are swelling with people moving in from the countryside in search of jobs, opportunities, and the prospect of a better life for them and their children.

In contemporary China, for example, there are approximately 150 million rural migrants in coastal cities, almost 11 percent of the total population. Large cities, especially in developing countries, are often fringed with the settlements of rural migrants.

Sometimes this shift takes place within one country, as in the case of China. In other cases, people move from rural areas in one country to cities in another country. For instance, rural migrants from Mexico move to cities in the United States as well as to cities inside Mexico.

Rural to urban migration involves not only a change in location but often a remaking of cultural identity as people move from the small-scale, intimate societies of the countryside to the larger, more heterogeneous mix of the big city.

An immigrant city was the backdrop to what has been termed the **Chicago School** of urban sociology. One of its most celebrated figures is Robert Park; its other major figure, E. W. Burgess, is considered in chapter 18. Robert Park (1864–1944) had a great love of the city. He was a newspaper reporter before he became a full-time academic, and his writings are infused with a journalist's acute sense of the city. In his 1916 essay, *The City: Suggestions for the Investigation of Human Behavior in the Urban Environment*, Park posits an **assimilationist model** in which new immigrants provide the dynamic raw material in the creation of a new urban society. New categories of people are created in the finer division of labor found in the city and the greater tolerance of difference. While the small community tolerates eccentricity, the great city rewards it.

Forced Migrations

Forced migrations occur when people are forced to move. One of the most significant cases of forced migration was the **slave trade**.

The Slave Trade

From the fifteenth to the nineteenth century, people were torn from their communities in West and Central Africa and transported to the coast to be shipped off as slave labor for plantations, mines, and homes in the Americas. There was a **triangular trade** pattern that linked port cities such as Liverpool, England with West Africa and then the colonies in the Caribbean and South and North America. Ships would carry trade goods to Africa, take on slaves there, sail across the Atlantic to the New World, and then return to Europe with commodities such as sugar and cotton for European markets and industries.

The human toll was enormous. Almost 4 million died during transportation within Africa and around 2 million died in the crossing as people were tightly packed together in appalling conditions. Around 11 million people arrived as slaves. Most went to South America; what is now Brazil received almost 40 percent. Around 7 percent ended up in the USA. The effects were enormous. In Africa the population stagnated and the small differences between the African and European economies widened. The profits from slavery were a basis for the accumulation of wealth in Europe and the Americas. The endowment of Brown University was built on the slave trade. The Catholic university of Georgetown made money by selling slaves. The evils of slavery permeated societies, permanently disfiguring their race relations and their very humanity. Slavery was the poisoned chalice that the Founding Fathers of the Republic such as Jefferson refused to challenge. It was only in the bloody Civil War of 1861–65 that slavery was ended in the USA. Struggles like black emancipation and equal rights continue to this day. In 1999 the city of Liverpool offered an official apology for its role in three centuries of slave trading. In 2007 the state legislature of Virginia also offered an official apology. More states, including Alabama, Maryland, and North Carolina, have made similar declarations. In 2008 the US House of Representatives passed a resolution condemning American slavery and subsequent discriminatory laws. Throughout Africa there are calls for tribal chiefs to account for and apologize for their role in the slave trade.

The enforced migration of people from Africa changed both Africa and the New World. It is impossible to understand race relations in the US without accounting for the history of slavery. Accumulated wealth in Europe and the Americas is built, in part, on the back of unwilling migrants forced into slavery.

Forced Migration Today

Today, more common causes of forced migration include war, political conflict, and social unrest. While international refugees garner most public attention, millions of people are displaced inside their own country. The Internal Displacement Monitoring Centre estimated that in 2015–16 there were 38 million internal refugees. The figure currently includes 6.6 million in Syria, 6 million in Colombia, 3.3 million in Iraq, 3.1 million in Sudan, 2.9 million in the Democratic Republic of Congo, 2.6 million in Nigeria, and half a million in Ukraine.

International refugees are forced across borders. Precise measurement is difficult but estimates suggest that currently almost 20 million people have been forced to leave their country. They include people escaping civil war in Syria and families escaping gang violence in Honduras. The vast majority are housed in developing countries. The largest refugee camps in the world are in the Horn of Africa

9.9 Refugee camps in East Africa

(Figure 9.9) where drought, civil war, and social dislocation have displaced many people seeking food, shelter, and security and who end up in refugee camps. Dadaab in Kenya, the largest refugee camp in the world, houses close to 330,000 people.

We can also identify economic refugees who are not forced because of war or social conflict but feel compelled to move overseas in search of economic opportunity lacking at home. The mingling of political and economic refugees was evident during the million-strong wave of refugees to Europe in 2015. They were escaping the murderous civil war in Syria as well as the lack of economic opportunities in Africa, Iraq, and Afghanistan. The flood of refugees was so large and so sudden that it overwhelmed the traditionally welcoming nature of German society; in its wake, barriers to refugees were raised across Europe and the rhetoric of nationalism and anti-immigration became more strident throughout the continent.

International Migration

Voluntary international migration is now an important form of population movement. Almost 192 million people live outside the country of their birth, around 3 percent of the total world population. Half of all international migrants are women. If we include illegal and undocumented workers, the level of international migration across the globe is even higher. These immigrants are either temporary migrants, perhaps sending money back in the form of **remittances**, or permanent migrants making a life in a new country. International migrants are the raw material of cultural globalization as they bring their culture with them, adding a new cultural dimension to their destination. It is rare that cultures are mechanistically transported; there is creative reworking as cultural forms are adapted and changed.

Compared to the rest of the population, international migrants tend to be younger and better educated, and are moving for employment-related reasons.

There are powerful push-and-pull factors at work as people search for better living conditions or religious and political freedoms in other countries. The level of migration depends on the range of factors pushing and pulling migrants as well as the regulatory regime that controls migration. Countries vary in the severity of their **immigration regimes**, which can be defined as the set of rules that govern the entry, settlement, and assimilation of foreign migrants. Regimes tend to be more severe during economic downturns and more relaxed when economic growth creates labor shortages. In the late nineteenth century, for example, the United States had an open-door policy that encouraged immigrants from Europe but severely restricted those from China. Today, most countries try to control immigration. Regimes vary from the more open, such as Australia's and Canada's, to the more closed and restricted, such as South Korea's.

Migration affects both origin and destination countries. Origin countries may lose some of the most talented population but receive remittances. Destination countries may receive skilled migrants, but there may be a depression of wage levels in sectors where the often cheaper immigrant labor force predominates.

International migration tends to increase during periods of increased globalization, such as in the late nineteenth and early twentieth centuries and over the past thirty years. Figure 9.10 plots the immigration history of the

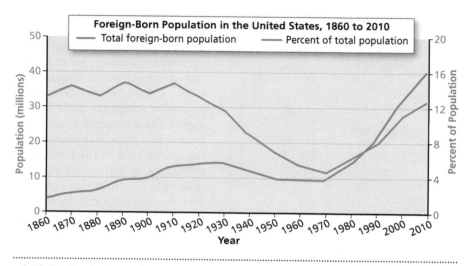

9.10 Foreign-born population in the United States, 1860 to 2010

United States. High absolute and relative numbers of immigrants were recorded from 1890 to 1930. During this period, there were few restrictions on immigration from Europe. People came from Ireland and Italy, Russia and Scandinavia, in the hundreds of thousands, changing the religious and ethnic composition of the country and leaving a permanent legacy. This period of large-scale immigration ended after 1930 as tighter immigration controls were launched at the start of the Great Depression. The number of foreign-born declined in absolute numbers and as a proportion of the total population. Between 1930 and 1960, the previous waves of immigrants were incorporated into the mainstream of US society. When the controls were lifted in the mid-1960s, a new wave of increased immigration ensued. The absolute and relative levels of the foreign-born rose to levels not seen for over a hundred years. Between 2000 and 2005, 8 million immigrants entered, the largest five-year total in the nation's history.

This second, more recent large wave of immigration has a wider source than the first, including Africa and Asia as well as Europe and Latin America, and gives the United States a new and different feel, making it more Hispanic, more Asian, and more cosmopolitan. While many immigrants in the first wave arrived in cities, in the second wave there has been a wider spread throughout greater metropolitan regions.

The current wave of international immigration across the world is now substantial. Resistance in destination countries is generated from perceptions of depression of living standards for native-born workers and feelings of local and national cultures being swamped by the foreign others. In the United States, for example, there is the added sense that between 12 million and 15 million people are in the country illegally. The description of this group is itself a battleground, with terms such as "illegals," which references the illegality of the people's permanent residence, competing with the

term "undocumented," which suggests an administrative mix-up that can be easily sorted out.

Many countries are now significant sites of foreign immigration. The Gulf States, for example, with their combination of wealth, small population, and large demand for labor, have some of the highest percentage rates of foreign-born, including the United Arab Emirates (88 percent), Qatar (75 percent), and Kuwait (74 percent). Among the larger developed nations, the highest rate is found in Australia (28 percent). In all these countries, the percentage of foreign-born has increased in the past thirty years, a result of increasing flows of migrant labor. In the United Kingdom, the percentage of foreign-born increased from 5.8 percent in 1971 to 12.4 percent in 2015—from over 3 million to just over 8 million in absolute terms. In the United States, the percentage of foreign-born has increased from 4.7 percent in 1970 to close to 15 percent in 2015. The USA has the largest absolute total of foreign-born, more than 47 million, myself included.

We should also note that there is also a wide range of countries with very low numbers of foreign-born. They include rich counties such as Japan (2 percent), poor countries such as Sudan (1.7 percent), Eastern European countries such as Poland (1.8 percent), authoritarian regimes such as China (0.1 percent), and South American countries such as Colombia (0.6 percent). A mixture of limited economic opportunities, more attractive alternatives, and restrictive immigration policies limits the number of foreign-born in these countries.

DIVERSE EXPERIENCES

There are multiple migrant experiences. Take the case of Dubai, where there are two very different streams of migrants. There is the encouragement of investors, technical experts, and those associated with **flight capital**, the practice of seeking a safe and opaque site for money made elsewhere. Then there are the construction migrants, often coming from rural communities in Pakistan and Bangladesh, who live many to a room and have tough working conditions. Their entry is tightly controlled and monitored. Luxury homes and migrant camps, foreign investors and building laborers, welcome guests and barely tolerated foreign workers: these are the extremes of divergent migrant experiences. There are also the expatriate communities of technical workers and experts—English financial specialists in Singapore, Scottish engineers in Saudi Arabia, European aid workers in Nairobi, Indian doctors in the United States. And there are the unskilled and often exploited workers—the female

Filipino domestic servants in Hong Kong, the Mexican gardeners in California, and the vast armies of undocumented and illegal workers in shadow urban economies around the world.

GATEWAY CITIES

International migration channels are often routed very largely, although not exclusively, through major metropolitan areas. In cities around the world, the number of the foreign-born has increased. In the United States, for example, a third of all the foreign-born are located in the five largest metropolitan areas. Marie Price and Lisa Benton-Short identify immigrant gateway cities around the world. They include

- established gateways such as Amsterdam, Birmingham (UK), New York, Sydney, and Toronto;
- emerging gateways such as Dublin, Johannesburg, Singapore, and Washington, DC; and
- exceptional gateways such as Riyadh and Tel Aviv.

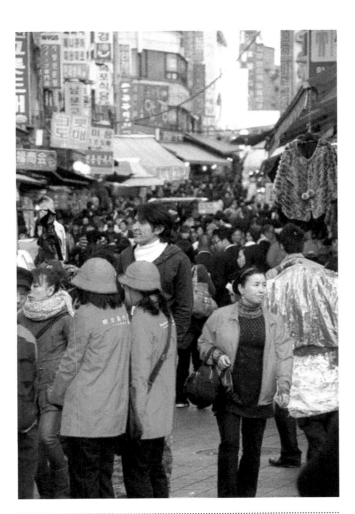

9.11 Ethnic homogeneity in Seoul, South Korea

Global and globalizing cities are now places of minority populations, **diasporic communities**, and **hybrid identities**. The metropolis is now the place of the Other, who, with their distinctive languages, cuisines, and cultures, now add to the cosmopolitan mix of the big cities. There is undeniable global mixing in the large metropolitan regions, but not in all big cities. In countries with restrictive regimes of tight regulations on the entry of foreign-born immigrants, even very large cities may remain remarkably homogenous. Seoul, in South Korea, has a population of more than 10 million, but less than 1 percent of them are foreign-born. Seoul is a closed gateway city.

Immigrant places are where otherness is renewed and celebrated, performed and remembered in creative acts of representation and memory. And this representation adds to the cosmopolitan mix of the city.

Models of Population Movement

There are numerous models of population movement.

Ravenstein

One of the earliest was developed by E. G. Ravenstein (1834–1913), who was born in Germany but moved to Britain, eventually becoming a professor of geography. He arrived at a time of intense population movement associated with rapid urbanization and industrialization. Spurred by a comment that the migration process was random, he analyzed UK census data in 1871 and 1881 to propose a number of laws of migration. These included:

1. Most migrants move only a short distance, toward the nearest city.
2. Growing cities attract migrants from nearby rural areas that depopulate rapidly; in turn, migrants from more distant areas fill the "gaps" left in the rural population. Migration is stepped.
3. Each main current of migration produces a compensating countercurrent.
4. Long-distance migrants tend to move to major cities.
5. Rural people have a higher propensity to migrate than urban people.
6. Women have a higher propensity to migrate than men within countries, but more men tend to move in international migration.
7. Migration is driven in largest part by economic forces.

These laws are not physical laws but generalizations made at a particular time in a specific context. While some of them (3, 4, and 7) stand up in many different circumstances, the others can vary depending on the context. There are more female than male Filipino international immigrants, undermining law 6. In the United States, more than one-third of all Filipino migrants live in just three metro areas, tending to support law 4.

Gravity Models

Gravity models suggest that population movement between two areas is a function of their population size and an inverse function of the distance between them. It is sometimes presented as an equation

$$\frac{P1 \times P2}{D^2}$$

where P1 is the population of city 1, P2 is the population of city 2, and D is the distance between them. It a is direct application of Newton's law of universal gravitation, which states that the pull between two bodies is function of their combined mass divided by the square of the distance between them. While this model may work in a world of empty space, the distance between places in the real world is more complex; it is warped by space-time and distorted by political boundaries. Nevertheless the model does provide some useful predictors such as how larger places attract more people and places closer together generate more migrant flow.

The space between places is also the site of **intervening opportunities**; the gravity model thus needs to be tweaked in order to account for these.

Lee's Push-Pull Model

Moving voluntarily requires a comparative assessment of the origin area's characteristics with those of possible destinations. Everett Lee specifies a model with reference to specific push and pull forces between origins and destinations. Typical push forces include limited economic opportunities, social unrest, and political instability, while pull factors include readily available employment opportunities, presence of friends and family, and social and political stability.

Origin and possible destination areas have a range of positive and negative factors. For example, positive factors for origin areas include connection to family, existing social networks, and a feeling of belonging. Negative factors may include lack of jobs, restricted social opportunities, and poor health and educational facilities. The main factors that push and pull migrants are employment and social opportunities. Migrants assess or sometimes are forced to assess the relative weight of different factors. When the pull or push factors reach a threshold, or when the positive factors in the destination outweigh the positive factors in the origin areas, then people move.

Consider the case of the great migration of African Americans. More than 6 million African Americans moved

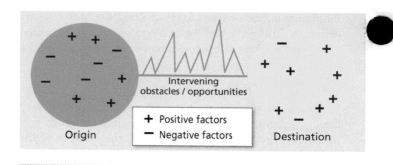

9.12 Lee's push-pull model

between 1916 and 1970 from the rural South to the towns and cities of the North and the East. They were pushed out by racial discrimination, economic marginalization, and restricted educational and social opportunities. They were attracted by the jobs and less explicit racial discrimination outside the South. African Americans could triple their wages by leaving the rural South for the urban industrial North and be free from the injustice of institutionalized racism. The **Great Migration** changed the identity of African Americans, making them a predominantly urban rather than rural population and in the process changing the racial balance of cities across the county.

Lee also identified intervening difficulties of distance, cost, and accessibility. The model is summarized in Figure 9.12 where the plus and minuses refers to the push and pull factors. Thus a migrant leaving Syria may decide to head for Austria since it is closer and perhaps easier to get to than Sweden. And while many migrants in North Africa and the Middle East would like to get to the USA, the distance, cost, and difficulties are high and the intervening opportunities are many.

The model assumes that people can rationally evaluate all the positive and negative factors. In most cases people act in a suboptimal manner rather than optimally. They have limited knowledge, limited resources, and often too little time to make complex and difficult decisions. They are strongly affected by the information and advice from friends and relatives. This explains the existence of **chain migration** where people move to a new destination and then convince friends and relatives to join them. This form of chain migration is a common characteristic of migration flows; people tend to move along channels and to destinations followed by friends and families. One chain for that links domestic female workers from the Philippines to jobs in Hong Kong and the Gulf States, another links male manual workers from South Asia to to working on construction sites in the Gulf States, and yet another links young men and women in Goa, India, with work on cruise ships. Once a strong link is established, a stable chain of migration links people with specific skills from specific origin areas with particular destinations.

9.2 The Zelinsky and Metz Models

We have already touched on the demographic transition model in chapter 3. This model was the inspiration for the geographer Wilbur Zelinksy, who wanted to develop a transition model for mobility. The **Zelinsky model** consists of three phases. In the premodern traditional society, residential movement and personal mobility are very limited and circumscribed. In the advanced society, residential mobility increases; migrants are on the move, especially in the form of rural-to-urban migration; and circulatory movement such as tourism begins to increase. In the super-advanced state there is an increasing acceleration of movement both in terms of residential and personal mobility. In this final stage Zelinsky suggests that there might be limits to increasing mobility; he gives the example of traffic lights that limit speed, despite the ability of vehicles to drive fast. Speed is reduced in the interests of traffic management and public safety.

Transport specialist David Metz employs a similar model, though he identifies four rather than three phases. In the first era of human travel, our hunter-gatherer ancestors walked out of Africa and populated the rest of the Earth. In the second era, they settled in agricultural communities and towns, where travel was generally limited to about an hour a day on foot. The third era began early in the nineteenth century with the coming of the railways, when the energy of fossil fuels could be harnessed to achieve faster travel through a succession of technological innovations, culminating in mass mobility made possible by the motorcar. There is now emerging evidence that we have passed peak car use. We are entering a fourth era in according to the **Metz model** in which aggregate travel times, trip rates, and distance traveled are holding steady. The "peak car" phenomenon, whereby car mode share in cities like London reached a peak and then recently declined, marks the transition from the third to the fourth era.

REFERENCES

Cresswell, T. (2010) Towards a Politics of Mobility. *Environment and Planning D: Society and Space* 28: 17–31.

Metz, D. (2013) Peak Car and Beyond: The Fourth Era of Travel. *Transport Reviews* 33: 255–270.

Zelinsky, W. (1971) Hypothesis of Mobility Transition. *Geographical Review* 6: 219–249.

Cited References

Abound, T. (2007) National Boundaries, Colonized Space: The Gendered Politics of Residential Life in Contemporary Jerusalem. *Anthropological Quarterly* 80: 997–1034.

Anderson, K. J. (1987) The Idea of Chinatown: The Power of Place and Institutional Practice in the Making of a Racial Category. *Annals of the Association of American Geographers* 77: 580–598.

Glazer, N. and Moynihan, D. (1963) *Beyond the Melting Pot.* Boston: MIT Press.

Hägerstrand, T. (1970) What about People in Regional Science? *Papers of the Regional Science Association* 24: 7–21.

Hanser, A. (2005) The Gendered Rice Bowl: The Sexual Politics of Service Work in Urban China. *Gender and Society* 19: 581–600.

Lee, E. S. (1966) A Theory of Migration. *Demography* 3: 47–57.

Park, R. E. (1952) *Human Communities: The City and Human Ecology.* New York: Free Press.

Price, M. and Benton-Short, L. (eds) (2008) *Migrants to the Metropolis.* Syracuse, NY: Syracuse University Press.

Ravenstein, E. G. (1885) The Laws of Migration. *Journal of the Statistical Society of London* 48: 167–235.

Sethi, A. (2012) *A Free Man: A True Story of Life and Death in Delhi.* London: Jonathan Cape.

Wilson, E. (1991) *The Sphinx in the City.* London: Virago.

Select Guide to Further Reading

Anderson, K. (1991) *Vancouver's Chinatown: Racial Discourses in Canada. 1875–1980.* Montreal: McGill University Press.

Castles, S., de Haas, H., and Miller, M. J. (2014) *The Age of Migration: International Movements on the Modern World.* 5th ed. New York: Guilford.

Cordey-Hayes, M. (2012) Migration and the Dynamics of Multiregional Population Systems. *Environment and Planning A* 7: 793–814.

Goldscheider, F., Bernhardt, E., and Lappegård, T. (2015) The Gender Revolution: A Framework for Understanding Changing Family and Demographic Behavior. *Population and Development Review* 41: 207–239.

King, R. and Skeldon, R. (2010) "Mind the Gap!": Integrating Approaches to Internal and International Migration. *Journal of Ethnic and Migration Studies* 36: 1619–1646.

Laslett, P. (1989) *A Fresh Map of Life: The Emergence of the Third Age*. London: Weidenfeld and Nicholson.

Pablos-Mendez, A., Radloff, S., Khajavi, K., and Dunst, S. A. (2015) The Demographic Stretch of the Arc of Life: Social and Cultural Changes That Follow the Demographic Transition. *Global Health: Science and Practice* 3: 341–351.

Strulik, H. and Vollmer, S. (2015) The Fertility Transition Around the World. *Journal of Population Economics* 28: 31–44.

Websites

Internal Displacement Monitoring Centre
http://www.internal-displacement.org
International Organization for Migration
http://www.iom.int
International Migration Statistics
http://www.migrationpolicy.org/programs/data-hub/international-migration-statistics

Minnesota Poulation Center
www.popdata.org
UN migration issues
http://unstats.un.org/unsd/demographic/sconcerns/migration
UN Refugee Agency
http://www.unhcr.org/cgi-bin/texis/vtx/home

Interactive Map Websites

http://www.citylab.com/housing/2015/10/mapping-the-frenzy-of-the-europes-migrant-crisis/412396/
http://globe.chromeexperiments.com

http://www.census.gov/geography/interactive-maps.html
https://www.census.gov/2010census/popmap/

Learning Outcomes

- The bulk of the world's population is in the Northern Hemisphere, centered on the Tropic of Cancer.
- The world's population is distributed and demarcated in cultural, political, and economic space.
- More than a third of the world's population lives in just two countries, China and India. Half of the world's total population lives in just six countries. And the ten largest countries combined contain almost 58 percent of the world population.
- Population size is an important component of national power. Very small countries tend to wield less power than larger countries.
- Population distribution often connects with national imaginaries.
- There are disparities between population distribution and political representation in a number of countries across the globe, including in the major global powers of Japan and the US.
- In a process of cumulative causation, movement of people to areas of greater economic opportunity can widen regional disparities.
- Gender roles in an economy vary over time and space, partly influenced by rarely stable and unfixed gender identities.

- Although there are definitive biological definitions for various ages, what it means to be at different stages in life varies across the world and throughout history.
- Race and ethnicity are neither natural categories nor are they unchanging. The very definitions of race and ethnicity are social constructions, not biological processes.
- Population movement takes place on a daily, weekly, seasonal, and annual basis following circular and noncircular patterns, sometimes involving permanent relocations.
- Sometimes people are forced to move, as was the case in the slave trade. Today political conflict and lack of economic opportunities are powerful drivers of forced movement.
- International migration channels are often routed very largely, although not exclusively, through major metropolitan areas known as gateway cities. Examples include New York, Amsterdam, Dublin, Washington, DC, and Riyadh.
- There are numerous models of population movement: Ravenstein's model; the gravity models that focus on size and distance between places; and Lee's push-pull model, which focuses on a comparative assessment between locations.

10

The Geography of Religion

Religion is an important element of culture and an important subject for cultural geography. Many people define themselves primarily in terms of religious affiliation. Religious beliefs order society. In this chapter we will look at a number of elements in the geography of religion and how religion and religious beliefs structure the cultural organization of space.

The Geography of the Major Religions

Three broad types of religion can be identified: the **animist religions** of hunting-gathering societies, the great religions of Asia, and the religions of **monotheism** that emerged from the Middle East. We will look at each in turn but with a gentle reminder that there are similarities and connections as well as differences and ruptures between the three categories.

The Animist Tradition

The cosmologies of hunting-gathering societies are rooted in an animistic conception of the world. The essence of the animistic cosmologies of hunting-gathering societies is that the Earth is alive and nurturing. Nature is a spiritual resource as well as an economic entity, part of the fabric of connections between all things. Space is sacred, and territory is the fundamental source of meaning and identity.

In Cherokee tradition the word for Earth, *elohi*, also means history, culture, and religion. All living things and all places are bound up in a web of material and spiritual connections. Collective and individual identity is bound to specific places, plants, and animals. There is a kinship between

LEARNING OBJECTIVES

Describe the three types of religion their origins and global distributions.

Explain the differences that exist within each major religion.

Consider the ways that religion is used to interpret the environment, to organize and regulate spaces.

Discuss the complex reasons why religious conflict occurs over space.

Reflect on how religious beliefs and practices inform social dynamics and affect social relations.

animals and humans, a bond between the hunted and the hunter, planter and planted. A deep and profound religiosity imbues the economic relations of everyday life. Ritual and performance in dance and song, as well as material culture such as shelter, clothing, and decoration, link people, places, and living things in a tight nexus of material and spiritual connections.

Consider the case of the indigenous peoples of the central Australian desert. Before the coming of the whites in the late nineteenth century, the total Arrernte pre-contact population is estimated to have been between 8,000 and 10,000. Food was scarce. The landscape was a complex system of land titles based on lineage and family ties. Responsibility for "managing" specific sites often lay in the hands of senior elders, both men and women (Figure 10.1). The complex, interlocking titles were less monopoly controls, as we understand private property under a capitalist system, and more relational and totemic, allowing individuals and groups a broad range of claims, responsibilities,

and bargaining options that enabled long-term occupancy of a harsh environment with irregular water and variable food sources. The cosmology bound people to the land in intricate webs of meaning that also sustained long-term economic usage.

The cosmologies of these early societies did not prevent major ecosystem transformation and species collapse. Calvin Martin looked at the effect of early contact with Europeans on the Micmac tribe in the Gaspé Peninsula of Canada. Prior to the arrival of Europeans, the beaver was hunted in a sustainable way; there were millions of beaver, and their hunting was bound by rules, regulations, and religious invocations that severely limited the hunting of "relatives in fur masks." With the coming of the French and the Dutch, a vigorous trade was promoted. Europeans wanted furs from the Micmac and other tribal groups, for which they traded knives, axes, guns, and alcohol. The increasing competition for furs, the decline of traditional constraints caused by the deadly epidemics of the Columbian Encounter, and the waning influence of the traditional shaman led to a wholesale slaughter of beaver by Micmac hunters, who no longer hunted for immediate needs but instead without restraint. It became an unrestrained slaughter as the beaver was turned from a relative in a fur mask to a commodity.

As hunting-gathering societies were overwhelmed by agriculturalists and later urbanization and industrialization, the world lost much of the rich variety of the animist traditions. Figure 10.2 show the main religions across the world and reveals how the traditional animists are now restricted to the more peripheral and marginal lands of desert, tundra, and rainforest. It is a fast disappearing world.

ANCIENT ANIMISMS AND MODERN ENVIRONMENTALISM

In a rapidly modernizing world, the original animist traditions are not unchanging but transformed and reinterpreted. Sam Gill, for example, examined the cosmologies of Native American animist traditions. He found little original evidence of a Mother Earth figure. The myth appears in nineteenth-century Anglo-American interpretations of Native American religions. It was read in rather than excavated out and drew upon Christian beliefs in the Virgin Mary and understandings of Old World Neolithic cultures. However, it became part of Native American writings as Native Americans used the myth to express their attachment to the land in a way that was understandable for the dominant Anglo-American audience. A product of Anglo-American cosmologies was incorporated into Native American beliefs and then projected back to influence Anglo-American perspectives.

Animist traditions, real and invented, are now used to inform the contemporary **environmental movement**.

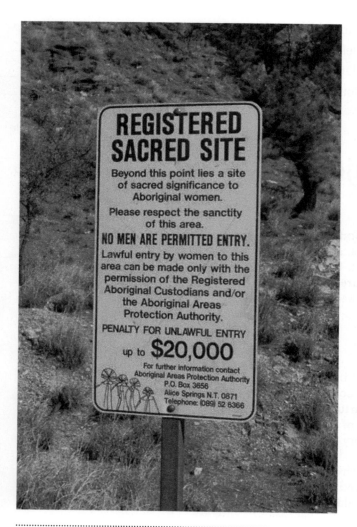

10.1 Sacred site in Alice Springs, Australia

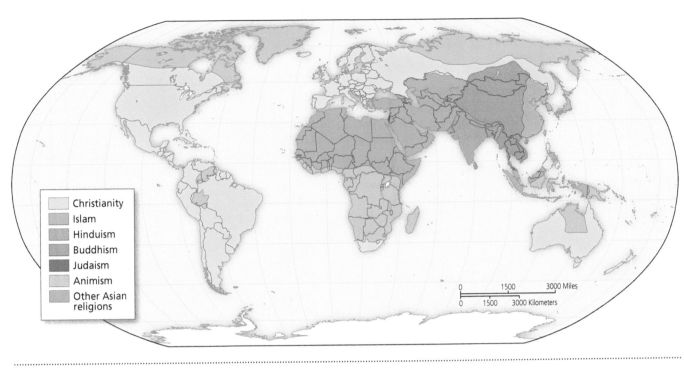

10.2 Main world religions

Here is the oft-cited 1854 speech of Chief Seattle of the Suquamish in response to the US government's offer to buy their land:

> But how can you buy or sell the sky? The land? The idea is strange to us. If we do not own the freshness of the air and the sparkle of the water, how can you buy them?
>
> Every part of the earth is sacred to my people. Every shining pine needle, every sandy shore, every mist in the dark woods, every meadow, every humming insect. All are holy in the memory and experience of my people.
>
> We know the sap which courses through the trees as we know the blood that courses through our veins. We are part of the earth and it is part of us. The perfumed flowers are our sisters. The bear, the deer, the great eagle, these are our brothers. The rocky crests, the dew in the meadow, the body heat of the pony, and man all belong to the same family.
>
> The shining water that moves in the streams and rivers is not just water, but the blood of our ancestors. If we sell you our land, you must remember that it is sacred. Each glossy reflection in the clear waters of the lakes tells of events and memories in the life of my people. The water's murmur is the voice of my father's father.
>
> The rivers are our brothers. They quench our thirst. They carry our canoes and feed our children.

> So you must give the rivers the kindness that you would give any brother.
>
> If we sell you our land, remember that the air is precious to us that the air shares its spirit with all the life that it supports. The wind that gave our grandfather his first breath also received his last sigh. The wind also gives our children the spirit of life. So if we sell our land, you must keep it apart and sacred, as a place where man can go to taste the wind that is sweetened by the meadow flowers.
>
> Will you teach your children what we have taught our children? That the earth is our mother? What befalls the earth befalls all the sons of the earth.
>
> This we know: the earth does not belong to man, man belongs to the earth. All things are connected like the blood that unites us all. Man did not weave the web of life, he is merely a strand in it. Whatever he does to the web, he does to himself.
>
> One thing we know: our God is also your God. The earth is precious to him and to harm the earth is to heap contempt on its creator.

It is a moving, eloquent speech. But it is an invention; we have no idea what the Chief actually said. The "speech" was rewritten at numerous times by different people with different agendas—some religious, some ecological, and some

10.1 Alternative Visions

The Huichol Indians live in a mountain area in western Mexico. Their total numbers are around 25,000 to 50,000. Although deeply impacted by the Columbian Encounter, they still retain many of their traditional belief systems. Huichol is still spoken, and there are two local governments: one answers to the local people while another represents the Mexican state.

The Huichol interact with the modern world. Their crafts cater to a tourist market. But they also practice a form of **animism** that imbues the area with primary spiritual significance. The landscape is a sacred geography whose protection is vital to the integrity of their worldview. The local mountains are not inert geological features but sacred portals to an understanding of the infinite, sites of cosmic significance. One mountain, Cerro del Quemado, known to the Huichol as Wirikuta, is considered the birthplace of the Sun.

Two mining companies, Canada's First Majestic Silver and the Mexican Real Bonanza, have a very different conception of the area. They see it is as a place to extract silver. In 2009 they purchased mineral rights from the Mexican government and will most likely use cyanide in open pit mining.

This is a form of struggle witnessed throughout the world. On the one hand, a small group of people with indigenous beliefs and uncertain land rights; on the other, powerful international corporations looking to extract commodities. The Huichol have sought to resist the mining proposals, while the mining companies have sworn to protect the sacred sites, maintain international environmental standards, and employ local people. Many of the non-Huichol local people support the mining proposal for the jobs it will generate. Behind the conflict are very different and competing ideas of the environment, as a commodity to be exploited and as a scared space to be revered. As one Huichol said of the mountain,

> this is the center of the world. It is our church. The mining company has promised that the mountain will look the same. But it won't be the same if you take away its soul.

REFERENCES

Booth, W. (2012) Where Cosmic and Commercial Collide. *The Washington Post*, February 14, pp. A1 and A11. Quote is from page A11.

Liffman, P. M. (2010) *Huichol Territory and the Mexican Nation: Indigenous Ritual, Land Conflict and Sovereignty Claims*. Tucson: University of Arizona Press.

a combination of the two. It was written into the script of a film made by the Southern Baptist Convention in 1972. The speech has entered the **religio-ecological** mainstream almost as a sacred text in its own right. It makes three important points:

> The environment is a spiritual resource more than an economic entity.
>
> There is web of connection between all things.
>
> Occupancy of the Earth is a religious bond with responsibilities and obligations.

The speech is less and more a lament for the present and hope for the future. The ecology of the animists is now an important element in the contemporary religio-ecological movement.

The Religions of Asia

The **Indic region** of northern India, centered on the Indus Valley, is the hearth source for Buddhism, Hinduism, Jainism, and Sikhism (Figures 10.3 and 10.4). The two largest are Hinduism and Buddhism. Hinduism spread across the Indian subcontinent into much of Southeast Asia. Later Buddhism spread along the trade routes of the Silk Road, finding expression and acceptance in China, Korea, and Japan and replacing Hinduism in much of Southeast Asia.

HINDUISM

Hinduism has more than a billion adherents and is the dominant religion in India, though not the only one. It is also found in Sri Lanka and Nepal, and is celebrated around the world as the Indian diaspora took their religion in their migrations to the Middle East, Australia, Western Europe, and North America.

Hinduism is one of world's oldest religions. It is a belief system that stretches back to the early civilizations of the Indus Valley over 6,000 years ago. It diffused south across India into Southeast Asia and the islands of what is now Indonesia. It is a monotheistic region, as it believes there is one life force, but it is also polytheistic with numerous deities. There are three male gods, Brahma, Vishnu, and Shiva, and a female deity, Devi, the mother goddess. There is a host of other gods such as Ganesh, a man with the head of an elephant who is a god of strength; Lakshmi, the goddess of wealth; and Hanuman, the monkey king, who is a god of cunning. Hinduism also has a strong belief in reincarnation.

With thousands of years of development, Hinduism has a long history, combining different strands of belief, from the earliest folk traditions of the worship of local deities, to the very recent associations with Indian Hindu nationalism. There is no one foundational text as, say, the Bible for Christians or the Koran for Muslims. Sacred texts include the Vedas, a large number of texts written around 3,500 and 5,000 years ago; the 700-verse *Bhagavad Gita*; and many texts of the Upanishads, written between 1000 BCE to 1500 CE. There is no single source of religious authority.

Hindus believe in an ultimate reality, Brahman. Numerous gods link this ultimate reality with the populace. The most popular include the four-faced Brahma, the creator god; the blue-skinned Vishnu, the preserver; and Shiva, the destroyer or transformer, with an eye in the center of his forehead.

Hinduism is not an unchanging set of beliefs; it has been touched by and transformed by historical changes. The increasing urbanization of India from 500–200 BCE, for example, challenged previous rituals. A Hindu modernism emerged in the nineteenth century that focused on social issues, while the contact with the British Raj impacted a wider understanding of Hinduism.

As one of the world's oldest religions, Hinduism provides the best case study of the plasticity of religious beliefs and practices. Religions have a tight core of belief but these are changed and transformed over time and in contact with new historical realities, changing social forces, and other beliefs and doctrines. Religions do not remain pure and untouched but rather are marked by their place in time and their time in place.

Hinduism is an ancient, complex religion that defies easy definition. It has changed and transformed over the millennia. However, there are a number of key beliefs. The core belief involves leading a righteous life, realizing wealth, fulfilling desires, and attaining liberation. The classic stages of life are student, householder, retired, and renunciation. The socially important household stage, when people take up their role in society, starts with a marriage that explains, in part, the significance and resources devoted to marriage celebrations in Hindu culture. The final stage is supposed to be marked by a detachment from material life to focus on spiritual matters.

Hindu temples are places of festivals, ritual, and community celebrations. The highest point in the temple represents Mount Meru, the sacred five-peaked mountain considered the center of the spiritual universe. In the temples of south India in particular, the outside walls are richly decorated. Rituals are also practiced at home and include worship at family shrines, yoga, meditation, and the chanting of mantras. There are also festivals celebrated throughout the year. Many of them are local and regional while a few, such as the annual festival of lights celebration known as Diwali, are found throughout the Hindu culture region.

There are a number of pilgrimage circuits that link holy cities and sites. One of the holiest is held every three years and the location rotates around the four cities of Allahabad, Haridwar, Nashik, and Ujjain. The River Ganges is worshipped as the goddess Ganga and devout Hindus believe that bathing in the river washes away sins and helps liberate them for the endless cycle of life and death. It is a problematic place for pilgrimage as it is one of the most polluted rivers in the world, a symbol of how sacred space is transgressed by the material world.

Partly in response to Western colonial presence, the loose collection of beliefs that constitute Hinduism became more codified. More recently, it has become the basis of identity and nationalist politics in India.

A significant social aspect of Hinduism, which was reinforced under the British, is the **caste system**. These are social categories that determine not only position in the social hierarchy but also appropriate occupation, behaviors, and social customs such as eligible marriage partners. There are many different castes, such as the high-caste Brahmins whose traditional occupations include priests and teachers. The Kshatriya formed the political and military elite. The Vaisya were involved in agriculture but now are landowners, merchants, and bankers. The Sudras are the "worker bees" of the system involved in many occupations such as barbers, gardeners, and potters. At the very bottom of the hierarchy are the Dalits, the untouchables, who number around 200 million people in India. They were restricted to the impure occupations such as leather workers, sweepers, manual laborers, and sewage workers. They were considered unclean and often physically segregated. Across India many Hindus became Muslims to avoid the debilitating and limiting caste designation. The conversion was almost universal in the rural parts of east Bengal, what is now Bangladesh.

The caste system is more complex than the simple categories outlined here. There are many more subcategories and further divisions of subcategories. There is no simple mapping of caste onto wealth. There are many poor Brahmins and rich Vaisya and affluent Dalits. The caste system is strongest at the village level. In Madhya Pradesh, for example, Dalits are prevented from entering most temples, and are often segregated at the edge of the village. As people move to the cities, the entrenched caste system that operates strongly at the village level is breaking down—not quite to the easier mobility of modern secular societies, but still, toward a greater fluidity. However, only 5 percent of marriages in India cross caste boundaries. Since 1950 the Indian government has actively worked to improve the employment opportunities and living conditions of the Dalits. They are now founds in the upper levels of the government and government-controlled enterprises and their living conditions are on par with the rest of

10.3 Hindu temple in Kuala Lumpur, Malaysia. **10.4** Buddhist monks in Yangon, Myanmar. **10.5** Jain Temple in Mumbai, India.

the population. They still face serious discrimination in public life, with high dropout rates from schools and more limited access to public services. Violence against Dalits continues, though it is on the decline. Urbanization and policies of affirmative action have eroded much but not all of the bias in the caste system. There is also variation across India. The caste system was more challenged in Kerala, for example, part of the more radical political tradition in a state that regularly elects communist state governments.

BUDDHISM

Buddhism emerged as a critical offshoot of Hinduism, and like Hinduism also developed in northern India, around the sixth century. It spread across East and Southeast Asia. The transformation of Hinduism into Buddhism is embodied in the life of Buddha, who was born a prince in northern India around 563 BCE. Spared the vicissitudes of life by his wealthy family, he instead chose poverty, ill health, old age, and eventual death. He distanced himself from the caste system of Hinduism and taught a middle way between the asceticism of a retreat from society and the indulgence of society driven by desire. Buddha preached the pursuit of enlightenment.

A fundamental belief of Buddhism is that desire in any form, whether for wealth, fame, or sensory pleasure, causes misery. Bliss is achieved through not having desire. Retreat from the material world through permanent or temporary periods as monk or nun is a common practice. Buddhism slowly lost influence in India but gained adherents in the Southeast Asia. There are two main forms: Mahayana Buddhism found in China, Japan, and Korea, and Theravada Buddhism, which ties more closely to the Buddha's teaching of the search for nirvana. Theravada Buddhism is particularly strong in Sri Lanka, Myanmar, Thailand, and Laos and in these countries most young men try to spend a portion of their time as monks learning to mediate while begging for food.

In parts of Southeast Asia Buddhism was influenced by British colonialism. After the British gained control of Burma in 1824, Christian missionaries began a conversion program. Buddhist monks, fearful that their religion would crumble, encouraged people to study Buddhist texts and to meditate, a practice formerly limited to monks. Buddhism was reshaped and meditation became a standard practice in response to the threat from Christian proselytization. And today most people in the West think of Buddhist meditation as a centuries-old tradition rather as a product of nineteenth-century resistance to British colonialism and its contact with modernism.

This provides yet another example of the constantly changing nature of religious practices and beliefs and their evolution as they come into contact and sometimes compete with other religious faiths.

OTHER ASIAN RELIGIONS

Hinduism and Buddhism are the two main religions of South and East Asia, but there are also others. Sikhism combines Hindu mysticism and Muslim monotheism. It is indigenous to this region and was founded in the fifteenth century in the Punjab. By 1800 a vast Sikh empire stretched over Pakistan and India with a capital in Lahore. The entry of the British undermined this empire, and by 1900 the Sikh heartland had become the Punjab region of British India. Today there are around 25 million Sikhs, the vast majority in the Indian state of Punjab, smaller numbers in Haryana, India, and a diaspora in Asia, North America, and Europe.

Jainism is an ancient religion with around 5 million followers. Its main teachings include nonviolence, a detachment from the world, and a commitment to honoring other forms of life. Jains are vegetarian to avoid inflicting unnecessary suffering upon animals (Figure 10.5).

Shinto is the largest religion in Japan, intermittently followed by around three-quarters of the total population. It is considered Japan's traditional religion and is constituted in thousands of public shrines and temples devoted to the worship of numerous gods. Shinto is the religion of Japan's traditional agricultural society, and many of the rituals were originally festivals to placate the gods to ensure a good harvest and to thank them for the bounty.

Shintoism provides a good example of an important trait of all religions: a hard core of true believers and regular adherents, and an often much larger periphery of occasional practitioners. There are, for example, Catholics who go to Mass and confession every week, and many others who may go to religious services only during Easter and Christmas, to bury someone, or to attend a wedding. They are all Catholics, but some are in the **core of believers**, while the others are in the periphery. In Japan only a very small minority of Shinto practitioners are in the core. Most use Shinto temples just for weddings and occasionally visit its shrines to make an offering for a successful business venture or social project; it is not an integral part of everyday life as, say, Buddhism is in Thailand or Myanmar. There is a wide continuum to religious beliefs from the ardent to the casual and from the committed to the irregular.

Confucianism is as much a philosophy as a religion. It developed from the teachings of Confucius (551–479 BCE). It stresses family loyalty, social harmony, and respect for hierarchy. It cultivates virtue, education, and social ethics with an emphasis on tradition and stability. It is a major influence not so much as a religion but as an organizing social worldview in China, Singapore, Taiwan, Korea, Japan, and Vietnam. Its influence is profound. The spectacular economic growth of East Asian countries in the past decade is due in no small part to the Confucian legacy of good government and social harmony. All the countries mentioned have experienced spectacular rates of growth starting with Japan and more recently in China and Vietnam. That is one reason why the success of East Asia has been difficult to replicate in other parts of the world that do not share the Confucian legacy. The social order of Confucianism does play an important part in their economic success, based as it is on hard work, family sacrifice, and a belief and trust in the role of government. When the Communists took control of China in 1949, they initially discouraged Confucianism, but now his teachings are widely taught and admired. As part of their projection of soft power, the Chinese government now supports Confucius Institutes in countries around the world.

The title of this section is misleading. While these religions developed in Asia, they have spread around the world. There are Buddhists in America, Hindus in Britain, and Sikhs in Canada. Along the way, the rituals and practices have subtly transformed. The practice of yoga, long an element of Hinduism, has now been severed from its religious roots in the West to become a physical exercise. So when you see people carrying yoga mats in the cities of the West they are less likely to be religious devotees and more likely to be people coming and going from physical exercise class.

The People of the Book

The **Semitic region** of the Middle East is the hearth area of three of the worlds' significant religions: in order of historical appearance, Judaism, Christianity, and Islam. They are the religions of the Book, where religious beliefs are encoded in written texts. Their texts often contains the same myths and stories. Some stories are sacred to all three **Abrahamic religions**: Adam and Eve, Abraham, Cain and Abel, Jonah and the whale, Moses, and Job. The stories helped to define each religion as both special in God's eyes and as supporting community identity in face of its other two rivals. Robert Gregg has looked in detail at some of the shared stories and their varied tellings. In the Jewish telling, Jonah was a cautionary tale of a complex figure: an angry prophet who did not repent, and later a rehabilitated agent of God. For the Christians, he was a rescued prophet who prefigured Jesus and forecasted Christ's victory over Death.

In the Muslim telling, he was an angry man, then a penitent one who excelled in glorifying God.

Let us briefly consider the geography of these three religions.

JUDAISM

Judaism emerges from the early city-states of Mesopotamia. Abraham and his family clan left the city of Ur and set out for Canaan around 3,000 to 2,500 years ago. Urban sophisticates, they left civilization to establish settlements in what was considered wilderness. Abraham eventually came to believe in a monotheistic religion, the foundation of Judaism. The Judaic theology has roots deep in Mesopotamian beliefs and writings. *The Epic of Gilgamesh* is echoed in the story of Genesis, including the idea of a great flood. The Jewish universe was one ruled by an all-powerful God, accessible by prayer and incantation and present in appearance and action. Abraham's faith was tested in the searing experience of his willingness to sacrifice his son. The Jews, who saw themselves as the chosen people of God, were continually tested in their faith. Famine stalked the land and they were taken into slavery in Egypt, to build the great cities and temples of the empire on the Nile. Led by their prophet Moses they managed to escape, as described in the story of Exodus, and displaced the Canaanites in a brutal campaign described in the Bible as a form of ethnic cleansing. The Jewish kingdom soon become rich as it was strategically located on the routes between major trading regions. Hebrew emerged as the dominant language, and a vast temple complex was constructed in Jerusalem.

And then a series of disasters tested the people's faith in the love of God. The country came under the rule of nearby empires, first the Assyrians and then the Babylonians. The Jews resisted full incorporation into the Babylonian Empire and when they revolted, the rebellion was crushed and, in 586 BCE, the Temple in Jerusalem was destroyed. Many Jews were taken off into exile until freed by the Persian king in 538 BCE. The Jews returned and rebuilt the Temple. Its rededication, centuries later, is the basis of the Hanukkah celebration. The Romans destroyed the temple complex in 70 CE, and the Jewish state disappeared for close to two thousand years until 1948.

The Old Testament Bible, then, is a historical geography of the Middle East albeit embroidered and tailored for specific purposes. Four distinct sources make up the Old Testament. The J source is the earliest and draws on Mesopotamian and Canaanite sources, the E source was written two centuries later, the D source is a book of laws for more centralized worship at a time when the priests were gaining power, and the P source was written at the same time as the Babylonian exile. Ecclesiastes, written around 400–200 BCE, draws upon ancient wisdom literature in which a character offers important life lessons, but also expresses the skepticism and muted joy and mournfulness that perhaps reflects the experience of exile. In keeping with the idea of the Middle East as a cauldron of shared and different cultural traditions, Ecclesiastes also contains Persian words and Greek themes.

After the destruction of the Temple in 70 CE, Jews dispersed throughout the Middle East and later Europe and the New World. Centuries later, from around 1940, Jews from across much of Eastern and Central Europe either left or were forced to leave as Nazi Germany conquered Europe, or were killed in the horror of the Holocaust. Jews in the Middle East largely escaped the Holocaust that blighted European Jewry. But their position changed dramatically after the creation of Israel in 1948. Wars between Israel and the surrounding Arab countries made life difficult for Jews of these countries as they were now associated with **Zionism** and Israel. Jews left them for Israel or the West, especially the US. Outside of Israel, the region lost a Jewish presence that had survived for thousands of years.

Judaism is not a universalizing religion such as Christianity. It does not seek out converts or promote expansion into other faith communities. Judaism is the religion that connects Jews with their God; it is not promoted as a universal religion open to everyone. It is the religion of a people, not a religion for all the people.

Jews and significant Jewish communities are found in many countries of the world, including the USA, often playing significant roles in local economies and political life. The "Israel lobby," for example, generally taken to mean a commitment to Israel, is an important element in US presidential politics, and plays a key role in crucial electoral states such as Florida. However, you can see from Figure 10.2 that Judaism is the dominant religion only in Israel. Israel is now the recognized homeland for Jews. In 1897 only 50,000 Jews lived in Palestine and they accounted for less than 1 percent of world Jewry. By 1950 it was 10.6 percent, and now, with 6 million Jews in Israel, it is 45 percent. Between 1989 and 1991, one million migrants came to Israel from Russia. Since Israeli Jews are younger and have higher birth rates than non-Israeli Jews, by 2050 the majority of all Jews in the world will live in Israel.

CHRISTIANITY

Jesus was a Jew claimed by Christians as the Son of God. He grew up and died in the Middle East, a region then situated on the edge of the Roman Empire. His short life and resurrection became the basis for the New Testament. This minor Judaic sect soon spread along the trade routes of the Roman Empire, first between seaports along the Mediterranean and then later inland. Armenia became the first Christian country around 303 CE. The Romans initially persecuted the Christians and suppressed their religion, but after 313 it was officially recognized, and in 380 became the official religion of the empire. The Catholic Church embodied the Roman Empire; it is

10.6 St. Joseph's Cathedral in Antigua, Guatemala, first built in 1594

still headquartered in Rome, and its faith and influence spread across the territories of the former Roman Empire. Christians were persecuted because their religious faith undermined civic responsibilities. Yet when the Roman emperor Constantine converted to Christianity in 312 CE, it had the backing of political power and became the dominant religion throughout Europe. It built upon the older, more place-based religions, often using their sacred sites as locations for places of Christian worship. Many important cathedrals in Europe were sited on pre-Christian sacred sites.

After the fall of Rome and the passing of the medieval period into the Renaissance, Christianity was exported around the world as part of European colonialism and imperialism. The Spanish took their Catholicism with them when they invaded the New World (see Figure 10.6). The British settler societies of North America and Australia brought their Protestant religion with them, self-consciously believing in its superiority over local religious beliefs and practices. Christianity was one of the more expansionist religions, associated as it was with the powerful colonial forces of the day. In some countries, it was resisted. British control in India and the Middle East did little to change the religious beliefs of the majority of the colonized people. In others, it was incorporated into complex blends of the old and the new. In parts of South America, the old gods still appear, suitably modified in Catholic worship; in Haiti, Catholic rituals are part of ceremonies that also draw upon older African practices brought there by slaves. In still others, it was enthusiastically adopted. In late nineteenth- and early twentieth-century Korea, Christianity

gave meaning and purpose to the lives of ordinary people denied both in a very hierarchical and feudal Confucian system.

ISLAM

Islam emerged from desert oases in an Arabia sandwiched between the competing powers of the Byzantine and Persian Empires. Around 610 a merchant, Muhammad ibn Abdullah, sleeping in a cave near Mount Hira received a revelation and wrote down, in what became the *Quran*, the words spoken to him by God. Muslims accept the Quran as the literal word of God; that is why devout Muslims recite the Quran in Arabic and are particularly sensitive to criticisms of it. In 622 Muhammad, with his followers, left Mecca for Medina; this date is taken as the start of Muslim calendar, much like the birth of Christ is the starting point of the Christian calendar. A persecuted cult quickly became a religious and military force, first by taking control of Mecca in 630 and then by extending its reach in one of the fastest imperial expansions in the premodern world. Like Christianity, Islam was a universalist doctrine, in which faith, not ethnicity, was important. And like early Christianity it combined secular and religious power. Whereas in the Christian world, there was a steady separation, especially in modernity, between the religious and political spheres—rarely a neat or clean break, and in some cases still a site of struggle and contestation—this distinction did not materialize in Islam, where the religious and the political were conceived as one and the same.

By 632, barely twelve years after the Prophet's death, an Islamic empire had taken control over former Persian and Byzantine territory; a century later, by 732, Islamic power had swept through the Middle East, across North Africa, and up through Spain and into France. Like all rapid and successful imperial expansions, it was fueled by an economic incentive to gain monopoly control over taxation, goods, and people as well by a religious fervor to spread the word of Islam to the entire world. The designation of a region known as the Middle East and North Africa reflects the geography of Muslim expansion from the heartland in Arabia.

As Islam spread from its core region, it was influenced by local traditions. Thus Islam in Southeast Asia lost some of its desert austerity as it was grafted onto local traditions of animism, Hinduism, and Buddhism. To this day, Islam

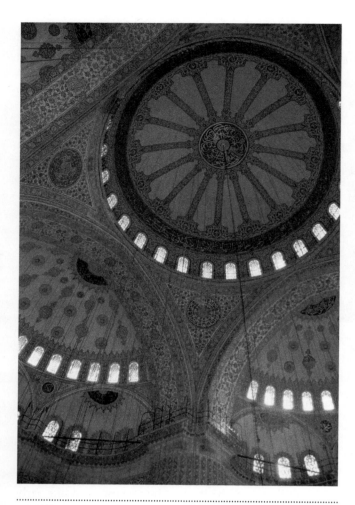

10.7 The interior of the Blue Mosque, Istanbul

in Indonesia is more socially moderate than the fundamentalist Islam of Saudi Arabia.

Early on, Muslim places of worship were forbidden to show human figures, which led to wonderful displays of Koranic writing and dazzling geometric forms as decoration.

The Geographies of Religious Belief

Figure 10.2 is a useful guide to the global distribution of religion. However, it is a broad-brush approach that assumes religious homogeneity subsumed under very wide categories. If we change the scale from the global to the world regional and more national level, then a different picture emerges and we can see the many varieties of Christianity, Buddhism, and Islam.

The Sunni–Shia Split in the Middle East

Sharpening our focus from the global to world regional allows us to see marked differences even within the same religion. Take the case of Islam in the Middle East. It is divided into two main sects, Sunni and Shia. Both share the basis tenets of Islamic teaching: a belief in Muhammad as the messenger of God, daily prayers, pilgrimage to Mecca, alms, and fasting during Ramadan. Their argument lies in different interpretations of the succession after the death of the Prophet Muhammad. The Shia believe that the succession should be hereditary and follow family bloodlines. The first caliphs after Muhammad were Sunni choices from outside his immediate family. The fourth was Ali, Muhammad's son in law, and thus more acceptable to Shiites. When he was killed at a battle in 661, a rift developed between the two branches that was further widened when Ali's son, Hussein, was killed at the battle of Karbala 680 CE, or 61 in the Islamic counting. It is now a Shiite sacred site.

Sunni Islam is the dominant form with more than 1.3 billion followers. Shiites are a minority with less than 250 million, concentrated in Iran, southern Iraq, southern Lebanon, and eastern Saudi Arabia and neighboring Bahrain (Figure 10.8). The split has grown in recent years, embodied in two major and competing powers in the region, Saudi Arabia and Iran. Saudi Arabia is one of the richest Sunni Arab countries, where fundamentalist teaching see Shiites as non-Muslims; and Iran has been the only officially Shiite state since the Islamic Revolution in 1979. There is conflict between the two countries, and their proxies in Iraq, Syria, Lebanon, and Yemen, for geopolitical dominance, especially in weak, dysfunctional, or collapsing states such as Yemen and Syria that provide space to project national interests. The rise of a more fundamentalist Sunni Islam across the Middle East has endangered minority Shia communities in countries such as Pakistan and Afghanistan. The Middle East remains a contested space between the two branches of Islam, their primary states, their proxies, and their terrorist offshoots.

The Case of the USA

Religious beliefs and practices are also expressed spatially at a national level. Figure 10.9 plots the majority church affiliation in every county in the United States. Distinct regional groupings are evident: Mormons in the mountain West, grouped around Salt Lake City; a Baptist heartland region south of the **Mason-Dixon Line**, the so-called **Bible Belt**; and a concentration of Lutherans in the upper Midwest. In part, these regional differences reflect the settlement patterns, as Mormons trekked west to establish their way of life far from mainstream America, the Scots-Irish poured over the Appalachians in the eighteenth

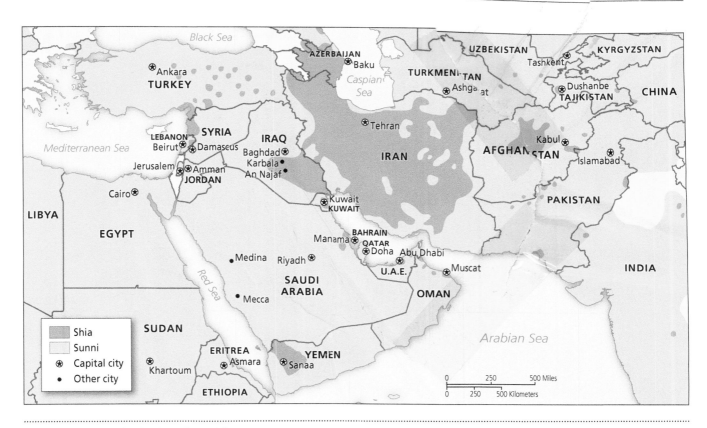

10.8 The Sunni–Shia divide in the Middle East

century, and Germans and Scandinavians later settled in the upper Midwest. There are also small outliers in select rural counties such as the Mennonites. Different groups coming at different times brought their religions with them. These religious affiliations have had wider impacts on the tone and practices of local communities in these regions. Using a slightly different religious classification, geographer John Bauer looked at regional changes in religious affiliation from 1980 to 2000. While there was great stability, there were also some marked changes as the strength of the Catholic region deepened and extended in the West, a trend associated with increasing size of the Hispanic population.

The Place of Religion

Traditional maps that plot the national distribution of the major religions are now inadequate to convey the complexity of religious devotion. When there are mosques in Detroit and ashrams in Memphis, the spatial geography of religion is more complex than can be summarized in simple maps that shade in entire areas with a single category of religious belief.

At the regional and local levels there are profound differences in religious affiliation and the degree of churchgoing.

In the US we can make a distinction between the Bible Belt, where fewer than 10 percent consider themselves irreligious, compared to 30 percent in New England and the Pacific Northwest. The Bible Belt is not just an area of Protestant churches but also of an explicit religiosity, compared to the rest of the country, often combining social and political views in a social conservatism that impacts political affiliations. The context of place is important in influencing religious practices. In more religious communities such as the Bible Belt, social pressures may make it more difficult not to go to church, whereas in the more irreligious Pacific Northwest, going to church, since it is not a widely practiced ritual, may be less frequent even for people of faith.

At the very local level, religion often reflects and shapes local communities. Local churches, for example are important community resources, providing sites of worship and witness, public spaces for activism, and places of devotion and celebration. In contemporary American Evangelicalism, the suburban megachurches also offer a space with meaning amid the dispersed character of suburban sprawl.

And it is often at the very local level, where different religious groups share the same public space, that both ecumenical cooperation and religious conflict can occur.

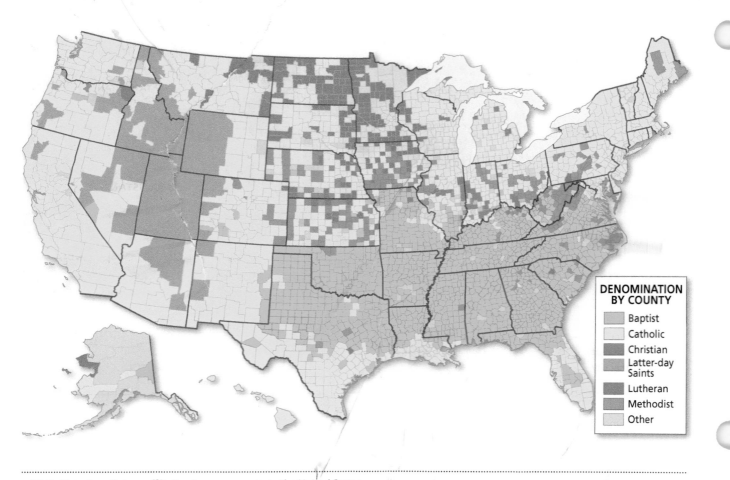

10.9 Majority religious affiliation in every county in the United States

DENOMINATION
BY COUNTY

- Baptist
- Catholic
- Christian
- Latter-day Saints
- Lutheran
- Methodist
- Other

In the wake of 9/11, Muslims in Europe and in some parts of the USA have felt threatened by a rise in anti-Muslim rhetoric. Controversies over building mosques and debates about the proper place for religious observances, such as dress codes, are all highlighted in local places shared by different religious practices.

Religion and Urban Public Space

One persistent theme is that religion declines with modernization. As societies become modern, more connected to the rationalities of science and technology, magic loses its hold, spirituality is replaced by materiality, and religion loses its appeal. Matthew Arnold's 1867 poem *Dover Beach* suggested that the retreat of religion and faith left us abandoned on

> *A darkling plain*
> *Swept with confused alarms of struggle and flight*
> *Where ignorant armies clash by night*

The city as the embodiment of the modern and the industrial and mechanical is often depicted as the locus for the breakdown of religion and faith. In fact, this trajectory is rarely so simple. We can consider three examples where faith within cities remains strong.

Tokyo is a very modern city, having been built and rebuilt twice in the last hundred years after earthquake damage and war bombing. Most Japanese report that though they think religious belief is important, their formal religious commitment is slight. Yet in this secular, modern city there are thousands of religious sites, from major temples to tiny shrines. The more popular ones are sites of festivals, holidays, and ritual occasions. Premodern religious sites are integrated into the urban landscape; Buddhist sites are associated with funereal rites, and Shinto shrines, drawn as they are from premodern harvest festivals, affirm life in weddings and other celebrations. The sacred spaces embody the tension between the continuing strength of ritual practice and the weakness of commitment to religion.

In Bangkok, Thailand, the many markets are filled with sacred sites, including shrines and spiritual markers (Figure 10.10). These religious places continue to exist not despite the growing modernization and commercialization of society but because of it. The sacred space of the vernacular is transformed into more contemporary prosperity religions that align traditional Buddhism to the pursuit of individual material gain. Just as the cityscape is transformed by modernization, so the city's sacred geography is reimagined. Bangkok is the rule rather than the exception, as the religious and the secular coexist in modernizing cities around the world.

Orthodox Christianity was the state religion of Ethiopia until 1975, when a religious plurality was introduced. Today, in the main city of Addis Ababa, urban public space is a contested site for interreligious rivalry and competition between Orthodox Christians, Protestants, and Muslims. Many taxicabs display overtly religious messages. Religious festivals provide the opportunity for **faith communities** to display their strength. During Epiphany the Orthodox Christians display banners of green, yellow, and red, the colors of the Ethiopian national flag, across the city. Posters of saints adorn the stress and public squares, and young men wear T-shirts with the slogan "Ethiopia, A Christian Land." The Muslim faithful take to the streets in large numbers during their two annual holidays of Eid al-Adha and Eid al-Fitr to show their strength and commitment. And Protestants mount loudspeakers on vans and patrol the streets. Urban public space in the city is not a secular place, but the arena for displays of religious beliefs.

cathedrals in Europe were constructed on pre-Christian scared sites, and the key Christian festivals of Christmas and Easter are built on the observances of pre-Christians sun worshippers, who celebrated the winter solstice, and the festival rituals of pre-Christian agriculturalists who celebrated the start of the growing season. The cult of the Virgin Mary developed from pagan beliefs in the Earth Mother.

Many of the myths and stories of the animist and sacred books refer to environmental changes. The recurrences of flood myths are less fantasies than references to the melting of the ice sheets and subsequent flooding at the end of the Pleistocene. The flooding of the Black Sea around 7,500 years ago may be the basis for the story of Noah. Many of stories of early religions are imaginative and elaborate retellings of actual environmental change and collapse.

The monotheistic religions of Judaism, Christianity, and Islam depict a sky-centered, universal God looking down over the entire world. The brutal defeat of the Canaanites by the Israelites, mentioned at various points in the Bible, is one of the first descriptions we have of the destruction of an Earth-based **polytheism** by a monotheistic religion. The Israelites' basis for this holy war was the fundamentalist religious doctrine that people should worship only one God. The Canaanites were depicted as corrupt and evil, and their fertility-based polytheism was replaced by the stringent demands of adherence to a single universal God. The war between the Canaanites and the Israelites played out again and again across time and over space as monotheistic religions displaced the more local, place-based belief systems, whether it was in the

The Religious Organization of Space

Religion and the Environment

The very first religions were animist, firmly based in place, with sacredness located in specific sites. Some of these belief systems grew into more universal religions that effectively lifted the religious gaze above particular peoples and places to an all-seeing, all-knowing, single God on high. The transition was a gradual grafting rather than a sharp break. Most of the major

10.10 Shrine in downtown Bangkok

destruction of local religious traditions in South America by the Catholic Church or the marginalization of indigenous Native American beliefs by Christian missionaries in North America. Across the world, it is a similar tale of sky-based monotheistic religions displacing more Earth-based polytheistic beliefs. Barbara Ehrenreich makes a convincing case that many religions began more as festive events, ceremonies filled with music and dance, what she terms "**ecstatic worship**." As the religions became connected with establishment power, the authorities sought to control and reduce these forms of ecstatic worship. Christian worship, for instance, was made more formal over the medieval period. When Christian missionaries arrived in the wake of imperialism and colonization, they similarly devalued the ecstatic forms of worship that they encountered.

Organized religions, with their gods and priests and rituals to make sense of the mysteries of life, also grew out of the need to justify and legitimize the social hierarchy in the early city-states. When these empires expanded and came into contact with one another, creative and destructive cultural collision ensued. Karen Armstrong updates the idea of a pivotal "**axial age**" that lasted from 900 to 200 BCE, during which the great religious traditions of the world came into being across the globe—Confucianism and Daoism in China, Hinduism and Buddhism in India, monotheism in Israel, and philosophical rationalism in Greece. Their universal messages of empathy and compassion grew out of the collective experience of living in cities as well as their awareness of people in other cities and regions.

In recent years there has been a renewal of interest in the idea of sacred space as part of modern environmentalism. Such environmentalism draws upon the notion of sacred space to promote environmental protection and a sacred duty to protect and save the environmental bounty that the Creator provided. There is now a significant faith-based approach to environmental issues.

Sacred Sites

In place-based animist beliefs, the landscape was filled with spiritual significance. In the more universal religions, there is a more marked categorization of sacred and profane space. Most of the world was profane, a word whose etymological meaning is "in front of (i.e., outside) the temple." A small number of places were considered sacred sites: birthplaces of prophets, scenes of revelations, centers of religious authority and worship—places where the veil between the material and spiritual world was only lightly drawn. Jerusalem, for example, is a sacred site for the three religions of the Book: Christianity, Judaism, and Islam. The sharing of this sacred space by different religions is part of the political tension in the city (see Box 10.2). Sacred sites can be places of retreat as well as sites of engagement.

Different sects of even the same religion have their own sacred sites. Catholics visit the shrine at Lourdes in search of cures for illness, but Protestants do not. The profane treatment of sacred sites can become a source of friction. In 1984 Indian troops attacked militants hiding out in the Golden Temple in Amritsar, one of the holiest places for Sikhs. The damage incensed the Sikh community and led to sectarian violence and the assassination of the prime minister who ordered the attack. Sectarian rivalry is expressed most vividly in the destruction of these special sites. In 2004, when terrorists exploded a bomb near the golden-domed mosque of Karbala in Iraq, they desecrated one of the most revered sites for Shia Muslims. The explosion triggered a bitter round of sectarian violence in Iraq between Shia and Sunni.

Pilgrimages

Some sacred sites become places of pilgrimage. Catholics in medieval Europe made pilgrimages to the tomb of St. James in Santiago de Compostela in Spain. Pilgrims continue to follow these trails across the Pyrenees and through northern Spain.

Part of the appeal of the pilgrimage is the search for spiritual revelation through physical endeavor. Pilgrimage involves a journey, a move into greater spiritual awareness through physical exertion. In many societies we have retained the sense of the efficacy of physical exertions in the gym and and health club but disconnected it from the spiritual dimension.

The world's largest pilgrimage is the hajj. One of the principal duties of every devout Muslim is to make a pilgrimage to Mecca at least once in their life. Millions converge on the city each year. They circle the Kaaba, a giant rock, seven times counter-clockwise, walk four miles to Mina, then a further eight miles to Mount Arafat where they pray and then collect pebbles. The next day they walk back to Mina, where they throw their pebbles at the pillars that represent the devil, then return to circle the Kaaba seven more times. Because of increased accessibility and cheaper air travel, many more Muslims can now visit the city and make the pilgrimage. Crowds are huge, numbering close to 2 million at any one time and presenting major problems owing to the large number of people in a restricted area over just a six-day period. During the pilgrimage of 2015 more than 760 people were killed and almost a thousand were injured when crowd control measures failed to stop a stampede after crowds coming from different directions collided.

Religious and Secular Space

There is a spatial variation in the importance of religion in people's lives. When people across the world were asked in 2016 what was the most important criteria of self-identity,

10.2   Jerusalem

Jerusalem has a long history. With a fresh water spring making it an oasis in a dry land, and Mount Moriah forming a high point on the landscape, it became a sacred site for early peoples and religions. To this day it remains at the very heart of three of the world's great religions.

Jerusalem figured greatly in early Jewish tradition when Abraham took his son to Mount Moriah to sacrifice him. Under King David it became the political and religious center of a Jewish state. He chose the site because of its sacred appeal but also for geopolitical reasons; it was centrally located among the twelve tribes of Israel but not directly linked to any one of them. King Solomon built a great temple around 1000 BCE. It was destroyed by the Babylonian empire in 587/586 BCE, whose forces not only torched the city and Temple but also carried off its wealthy citizens and many of its skilled craftsmen. Most of the Jewish elite was transported to Babylon where, according to Psalm 137,

> By the rivers of Babylon we sat and wept
> When we remembered Zion.

The Psalm goes on to promise,

> If I forget you, Jerusalem,
> May my right hand forget its skill.
> May my tongue cling to the roof of my mouth
> if I do not remember you,
> if I do not consider Jerusalem
> my highest joy.

The Jews had more luck with the next emerging empire, the Persians, whose great king Cyrus transported them back to Jerusalem and even gave them funds to rebuild the Temple around 520 BCE (something to remember as the Israel–Iran conflict in the present era deepens). Jewish religious writings after this event are imbued with some of the beliefs of the Persian state religion of Zoroastrianism and its cosmology of a world divided into forces of good and evil.

The Temple was refurbished and expanded, and Jerusalem grew, drawing pilgrims from all over Judea. The city came under the rule of the rising empires of the region, the Greeks under Alexander and then the Romans. When the Jews revolted against Roman rule, the city was demolished in 70 CE, the Second Temple destroyed, and the survivors exiled or sold into slavery. The Roman emperor Hadrian rebuilt the city but banned Jews.

Jerusalem also played a huge role in the Christian tradition as the place where Jesus was tried, crucified, and resurrected. When it came under Byzantine control around 313 CE the Christian element was emphasized; the Church of the Holy Sepulchre was constructed, and the city became a place of Christian pilgrimage. In the medieval maps of Europe Jerusalem is often depicted as the center of the world.

The city plays an important role in Muslim theology, too. In 638 CE it came under Muslim control, and in 691 Abd al-Malik built the Dome of the Rock, one of the earliest mosques ever built, on Mount Moriah. Jerusalem is considered the third most holy site after Mecca and Medina, and for a while devout Muslims prayed in the direction of Jerusalem before the orientation was shifted to Mecca.

The city went through a cycle of destruction and rebuilding as different Muslim dynasties changed their attitude to

Sacred sites in Jerusalem

churches and synagogues. Apart from a brief period of Crusader rule, the city came under centuries of Muslim rule and was then part of the Ottoman Empire from 1517 to 1917. Suleiman the Magnificent 1538 rebuilt the city walls, still extant, that define the Old City, divided traditionally into four quarters (Latin Christian, Armenian Christian, Jewish, and Muslim) and containing the Temple Mount, also known as the Noble Sanctuary.

In 1947 the UN voted to create a Jewish and Palestinian state in Israel. The State of Israel was declared in 1948, but Jerusalem sits on the border between Israel and Jordan. West Jerusalem was the capital of Israel while the east was part of Jordan. During the Six-Day War of 1967 Israel captured the Old City and east Jerusalem came under Israeli control. The city has become more distinctly Jewish as settlers moved into east Jerusalem and there is a marked segregation in housing and use of public services, with separate Arab and Jewish bus routes.

Jerusalem is not just a place of streets and buildings; it is also a place of imaginings, longings, hopes, and fears. For Jews it is the place where Heaven and Earth meet, a site of pilgrimage, the center of the universe. For the Christians the city is also a place of pilgrimage as it is intimately associated with the life, death, and resurrection of Jesus Christ—a city made holy. The city plays a central role in the sacred geography of Muslim belief as it ranks only just below Mecca and Medina as a place of pilgrimage and just after Mecca in religious significance.

These sacred geographies superimpose and interconnect in the tight quarters of the Old City. On the 36-acre site a platform, known as the Temple Mount or Noble Sanctuary, is the site of the Second Jewish Temple; the Muslim Dome of the Rock, modeled in part after the Christian Church of the Holy Sepulchre to commemorate Muhammad's ascension to Heaven; and the Al Aqsa Mosque, constructed in 705 on part of the old Jewish temple. At the edge of the platform is the Western Wall, also known as the Wailing Wall, all that remains of the Second Temple and one of the holiest sites for observant Jews.

Jerusalem is a holy place for three religions that have enjoyed periods of tolerance as well as interaction marked by conflict, strife, and outright hostility. The city remains at the center of conflict between Palestinians and the Israeli state.

---

43 percent of those interviewed in Pakistan said religion. This contrasted with only 4 percent in the UK, 5 percent in France, Germany, and Spain, and 6 percent in Canada. Among the richest countries of the world, the US had the highest percentage, 15 percent. And these differences have impacts. An explicitly Christian politician has a hard time being elected prime minister in the UK, while a committed atheist would have a similarly difficult time in the USA. But notice that only 15 percent in the US interviewed cited religion as the main defining criterion. So even in a nation noted for its religiosity, the large majority of people use other criteria such as national citizenship, being a world citizen, local community, race, or culture to self-identify.

Societies and local places vary in their degree of religious tolerance, from fundamentalist societies that allow only one form of devotion to be openly displayed, to more multi-faith societies that show more tolerance of religious diversity. But even in more tolerant societies a tension exists between freedom to practice in private and limitations in the public sphere. All societies regulate religious practices in the public sphere, some more than others. In a number of European countries, restrictions are now being placed on the building of mosques and the public displays of Muslim religious practices. In Egypt Muslim men can marry Christian women but Christian men cannot marry Muslim women unless they convert to Islam.

In societies dominated by one religion, there is a conjoining of public and private space, a centrality of religious space in social and spatial organization. In medieval Europe, the tallest and most central buildings were churches and cathedrals. They towered over the surrounding city as a constant symbol of both religious expression and the central importance of religiosity in organizing human affairs. Myanmar remains a predominantly Buddhist society, and in its capital city of Yangon the most conspicuous building, set high on a hill and easily visible from all parts of the inner city, is the Shwedagon Pagoda, the most sacred site in the country. It was constructed over 1,500 years ago and rebuilt numerous times after earthquakes. It is the defining landmark of the city and the center of Buddhist sacred geography in the entire country. Its centrality also attracts political demonstrations. In 2007 Buddhist monks critical of the junta gathered in the temple as s symbol of resistance to military rule.

The rise of a secular public space is most evident in the founding of the USA, where the constitutional separation of church and state in effect means an explicit secularization of public space. In America and elsewhere, this division is still an arena for debate and conflict. Some, like fundamentalists, still call for the religious control of public spaces, such as Orthodox Jews in Israel asking for separate buses for men and women in public transport. Others, including avowed secularists, demand a more complete erasure of religion from public space, such as calls to ban religious dress in French public schools. The precise separation and demarcation of public and private space, and how it maps religion and religious observance, can be divisive issues in multicultural, multi-faith societies.

Spatial Organization of Religion

There are also the territorial administrative units of religion. The Catholic Church, with its center in Rome, is divided into archdioceses, dioceses, and parishes. In more multicultural societies, the spatial organizing role of religion may be more opaque. An *eruv* is a Jewish ritual enclosure that marks out a space in which Jews observing the Sabbath can carry things and move between public and private space without transgressing Orthodox beliefs. In the Upper West Side of Manhattan, an area of 200 city blocks was separated out using existing walls and fences and specially made cord and wire. Figure 10.11 shows the

10.11 An eruv in New Rochelle, New York

map of another eruv in New Rochelle, New York, centered around the Congregation Anshe Sholom. An eruv is a silent presence, a marker for only some members of the community.

Religious Conflict

Religious conflicts occur for many reasons, but space and place are often important sites and reasons for religious conflicts.

Conflict can occur on the borders of faith communities. Figure 10.12 shows the gradual nature of the shift from Muslim to non-Muslim from north to central Africa. The liminal division is most obvious in nations that straddle this line such as the Central African Republic and Nigeria. The Central African Republic sits at the edge of the division with Chad and Sudan to the north and Congo to the south. Almost 80 percent of its 4.7 million people are Christian and the remainder Muslim. Christians are

mainly sedentary farmers inhabiting the south, while the more nomadic Muslims live in the north of the country. A civil war in 2013 was fought along this religious division, overlain with struggles over land and political power. Muslims in the south of the country moved to its north or overseas to escape violence. The Muslim population of Bangui, the Republic's capital, fell from 138,000 to 900. More than one million people were displaced by the civil unrest and violence.

There is also conflict when different religious groups share the same national space. Even in predominantly Islamic Pakistan there are Christians who officially make up 1.5 percent of the country's 180 million. Other estimates suggest closer to 5 percent. As the country has become more rigidly Islamic, the position of Christians has become more problematic. In 2013 more than 100 were killed by a suicide bomb attack in a Christian church. Christians are at the bottom of the socioeconomic hierarchy, living in slums and forced to work in menial occupations. One million live in Karachi, Pakistan's largest city. One of the largest Christian crosses in the world, 140 feet tall, was erected outside the entrance to the Christian cemetery in the city, where gravestones are often vandalized. The cross not only signifies belief but also is a symbol of hope for a beleaguered minority.

Conflict is heightened when religious differences are compounded by language and ethnic differences such as in the conflict in Sri Lanka between Hindu Tamils and Buddhist Sinhalese. The Tamils live in southeast India and northeast Sri Lanka. After Sri Lanka gained independence in 1948, the Tamils faced discrimination, and in 1983 a militant Tamil group, the Tamil Tigers, attacked a military base. The resultant civil war was a brutal affair with war crimes on both sides, 100,000 casualties, and hundreds of thousands displaced as the war zone moved across the country. Worst hit was the Tamil area in the north. The civil war officially ended only in 2009 with the final defeat of the Tamil Tigers by the Sri Lankan military. In the very last stages of the war, 40,000 civilians died and 300,000 more were displaced and moved into camps.

RATIO OF MUSLIMS TO CHRISTIANS

- 100+ times as many Muslims
- 2–100 times as many Muslims
- Equal numbers
- 2–100 times as many Christians
- 100+ times as many Christians

ATLANTIC OCEAN

INDIAN OCEAN

0   500   1000 Miles
0   500   1000 Kilometers

10.12 Muslims and Christians in Africa

Conflict can also occur between different sects of the same religion. In Pakistan terrorists from the country's majority Sunni Muslim sect (80–90 percent of the population) have waged a campaign against the Shia minority (10–20 percent). Between 1987 and 2007 as many as 4,000 people were killed as Shia were targeted by Sunni terrorist groups.

In Northern Ireland, a long-running conflict between Protestants and Catholics known as the Troubles lasted from 1960 to 1996. Over 3,500 people were killed by Republicans (Catholics committed to breaking away from Britain), Loyalists (Protestant committed to remaining in the UK), and British security forces.

Religious differences are rarely the only source of conflict, but when they are combined with economic differences and social and political marginalization, then religion can become the source of identity and mobilization. Religious differences are not monolithic, unchanging categories. Conflicts are not fixed in time or space, but emerge, rise, and fall in connection with internal and external forces.

Religion and Society

Religion is as much a social phenomenon as a spiritual one. Its practices and beliefs connect with wider social issues. Here we will consider just three: identity, globalization, and gender.

Identity

Religious belief is not a singular form of identity, but is also folded up with other forms of identity that have a spatial component. In Northern Ireland, for example, to be Catholic is often associated with those who support a connection with Eire rather than the UK, while to be Protestant is bound up with a commitment of being part of the UK.

Sometimes religion becomes part of national identity. Although Poland had a significant Jewish population before the Holocaust, it was primarily a Catholic country when the Communists controlled it from 1945 to 1989. Part of Polish resistance to the dominance of the USSR was in and through the practice of Catholicism. Being Catholic was part of Polish nationalism and an act of resistance to Soviet authority.

In some cases national identity is a contested site between competing religions. India, for example, has one of the world's largest Muslim populations: officially 177 million, a figure exceeded only by Indonesia (204 million) and Pakistan (178 million). Yet this Muslim minority is threatened by the rise of a Hindu nationalism that tends to imagine India as a Hindu society. Conflicts between the two communities continue to fracture the society at national and local levels, especially as India's main adversary, Pakistan, proclaims itself as an Islamic Republic.

Globalization and Religion

Religious beliefs are part of cultural identity—for some people, the principal source of their identity. So when they feel their beliefs are under attack or threat, from secularism or other religions, the political backlash can be profound. We are in an interesting time with powerful competing forces. On the one hand, we are in the middle of a major wave of cultural globalization and the expression of religious diversity in shared national spaces; on the other, we are also witnessing a rise of religious fundamentalism, whether in an Islam that promotes sharia law based on the Koran, a Hinduism strongly associated with Indian nationalism, strict Judaism that sees occupation of the West Bank as part of God's covenant, or Christians who argue for the Ten Commandments as the guiding light of civil law. Geographer Roger Stump provides geographical perspective on religious fundamentalisms with reference to the control of scared space—the attempt to conjoin public and private space with religious structures, and the association between territory and identity.

Religious belief provides many with a simple message in a complex world, an articulation of fixed primary identity when cultural change is everywhere, a platform and vehicle for political mobilization, and an anchor of meaning and relevance in a society of ever-accelerating transformation. Yet these permutations are also important generators of religious fundamentalism. At a more global scale, we are now in an era of both religious fundamentalism and religious diversity. The recent wave of globalization has spread and diffused different religions around the world.

Globalization also involves the movement of people and ideas, including religious beliefs and practices, around the world. Religions are diffused across the globe but become reterritorialized in specific sites. Diasporas are often constructed and maintained in and through church gatherings and religious festivals. For example, Filipino domestic workers in Italian cities are largely invisible in the public arena, but each year in the city of Padua, as well as in Berlin, Germany, and New Orleans, USA, they celebrate their own festival in the streets. The Filipino diaspora is maintained and Filipino identity is performed by the annual religious celebration known as Santacruzan. By appropriating public space through a religious ritual, Filipino immigrants make themselves visible to reconfirm their identity and renew their community. Likewise, immigrants from the Democratic Republic of Congo connect in foreign places in and through their churches. The Congolese diaspora in Atlanta and London have through their own Pentecostal and Catholic churches. The religious landscapes of both cities, tweaked by Congolese immigrants and their religious practices, are shaped by local circumstances. David Garbin describes the spatial appropriation and re-enchantment of the urban landscape as the Congolese often situate their churches at the margins, in former industrial areas and warehouse districts.

Diasporas are also recast as religious identities if the religious practices of immigrants become their defining feature, either in their own eyes or those of the host society. When Turks living in Germany practice Islam, they are seen—and start to see themselves—primary as Muslims rather as Turks or Turkish Germans. They become less Turkish and more Muslim. Distinctly "Muslim" spaces are created by immigrants from many different counties. In British cities, immigrants from a variety of Muslim nations such as Algeria, Afghanistan, Bosnia, Tunisia, and Iraq find collective religious homes and identity in the mosque and the madrassa, their shared religious identity now trumping their different national origins.

Social Differences: Gender

Religion embodies, defines, and reflects social differences. Religion can play an enormous role in differentiating men and women, as the social differences are often codified in religious beliefs and practices. Many religions promote segregation between men and women and institutionalize gender differences in leadership roles, ordination, spatial behaviors, dress, and comportment (see Figure 10.13). The Catholic Church, for example, does not allow women to be priests. In effect, men run the church and women play a secondary role. Men fill all the positions of real power, from pope, cardinal, and archbishop to bishop and priest. In traditional Islam women's lives are shaped by religious strictures on dress and public behavior, codified in states such as Saudi Arabia where women must remain veiled in public and are not allowed to drive.

Many feminists decry formal religions as patriarchal structures that can lead to the oppression of women. The monotheistic religions do have a patriarchal character, often assigning women restricted social and religious roles, defining their functions for a private rather than public space.

Gender is perhaps the most obvious example of how religious beliefs and assigned spatial behaviors reinforce gendered social differences. Can you think of others?

Witness to Power

As social institutions, religions embody and reinforce, yet sometimes overturn and question, social hierarchies, gender differences, and socioeconomic power relations. The antislavery movement in the UK and the USA was in large part driven by Christians appalled at slavery and its exploitation of fellow humans. Abolitionists felt compelled by their religious beliefs to work to end the slave trade. The more recent civil rights movement in the USA was in large measure advanced by religious groups, especially by African American religious institutions and leaders who felt that racial discrimination was evil. Religion has reinforced existing power arrangements but has also been a vehicle for undermining, contesting, questioning, and sharply criticizing authority. The Catholic Church throughout much of Latin America's history often buttressed the existing social order. But the region also gave rise to progressive liberation theology that speaks to the plight of the poor and the dispossessed. The movement has its martyrs; Oscar Romero, the archbishop of San Salvador, and a critic of El Salvadorean elites and US-backed military regimes, was gunned down at the altar in 1980 while celebrating Mass. He was made a saint by the Vatican in 2015.

From Buddhist monks in Myanmar protesting against the military junta to Catholic priests protecting the poor in the slums of South America, and Muslim feminists in Iran fighting for women's rights, religious believers are using their faith as a guide and justification for profound social criticism.

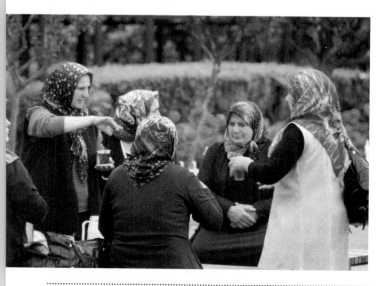

10.13 Women in Istanbul, Turkey

Cited References

Armstrong. K. (2006) *The Great Transformation: The Beginning of Our Religious Traditions*. New York: Knopf.

Bauer, J. T. (2012) U.S. Religious Regions Revisited. *The Professional Geographer* 64: 521–539.

Ehrenreich, B. (2006) *Dancing in the Streets*. New York: Metropolitan.

Garbin, D. (2014) Regrounding the Sacred: Transnational Religion, Place Making and the Politics of Diaspora Among the Congolese in London and Atlanta. *Global Networks* 14: 363–383.

Gill, S. (1987) *Mother Earth: An American Story*. London and Chicago: University of Chicago Press.

Gregg, R. C. (2015) *Shared Stories, Rival Tellings*. Oxford: Oxford University Press.

Kuppinger, P. (2014) Flexible Topographies: Muslim Spaces in a German Cityscape. *Social and Cultural Geography* 15: 627–644.

Martin, C. (1978) *Keepers of the Game: Indian-Animal Relations and the Fur Trade*. Berkeley: University of California Press.

Saint-Blancat, C. and Cancellieri, A. (2014) From Invisibility to Visibility? The Appropriation of Public Space Through a Religious Ritual: The Filipino Procession of *Santacruzan* in Padua, Italy. *Social and Cultural Geography* 15: 645–663.

Stump, R. W. (2000) *Boundaries of Faith: Geographical Perspectives on Religious Fundamentalism*. Lanham, MD: Rowman and Littlefield.

Select Guide to Further Reading

Becci, I., Burchardt, M., and Casanova, J. (eds) (2013) *Topographies of Faith: Religion in Urban Space*. Leiden: Brill.

Bergmann. S. (ed) (2014) *Religions, Space and the Environment*. London: Transaction Publishers.

Collins-Kreiner, N. (2010) The Geography of Pilgrimage and Tourism: Transformations and Implications for Applied Geography. *Applied Geography* 30: 153–164.

Gottlieb, R. S. (ed) (1996) *This Sacred Earth: Religion, Nature, Environment*. New York: Routledge.

Heine, S. (2012) *Sacred High City, Sacred Low City: A Tale of Religious Sites in Two Tokyo Neighborhoods*. Oxford: Oxford University Press.

Hopkins, P., Kong, L., and Olson, E. (eds) (2013) *Religion and Place: Landscape, Politics and Piety*. Dordrecht: Springer.

Ivakiv, A. (2000) Toward a Geography of "Religion": Mapping the Distribution of an Unstable Signifier. *Annals of the Association of American Geographers* 96: 169–175.

Krech, S. (ed) (1981) *Indians, Animals, and the Fur Trade: A Critique of* Keepers of the Game. Athens: University of Georgia Press.

McMahan, D. (2008) *The Making of Buddhist Modernism*. Oxford: Oxford University Press.

Morin, K. M. and Guelke, J. K. (eds) (2007) *Women, Religion and Space*. Syracuse, NY: Syracuse University Press.

Orsi, R. A. (ed) (1999) *Gods of the City: Religion and the American Urban Landscape*. Bloomington; Indiana University Press.

Shelton, T., Zook, M., and Graham, M. (2012) The Technology of Religion: Mapping Religious Cyberscapes. *The Professional Geographer* 64: 602–617.

Sopher, D. E. (1967) *Geography of Religions*. Englewood Cliffs: Prentice-Hall.

Stump, R. W. (2008) *The Geography of Religion: Faith, Place and Space*. Lanham, MD: Rowman and Littlefield.

Watson, S. (2005) Symbolic Spaces of Difference: Contesting the Eruv in Barnet, London and Tenafly, New Jersey. *Environment and Planning D: Society and Space* 23: 597–613.

Wellman, J. K. Jr., and Corcoran, K. E. (2013) Religion and Regional Culture: Embedding Religious Commitment Within Place. *Sociology of Religion* 74: 496–520.

Wilford, J. (2012) *Sacred Subdivisions: The Postsuburban Transformation of American Evangelicalism*. New York: NYU Press.

Wilson, A. (2008) The Sacred Geography of Bangkok's Markets. *International Journal of Urban and Regional Research* 32: 631–642.

Websites

Religion in Everyday Life in the USA
http://www.pewforum.org/2016/04/12/religion-in-everyday-life/

The Future of World Religions
http://www.pewforum.org/2015/04/02/religious-projections-2010-2050/#projected-growth-map

The Gender Gap in Religion
http://www.pewforum.org/2016/03/22/the-gender-gap-in-religion-around-the-world/

Interactive Map Websites

http://www.pewforum.org/2015/04/02/religious-projections-2010-2050/pf_15-04-02_projectionsoverview_worldgrowthrate640px/

How Religious Is Your State?
http://www.pewresearch.org/fact-tank/2016/02/29/how-religious-is-your-state/?state=alabama

Visualizing Gender Differences in Religious Commitment Around the World
http://www.pewforum.org/Interactives/Religion-Gender-World/

Learning Outcomes

- There are three types of religion: animism, religions that emerged in Asia, and monotheistic religions that developed in the Middle East.
- The major religions have diffused out from their origin areas.
- The "axial age" from 900 to 200 BCE saw the creation of the great religious traditions of today, including the two main hearth areas of northern India (Buddhism, Hinduism, and Sikhism) and the Middle East (Judaism, Christianity, and Islam). All of these spread throughout the world.
- There are substantial differences within the major religions such as the Sunni–Shia divide in Islam or the Catholic–Protestant divide in Christianity.
- There are also major differences within certain countries in terms of religious belief. The US, for example, has a rich range of religious diversity.
- Religious beliefs and worship remain strong in cities around the world.
- The environmental movement is informed by religious beliefs.

- The profound relationship between religion and space is expressed in the notion of sacred sites, the religious organization of space, and the spatial expression of beliefs and practices.
- All religious have special areas, sites considered sacred. These are honored and visited in pilgrimages.
- The division between religion and secular space can be a contentious issue in many countries and cities.
- Religious conflicts can occur over space and about place.
- Religious conflicts are most severe when they overlap with differences in wealth, access to power, race, and ethnicity.
- Religion is an important source of personal identity, although its importance varies around the world.
- Globalization is an crucial background to the rise of religious fundamentalism around the world.
- Religion plays an important role for immigrants in securing community ties and maintaining identity.
- Religion can play an important role in assigning and reinforcing social roles such as gender differences.
- Religious beliefs are used to legitimize but also to question existing social relations.

The Geography of Language

Language is an important part of culture that is intimately connected to place and space. We make sense of the world by naming places and spaces. Language is rooted in place but also emerges from **spatial diffusion** and contact. Language can be a hard medium in which to lock up, in sounds and words, our understanding of the world, but it is also fluid enough to change and reflect changing circumstances. Language embodies the fixity as well as the flow of cultural geographies.

Language of the Local

Language is connected to specific places and local geographies. Indigenous peoples of the northern latitudes have a rich variety of words for snow. Igor Krupnik and colleagues at the Smithsonian studied the vocabulary of ten Inuit and Yupik dialects. They found that Inuit speakers in Canada's Nunavik region have fifty-three words for different types of snow, including a word for wet snow that can be used to ice sled runners. Speakers of a local dialect around Wales in Alaska have seventy different terms for sea ice. When life is firmly based on local conditions, language has to become a very fine filter of environmental understanding. The Sami people of Finland and surrounding countries have 180 words related to snow and ice and a thousand separate words to describe reindeer. Indigenous languages are rooted in place and highly sensitive to the nuances of the local environment.

Local places are the context and setting for a multitude of languages. In the premodern world, there were many more languages spread widely across the surface of the habitable globe. Figure 11.1 depicts the complex linguistic map of indigenous languages of the Pacific Northwest. Notice the mosaic of many language families across this one single region.

We can focus in on one particular group, known to themselves and others as the Nooksack. They are part of a larger group of people known

11

LEARNING OBJECTIVES

Summarize the various influences of culture and locality on the development of language.

Consider how languages get distributed through global political and economic interactions.

Describe the differences that emerge in spoken languages and the variation in usages across assorted spaces.

Discuss the important role of language in national identity and politics, with consideration of the spoken and written word.

Reflect on the reasons for contemporary losses in linguistic diversity and the ways that populations are responding.

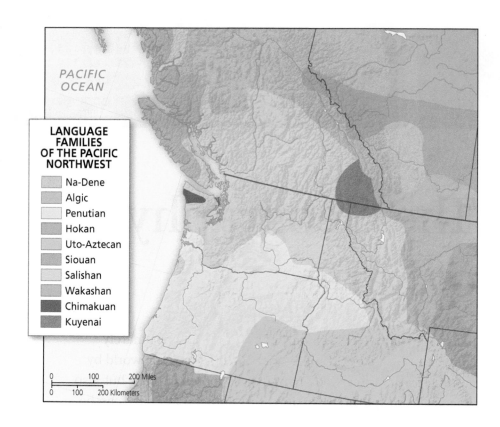

11.1 Indigenous languages of the Pacific Northwest

LANGUAGE
FAMILIES
OF THE PACIFIC
NORTHWEST
- Na-Dene
- Algic
- Penutian
- Hokan
- Uto-Aztecan
- Siouan
- Salishan
- Wakashan
- Chimakuan
- Kuyenai

as Salish, part of the Salishan language group shown in Figure 11.1 The Nooksack were a **speech community** comprising a group of villages located around the Nooksack River. The population was estimated at around 1,125 in 1800. In 1855 they agreed to give up the title of their land in return for fishing and hunting rights and a guarantee of government services. Government attempts to keep them in reservations were fiercely resisted. Some of the Nooksack eventually gained formal title to their traditional land. In recent years, as part of their cultural survival, the tribe created an inventory of cultural sites. Allan Richardson and Brent Galloway uncovered the rich connection to the land through patient documentation and discussion with Nooksack elders about place names.

There are profound differences between the names given by English settlers/speakers and the original Nooksack. Many of the English place names refer to people. Thus Johnson Creek was originally Koy-yoti meaning "former" or "in the past." Many of the English names sometimes suggested positive feelings, in order to encourage settlement, such as Happy Valley and Eden Valley. The original Nooksack names, in contrast, focused on environmental factors and the flora and fauna used for food. Thus the Nooksack names for what were called Ryder Lake and Ryder Creek in English referred to "many dead trees." Fairhaven Creek was originally Xwsisel7echem (yes, that's the name and not a typo) and it means, literally, "place of always finding salvage in the back" (as of finding a drifted canoe). To recover the original Nooksack names is to reveal a complex and sophisticated way of understanding and describing the environment.

The Distribution of Languages

Roughly 6,000 to 7,000 languages are spoken around the world. Their distribution is uneven. The highest level of linguistic diversity is found in the equatorial region of the world, such as Papua New Guinea and southern Nigeria, which respectively have 820 and 516 distinct languages. In the tropical areas, climate allows year-round agriculture. Since there is less need for trade to secure food supplies, it is easier for groups to maintain their languages. Papua New Guinea is particularly diverse in languages not only because it is tropical, but the rugged terrain and steep mountains make movement difficult, allowing very small and localized speech communities to remain intact.

Where places are less self-sufficient and/or engaged in more extensive trade, languages of dominant groups are diffused across a wider area. The Hausa language, for example, spread across the savannah areas of West and West Central Africa. Today more than 39 million people speak the language.

Trade links spread particular languages and enabled the emergence of multilingualism as people needed to communicate with those of other language groups. Even in the more isolated villages of Papua New Guinea, other languages are spoken and understood. A study of one village showed that everyone over the age of ten spoke two languages and could understand at least one more. In other words, multilingualism was probably a very common feature of the premodern world.

Table 11.1 indicates the main language groups. Half of the world's population speak only ten languages. The table also shows the second language usage for these dominant languages. Notice how some languages have little traction outside their core **cultural region**. The number of people who speak Japanese as a second language is less than 1 percent of those who use it as a first language. Similarly, while 207 million people speak Bengali as their first language, only 4 million more speak it as a second language. Compare this with English, which is more widespread outside of the core area of England. We can make a distinction between dominant languages with large numbers of speakers and **global languages** that reach outside their primary language area. While both English and Spanish have similar numbers of primary speakers, making them a global language in terms of size, more people speak English as a foreign language than Spanish, making English an extensive global language. We will consider the practice of global English later in this chapter.

Figure 11.2 shows the current distribution of dominant languages in the world today. Notice how certain languages are highly localized in particular countries: Chinese in modern-day China and Hindi in India. Others are more diffuse, such as Arabic, spoken all throughout the Middle East and across North Africa.

TABLE 11.1 Ten Most Spoken Languages in the World		
	NUMBER (MILLIONS)	
	AS FIRST LANGUAGE	**AS FIRST AND SECOND LANGUAGE**
Chinese	1,146	1,343
Hindi	366	487
Spanish	358	417
English	341	573
Arabic	280	n/a
Bengali	207	211
Portuguese	176	191
Russian	167	277
Japanese	125	126
German	121	128

Language and Power

Language is a good marker and a permanent legacy of the spread of empires, which diffused language use around the world. The Roman Empire created the first pan-European language of Latin, the language of the dominant elites, which became the accepted linguistic current of the expanding empire. As the empire declined, flow and spread was then replaced by place and stasis. Local variations became more important, and we can make a broad distinction between the Romance languages at the center of

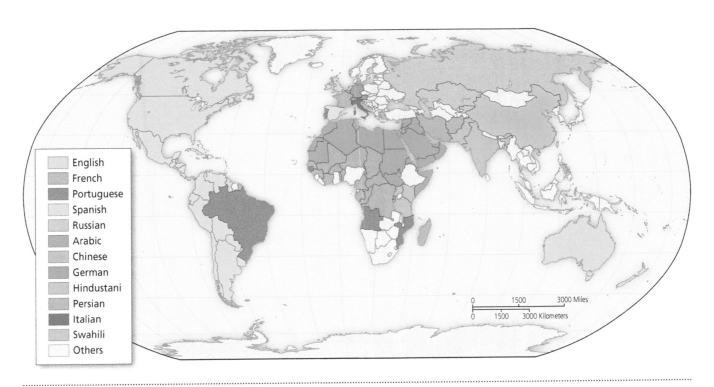

11.2 Dominant world languages

English
French
Portuguese
Spanish
Russian
Arabic
Chinese
German
Hindustani
Persian
Italian
Swahili
Others

the empire and the languages not descended from Latin, such as Old Norse, German, and English, that emerged at the periphery. We retain this linguistic legacy in the names of the days of the week. Days in the Romance languages have names that derive from the name of the Roman god Jupiter: *jeudi* (French), *jueves* (Spanish), and *giovedi* (Italian). In contrast, English still draws on the Old Norse name of Thor to designate the fifth day of the week as Thursday. The day is similarly named *torsdag* in Danish, Norwegian, and Swedish. The closer to the heart of the old Roman Empire, the stronger the linguistic legacy of Latin.

Empires spread language. The language areas shown in Figure 11.2 are in large part maps of imperial expansion. Arabic is spoken across the Middle East and North Africa in a direct mapping of the spread of the Muslim empire out from its core region in Saudi Arabia. French, English, Spanish, and Portuguese are European languages that diffused to appropriated regions around the world. Spanish spread into South America and Portuguese into South America and Africa.

Language and language use reflects colonial expansion and **imperial incorporation**. We can imagine a three-level linguistic model of colonization of increasing linguistic impact. First, in small trade colonies such as the Dutch trading port in Japan, there was limited linguistic contact outside the enclave. Second, in a situation of colonial exploitation such as mining and planation economies (e.g., the French in Indochina or the British in the Caribbean), segregated elites spoke the language of the imperial center. However, as locals learned the language of the powerful, the dominant language percolated down through the local society. Last, in settler colonies, immigrants from the home country brought their own languages, which became dominant and replaced a range of local ones. This is what happened with English in Australia, Canada, and the USA.

When the language diffused out from its home base it was adopted but also tweaked. Many settler societies spoke English, but subtle differences evolved in English usage between Australia, England, and the USA. Similarly, as the French expanded their commercial and political empire and gained colonies, local elites soon learned to speak the language of the dominant. But the French of the center was used alongside the local languages, and the result was the emergence of various forms of spoken French (see Figure 11.3).

Naming the World

The naming of things is how we humanize the world. Naming gives shape and meaning to our world; it turns space into place. Naming is never innocent of politics.

The naming of the Earth's surface is shaped by three basic toponymic eras, indigenous, colonial, and postcolonial. In the first, names are given by indigenous people. Consider the case of Australia, where there is still legacy of Aboriginal names. Cities such Wollongong remind us of an Aboriginal past. Then there are the colonial names. In Australia the principal cities of Sydney, Melbourne, Brisbane, Adelaide, and Perth all refer to names from English aristocracy. We live in a postcolonial period in which we are aware of the indigenous legacy as well as the

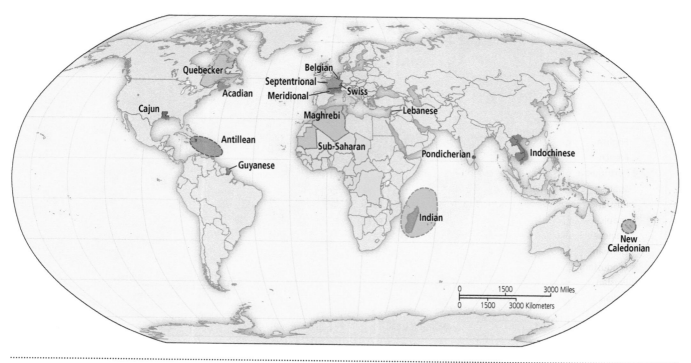

11.3 Dialects of the French language in the world

colonial rewritings. In the central Australian city of Alice Springs, the long-silenced Aboriginal name of Mparntwe is now used.

Naming the Seas

Three things make naming the seas a much more distinctive element than the naming of land features. First, large bodies of water are shared spaces, often surrounding or surrounded by different national territories. The east coast of China faces the west coast of the US; England's south coast is France's northern coast. Thus there is no simple hegemony over naming rights as in the case of land surfaces, which can make this a contentious issue.

We are more concerned with things closer than farther away; it is easier to agree with other countries when things are far away. Thus both Germany and the UK both refer to the Indian Ocean (*Indischer ozean* in German); it is far away from either country and both share the same toponymic history of the area. However, closer to home, the Baltic Sea in English is referred to in German as the East Sea (*Ostsee*). The English name probably draws upon Russian (*boloto* = marsh) or Lithuanian (*baltas* = white), while the German refers to its location relative to Germany.

For very large bodies of water with numerous landmasses and hence a variety of different nation-states involved, the indigenous names can be so varied and so many that colonial names become the standard. Take the case of the large body of water we call the Pacific. It probably had a rich variety of names as indigenous people named it in their own languages. After the sixteenth century it was opened up to European colonial trade and mercantile interests. In English it was named with reference to Europe and was originally called the South Sea. When Magellan crossed the ocean in 1520–21, he encountered no storms and named it Mar Pacifico. As the dominant global power of the time, the Spanish name displaced all the indigenous names and thus persists to this day. It persisted both because of the continuing Spanish legacy in the region and also because it was an easy solution to the complexity of many competing indigenous names. Colonial namings often replace a myriad of indigenous names. The larger the number of divergent indigenous names, the greater the force of singular colonial namings.

Second, territories try to exert greater influence on the naming of seas closer to them. National imaginaries cast a stronger shadow over more proximate territories. States rarely argue over the names of seas far from their shores. But seas close to their coastline are part of the national imaginary and the nation's sense of itself.

Third, conflict is more pronounced when there are only a few indigenous names. We can compare the Pacific Ocean, a vast sea body with many different indigenous communities and nation-states, with smaller bodies of water surrounded by fewer nation-states. The East Sea/Sea of Japan is a relatively small body of water surrounded by only three countries, Korea (North and South), Japan, and Russia, for whom it was at the edge of empire. The principal proximate interests are thus only Japan and Korea. The Japanese government prefers to use the name Sea of Japan while the South Korean government actively pushes for a change to a double naming system of East Sea/Sea of Japan to acknowledge the fact that the sea was known to Koreans as the East Sea for centuries. That a colonial relationship exists between the two countries makes the naming controversy all the more intense.

Two principles are important in naming seas: ensuring global intelligibility and a postcolonial sensitivity. We live in a postcolonial world, and the names of geographic features should reflect this fact.

11.4 Part of the Kata Tjuta range in central Australia

Toponymic Colonialism

We can identify a **toponymic colonialism** whereby colonial powers rename the landscape, replacing local words in a act of naming that is also an act of appropriation and control. Here are two examples.

First, consider the case of central Australia, long the ancestral home of the Arrernte people. In the late nineteenth century, white British Australians extended control into the region. The names of local features were replaced with English names. Table 11.2 shows the changes. The term "Gap" is quite literally a gap in the long MacDonnell Ranges, a dramatic feature of parallel red quartzite ridges that reach up to 3,000 meters at their highest point and stretch for 400 kilometers across the stark, eroded desert landscape. A white explorer, John McDouall Stuart, replaced the Arrernte description with the name of the governor of South Australia, Richard MacDonnell. Table 11.2 shows some of the other name changes that took place. The dramatic rock outcrops in the region were named Ayers Rock and the Olgas to commemorate white explorers and their wives. However, the indigenous terms Uluru and Kata Tjuta are now the preferred names (Figure 11.4).

Another example, this time from Asia: Korea became a Japanese protectorate in 1905 and came under formal Japanese control in 1910. Japanese colonial domination took many forms; among these, they suppressed the Korean language, especially in name usage. In 1939 Koreans were encouraged to change their surnames to Japanese forms. This linguistic colonialism was also evident in maps. Japanese place names appeared in maps of Korea from 1900 onward. Maps of Seoul showed the name Keijo or Kyongsong. From 1914 a new Japanese naming system was applied to maps. New administrative place names were adopted after annexation that added Japanese generic names such as Tong, Jeong, and Jeongmok or directly copied Japanese names

such as Bonjeong, Gilyajeong, Hwang Gumjeong, and Taepyong Tong. Roads and natural features were renamed in Japanese style. As well as Japanese inscriptions, there were Korean erasures. In Seoul, for example, 186 districts with Japanese names replaced 661 administrative districts with Korean names. Larger regional features were also renamed. The sea between Japan and Korea had multiple names through the centuries, predominantly named by the Koreans as East Sea. After annexation, as part of the new wider Japanese empire, it was named the Sea of Japan.

The process of **decolonization** also involves a renaming. As colonies became independent, one of the their first acts was to rename the landscape, erasing the colonial heritage and replacing it with more local names. We have already noted in chapter 6 the example of Tunis, where the main avenue originally named by the French as Promenade de la Marine was renamed Avenue Habib Bourguiba after Tunisia achieved independence.

After Vietnam achieved independence from France in 1954, the French-named streets were Vietnamized. Rue Catinat, named by the French authorities after a warship that consolidated their power in the region, was renamed as Freedom Street in Vietnamese. Name changes also occur

TABLE 11.2 Toponymic Colonialism in Central Australia	
ARRERNTE NAME	**ENGLISH NAME**
Mparntwe	Alice Springs
Lhere Mparntwe	Todd River
Akeyulerre	Billy Goat Hill
Anthwerrke	Emily Gap
Atnelkentyarliweke	Anzac Hill
Ntaripe	Heavitree Gap
Ntyarlkarle Tyaneme	unnamed ridge
Uluru	Ayers Rock
Kata Tjuta	The Olgas

11.5 Commercial street sign in Hoi An, Vietnam

with a change in regime. After the Communists took control of all of Vietnam in 1975, Saigon became Ho Chi Minh City and Freedom Street was changed to Uprising Street in Vietnamese. However, in Vietnam and across the world today, there is a commercial bilingualism as street signage displays things in English for foreign travelers (Figure 11.5). Although countries mark their independence by renaming places, streets, and cities in their own language, they also connect with global consumers by using English.

There are some holdovers. Georgetown remain a popular name for towns across the world, including an affluent neighborhood and a university in Washington, DC, a distant reminder in an independent country of when King George III ruled over the American colonies.

Contact Languages

Trade and especially the spread of commercial and political empires brought different languages into contact. New language forms emerge in the interaction of people with different languages. Various forms of **contact language** can be identified. A **pidgin** is a simplified language for communication between different languages. Pidgins developed as the British Empire extended its trade and political links around the world. West African Pidgin English, for example, was the contact language along the West African coast. It was also known as Coast English. Remnants of the language can be found in the more mountainous parts of Jamaica.

Creoles are formed in the contact between dominant and local languages and are adopted permanently by a group of people and passed on to their children. Various creoles can be identified. Consider two cases from the Caribbean—a melting pot of African, American, and European languages.

Papiamento is one of the two official languages of the island of Aruba; the other is Dutch. Papiamento is a creole language that draws on Amerindian, different African languages, Portuguese, Dutch, Spanish, and English. The pidgin stock of Papiamento is the simplified language that developed in the interaction between Portuguese slavers and African slaves and slavers. Later additions to the language embody a pre-Columbian linguistic heritage, Spanish control of the island for a hundred years and, since 1636, Dutch control. With its rich mix of languages from three continents, Papiamento embodies the complex history of the region.

The Garifuna are a distinct ethnic and language group emerging from the slave trade. West Africans, shipwrecked off the coast of Saint Vincent, intermarried with indigenous Caribs who had been shipped to the island by the French. In 1763, the British gained control over Saint Vincent and again the Caribs resisted. The British defeated them in 1796 and deported around 5,000 people, separating the more African-looking Caribs from the Amerindian-looking ones and banishing them to an island off the coast of South America. Only around 2,500 survived the voyage. The people soon moved off the island and settled along the Caribbean coast. They are now known as Garifuna and number close to 600,000. They are found in Honduras and Guatemala and, with subsequent migrations, also in New York City and Los Angeles. Their language derives from the Amerindian languages of Arawak and Carib as well as English and French, and combines Arawak grammar with African, English, and Spanish loan words.

Rather than being unitary, creoles are probably more accurately considered as a continuum where people can register different dialects according to the setting. All languages, not just creoles, have a fluid quality. Language use varies across space and place.

The Language of Place

Figure 11.2 is deceptive. It shows dominant languages as singular phenomena, covering entire spaces in a homogeneous categorization. In reality, of course, language is far more complex and interesting. There are even differences in the way we speak the same language. Although Table 11.1 uses the term "Chinese" as a singular category, different types of Chinese are spoken including Mandarin (1.052 billion as first and second language) Wu (77 million), and Yue (71 million). The category of English covers a rich variety of different types of spoken English from Scotland to Australia and from Jamaica to the USA.

Let us look at some of the variation in more detail.

The Regional Geography of Language

Spoken language varies by place. Linguists identify the variation by mapping the enunciation of the same word or plotting variant descriptions of the same phenomenon. For example, small lobsters are known as crawfish in the lower Mississippi Delta but as crayfish in New England and crawdads in the middle Mississippi Region. In a similar vein, pop, soda, and Coke describe the same thing in different places in the USA.

From this work it is possible to generate **dialect maps** that show how a single spoken language varies over place. In countries with centuries of settlement such as China there are very distinct regional dialects even within Mandarin or Cantonese Chinese. British dialects have evolved over close to five hundred years. You can take a fascinating dialect a tour of UK by visiting this website: https://www.youtube.com/watch?v=-8mzWkuOxz8. The accents are enunciated as the cursor moves over a map of the country. It is fascinating to see how speech changes over space, giving places a distinct dialect. Dialects are the localization of language. In Scotland, for example, the accent of Glasgow has much harder consonants than the softer, singsong dialect of the Highlands. My own accent is from the rural central region of Scotland. When I was a young boy and visited the big cities of Edinburgh and Glasgow, my accent marked me out as someone from the rural world. Figure 11.6 shows a sign in the written form of the spoken dialect of the Shetland Islands in the very far north of Scotland. Can you guess what it says?

In the UK, dialects are often compared with standard speech, effectively the speech of the elites of London and Southeast England. The term "Queen's English," for example, describes the standard form and hints at its elite status. Dialects are not innocent of social connotations. Indeed, what is considered a dialect and what is considered the standard form is a political act of categorizing people in a social hierarchy based on how they speak. The definition of dialect and standard speech codifies social differences and political power.

As a young boy growing up in Scotland, when I listened to the national radio or television I never heard anyone who sounded like me, my family, or any of my friends and neighbors. A more formalized English was spoken, called "Received Pronunciation." Scots' accents were definitely

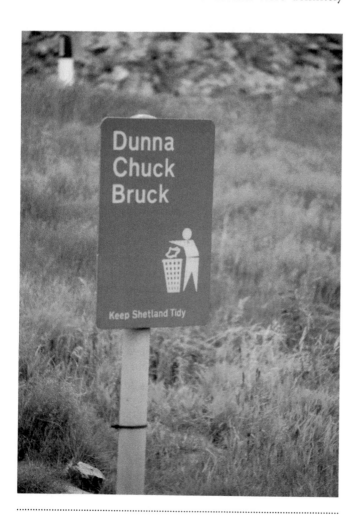

11.6 A sign in the Shetland dialect of Scotland

restricted to the periphery and the marginal and were denied expression in the national media. More recently, however, there has been conscious acceptance of regional dialects and different forms of spoken English. When you listen now to the BBC, you hear a wider array of accents.

Even in the USA, which is considered more linguistically homogenous, at least twenty-four different dialects have been identified among English speakers. The region depicted as the Southwest for example, contains elements of the Mexican dialect of Spanish. Along the Carolina and Georgia coastline, Gullah is identified in Figure 11.7 as regional dialect. It is in fact a creole language. More than a quarter million people speak Gullah, mainly descendants of slaves brought from Africa who developed a language based on African roots and the Coast English of West Africa blended with local English. It is language that reminds us of the slave trade that connected the USA with Africa.

There are also localized differences in how people use the same language. One study compared speech patterns of New Yorkers with non–New Yorkers. The former tended to exaggerate intonations and quickly uttered questions. The language of the New Yorkers was defined as high involvement while that of the non–New Yorkers was characterized as high considerateness.

Figure 11.7 assumes homogeneity by region. In fact there are variations within regions. We can identify speech communities composed of people who share a common language. One such community is composed of about 200,000 people in Brooklyn and the South Bronx who speak Garifuna, a language we have already encountered as a creole language from the Caribbean. Speech communities survive best with large numbers of speakers, dense internal social networks, and limited external connections. Speech communities with a small number of speakers are vulnerable. With few or weakened internal connections, as people die or move away, speech communities lose their linguistic oxygen. When members of a speech community have more contact with external institutions, people, and practices, the localized language can be overwhelmed and eventually displaced. Some physical environments, such as dense inner-city neighborhoods, can solidify speech communities. So Garifuna, with significant numbers of speakers in dense urban neighborhoods and strong internal connections, survives even as members of the community, and especially their children, interact with the wider community.

Language, Place, and Social Differences

Figure 11.7 is also deceptive. It implies variation only by location. Language variation occurs not only by location but also by social difference such as gender, age, and social class. Indeed, social differences are often embodied in different ways of speaking.

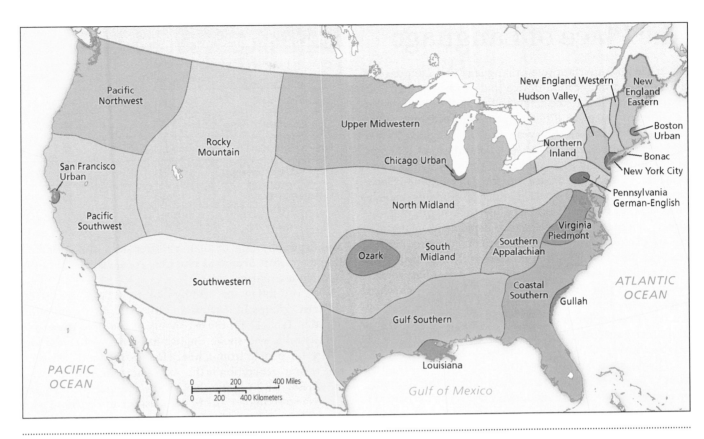

11.7 Dialect map of the USA

Place and socioeconomic status combine to produce social geographies of language. The higher socioeconomic groups tend to speak a similar national language despite their various locations, while further down the socioeconomic hierarchy, the national language can vary more by distinct regional dialects. Figure 11.8 shows the pattern for the UK. Regional variations in speech patterns are most marked at the lower levels of the hierarchy.

Speech patterns also reflect and embody socioeconomic differences, even within the same city. A pioneering study by the social linguist William Labov identified the different speech patterns of different socioeconomic groups in New York City. Language varied by pronunciation and in the range and type of words used. The work of sociolinguist Deborah Tannen shows how speech patterns also vary between men and women. Holding other things constant, women use less slang than men. Language also varies by age, with distinct youth cultures often having their own forms of speech.

We have a complex geography of language as people speak the same language differently depending on where and who they are. Despite the linguistic homogeneity implied by language maps such as Figure 11.2 and the official language policies of national governments, the linguistic reality is that most countries are populated by people speaking a variety of

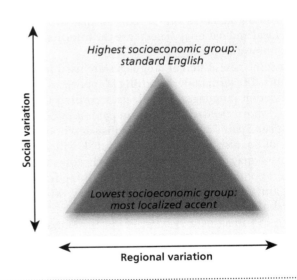

11.8 Social and regional accent variation

languages. People use different languages and even the same language in different ways in different contexts. A variety of distinct language domains can be identified, such as work, school, and home. Adding to this diversity are differences in language use between public and private, and by age, socioeconomic status, and gender.

The Place of Language

Place has an important role in language use. Language is a flexible communication device that can be tweaked and changed according to the context. We can speak differently in different places to different people in different social settings. Even people who speak only one language have a wide repertoire of language use, a continuum of different words, phrases, and intonations that can vary by place and domain. A **language domain** is defined in relation to place, role relationship, and topic. Language domains include such places as home, workplace, neighborhood, and the street. Language is a performance dependent on the role we are playing (parent, teacher, student, coworker, friend, stranger) and the domain. I speak differently with my friends compared to my colleagues, and my accent hardens when I telephone or visit with family in Scotland. It softens when I speak with my family in California. A distinction can be made between a formal domain, such as church sermons and official speeches, and a more informal domain of everyday conversation and e-mail between friends.

Disglossia refers to the existence of two or more varieties of language in a speech community. In the German-speaking parts of Switzerland, two varieties of German are used, often referred to as High German and Swiss German. High German is spoken by people from Germany; there are many German immigrants working in Switzerland, but local Swiss will often speak to each other in Swiss German. In much of the Arabic-speaking world, Classical Arabic is spoken by the elites and in more formal surroundings, while local and national dialects are the informal language of the home and street.

In some case different languages are used in different domains. The term **code switching** is applied when people use different languages or dialects according to context. Thus US-born children to Mexican parents in Los Angeles may speak Spanish at home but English at school. And the forms of Spanish and English they speak will vary. High schoolers may have a different speech code with their teachers compared to their English-speaking friends, and the Spanish they use with their parents may vary from the Spanish they use with friends.

We can imagine a rich variety of instances of different code switching. Turkish parents in Germany may speak Turkish to each other, but some blend of Turkish and German with their children. They may speak Turkish when they are at home but German in the public realm.

The spatial division between public and private, formal and informal, home and outside are often sites of code switching and changes in the way that language is employed.

Language Shifts

People move. They take their language with them, but if the move is to a different linguistic region, their language

TABLE 11.3 Language Shift in Australia		
ORIGIN (LANGUAGE)	**FIRST GENERATION**	**SECOND GENERATION**
Austria (German)	48	89
Chile (Spanish)	9	38
France (French)	37	77
Hong Kong (Chinese)	9	35
Lebanon (Arabic)	5	20

and especially that of their children may shift. Marked **language shift** occurs as children are socialized into the dominant language.

We can follow the shift by looking at language use by different groups in Australian cities. The first generation noted in Table 11.3 is the percentage of people born outside of Australia who spoke English at home. Notice the low figure for people from Chile, Hong Kong, and Lebanon. The second generation is the people born in Australia who speak English at home. Notice the marked shift toward English speaking for all children, especially for those with Austrian and French parents. Where there are substantial speech communities (there are lot more people in Australian cities who speak Arabic and Chinese than French and German), then the original language remains stronger. And if there are substantial cultural differences—wider between Australia and the world of Arabic- and Chinese-speaking immigrants compared to the average immigrant from Austria and France—then the language shift is less marked.

The Political Geography of Language

Language and National Identity

National identity is created, maintained, and reproduced in language. Graham Robb, in his wonderfully descriptive account of the historical geography of France, outlines the process whereby the national language of French finally replaced a myriad of different language and dialects. France encompasses at least five major languages, Catalan, Franco-Provençal, Gallo-Italian, Langue d'Oc, and Langue d'Oïl (now considered standard French), with at least ninety distinct dialects. At the time of the French Revolution, French as we know it today was spoken only in Paris and the surrounding area. Creating a national language was a major goal of successive republican governments. Even as late as the mid-nineteenth century, most of the people living south of the Loire did not speak French, and in Brittany most spoke Breton. The national education system fostered the exclusive use of French, so that French became the national language, a process heightened by the space-time convergence that Robb (2007) describes as the "contraction of space and the gravitational pull of Paris" (p. 248). But even today local variants persist, a testimony to the continuing connection between language and locality. Almost 2 million people speak Occitan, one and a half million Alsatian, and a half-million Breton. Local dialects persist, although the higher up the socioeconomic hierarchy, the more common the everyday usage of "proper" French (see Figure 11.8).

There is a complex relationship between language and national identity. A national language is more often than not a self-conscious creation. In 1783 Noah Webster (1758–1843) wrote of the need to "establish a national language as well as a national government." His *American Dictionary*, published in in 1828, was written to create a distinctly American English different from the English of the colonial power.

We have more recent examples. In 1947 the British Empire in India was partitioned into independent countries. Urdu was the dominant language of North India before Partition. It is filled with Arabic, Persian, and Turkish words, a legacy of the ebb and flow of empires. Hindustani was almost identical but written in Sanskrit. Urdu was in Arabic script. After 1950, the Indian government created a national language of Hindi from Hindustani with all the Arabic, Persian, and Turkish words replaced with fabrications from Sanskrit. The result was an invented national language expunged of any Muslim influence.

Hebrew as a language has an ancient history and a more recent emergence. It was the language of the early Israelites, but disappeared as an everyday language around 300 CE, surviving only as the **liturgical language** of the Jews. It was revived in the nineteenth century as part of early Zionism and is now one of the two recognized languages of Israel, the other being Arabic. Almost 90 percent of Israeli Jews are proficient in Hebrew, including 70 percent of Israeli Arabs.

Language policies vary from standardization onto a national language to a diffusion of standard languages into the vernacular. The official sign shown in Figure 11.6, for example, would have been unthinkable in the Shetland Islands fifty years ago, when the emphasis was on standardization. In new nations standardization is often the way to ensure national cohesion. Today, in a more postmodern world and more settled nation, vernacularization avoids bland uniformity and heightens the sense of place and regional identity.

The Written Word

While the spoken word varies, the written word is more standardized. Throughout the nineteenth and much of the twentieth century, the emphasis was on the construction of both written and spoken national languages. On closer inspection, many "national" languages arose from the promotion of one language at the expense of others. The promotion was aided by national education systems, print cultures of books and newspapers creating a national readership, and the marginalization and suppression of other languages within the national territorial space.

Language Policies

In reality, a variety of languages are used in most countries. Sometime this is reinforced and encouraged; other times it is suppressed in the pursuit of single language use. We can identity four different types of state language policy. *Domination/exclusion* refers to the continuing preference given to the dominant language and the suppression of minority languages. For example, the Hawaiian language was banned in public schools from 1896 to 1986. In the case of *assimilation,* the state encourages minority language speakers to learn the dominant language. Under a *pluralist* policy the state promotes tolerance of minority languages and the practice of multilingualism. Last, in a *confederation* languages are identified with specific regions of a country. The different categories are not mutually exclusive. The USA and individual states fall between the assimilationist and pluralistic approaches.

Some countries have an explicit policy of multilingualism with a number of official languages. Canada is a bilingual society with both French and English used in official records and documentation. Figure 11.9 shows signs at Vancouver International Airport. Notice that while the two official languages are used, so is Chinese, because of the many Chinese tourists and visitors to the city. Sri Lanka also has two official languages, Sinhalese and Tamil, although English is widely used, a legacy of colonialism and the dominant language when the country achieved independence (Figure 11.10). New Zealand has three official languages: English,

11.9 Sign at Vancouver airport

11.10 Language use in Sri Lanka

Maori, and New Zealand Sign Language. Switzerland has four: German, French, Italian, and Rumantsch. South Africa has one of the largest numbers of official languages, eleven: Afrikaans, English, Ndebele, Northern Soto, Soto, Swazi, Tsonga, Tswana, Venda, Xhosa, and Zulu.

The official languages do not cover all the languages spoken within a country. In Canada, for example, there are a large number of indigenous languages as well as new immigrant languages; at least eighty-three different other languages are spoken, and more than a million people speak Chinese.

Even in officially monolingual states, a multiplicity of languages is used on a daily basis. Although English is the official language of Australia, at least 388 different languages are spoken in the country, including many indigenous languages and the speech communities of recent immigrants. In Mexico, Spanish is the official language, but out of the total population of around 120 million, between 10 and 20 million speak an indigenous language. Indigenous speakers are disadvantaged in an officially unilingual society. While Amharic is the only official language of Ethiopia, it is spoken by only around 30 percent of the population. Around 34 percent speak Oromo, and in the marketplaces of Addis Ababa, merchants freely discourse in Amharic, English, and Tigrinya. At least 188 languages are spoken in the USA, 44 in the UK, 42 in Germany, and 39 in France, all polyglot societies with a large number of immigrants from around the world.

Official language policy can change abruptly with political ruptures. In Spain, at the fall of the Franco dictatorship in 1975, the promotion of Castilian Spanish was replaced by a greater multilingualism. In the region of Catalonia around Barcelona, public signs now display Catalan as well as Spanish words. The Basques were also given greater freedom to speak and write in Basque. During the apartheid era in South Africa, Afrikaans and English were the official languages; Afrikaans, a creole of English and Dutch, was the language of the white supremacists, but was also spoken by black and colored South Africans. After the fall of the regime, English became the preferred official language, while the state proclaimed eleven official languages. However, speaking the dominant language of English confers advantages in the labor market and in civic organizations.

Sometime the change occurs without political rupture. Around the world there is increasing recognition that the language rights of minorities should be protected. As an example, in 1977 the BBC launched a Welsh radio station.

Language Loss, Replacement, and Rebirth

We are losing linguistic diversity. Approximately 6,000 languages are spoken today, but by century's end it is estimated that only 600 are sure to survive. We are losing a language every four months. In a world dominated by fewer languages, we gain global communication but lose diversity, linguistic subtlety, and something that is almost impossible to resuscitate. Linguistic variety, like ecological diversity, is a sign of health and vitality. There is a strong correlation between areas of biodiversity and linguistic diversity at numerous scales of analysis. A loss of biodiversity is associated with a decline in linguistic variety. The loss of both impoverishes our world.

Languages are not just reflectors of the external world; they embody it. How we describe the world is crucially dependent on where we are and how we speak and write. Daniel Nettle and Suzanne Romaine drew attention to the extinction of language in their evocatively named book *Vanishing Voices*, and K. D. Harrison made a similar point with his book *When Languages Die*.

The consequences of the loss of language are profound. Indigenous languages of the Central Desert region of Australia, for example, are the living bond between people and the country given to them by their creators. The loss of the language cuts the cord with the cosmological understanding of the world, destroys the empathetic rendering of the environment, and undermines the cultural framework that gives social meaning and religious purpose. Without language, sacred space is rendered secular and meaningless.

At the point of contact, original languages of places are modified by new languages. In some cases they are replaced. Language replacement can occur when small speech communities are simply overwhelmed by the dominant language. They can also occur through the formal policies and naming practices of the dominant linguistic group.

Thousands of different languages were spoken in the Americas before the Columbian Encounter. Consider the case of the USA, where English became the dominant language and replaced a variety of local languages deeply rooted in place. Small tribal languages were particularly vulnerable because of the small number of native speakers and the overwhelming power and influence of the non-native language. In some cases the replacement was long part of official educational policy.

In the USA, only about 170 indigenous pre-contact languages are still spoken. Navajo is one of the largest speech communities, spoken by 250,000 people, but fluency is declining. While 25,000 people describe themselves as Mohawk, only 15 percent can speak the language. More than 74 languages in the USA are on the brink of extinction, including Quechan, whose speakers number no more than 2,500. They live in Arizona, and while elders can speak the language, it is losing its appeal to the younger generation. Only around 100 people know the language well. With such a small base, one generation's indifference can doom the language. A further 58 languages have fewer than 1,000 speakers and 25 have only between 1,000 and 10,000 speakers. With such small numbers, language revival is precarious.

Minority languages survive if they have large numbers of speakers. In Central America, for example, Mayan is still spoken by more than 7 million people. In South America, 8 million people or more speak Quechuan, the language of the Incas. In the Andes, the Inca Empire spread along an extensive road system, diffusing Quechua throughout a wider region. Today, between 8 and 10 million people speak it as their first language in the Andean belt of Bolivia, Peru, Ecuador, Colombia, and Argentina (Figure 11.11).

In some cases, formal recognition of a language can improve the chance of survival. Quechua, for example, is the official language of Bolivia, Ecuador, and Peru. Its recognition is part of the growing political power of indigenous people. The election of Evo Morales in Bolivia, for example, marked the first time that an indigenous person became president of that country; there had been seventy-nine presidents before him.

It is no surprise that Mayan and Quechua survive; they were the languages of sprawling empires that spread a common language across a wide area. Figure 11.11 is in effect a language map that corresponds to the reach of the Incan Empire along the roads of its imperial system. By contrast, the languages of isolated groups in Amazonia, with their small population bases, are more vulnerable to extinction. However, if they retain their internal coherence and have little contact with the outside, then, despite their small population base, they may survive.

Despite (or perhaps because of) the replacement and death of languages, there is evidence that people are aware of the need for linguistic revival and protection. The Nooksack language, for example, effectively died out from 1977 to 2001 with the death of the last native speakers. However, a cultural rebirth, including the place name inventory that we discussed at the beginning of this chapter, as well as people using recordings of native speakers,

11.11 Quechuan language

taking language classes, and recognizing that their language is an important part of their cultural legacy, has given Nooksack new life.

Globalization of Language

The globalization of certain languages is a recent trend. Spanish, for example, was one of the first European languages to be exported as part of the expanding Spanish Empire. Spanish became the language of the elites throughout Central and South America. The emigration of Chinese speakers spread the language along the migration routes to diasporic communities across the world. But the most global language in terms of its widespread diffusion and adoption as a second and third language is English, now the hegemonic language of the contemporary world. Waves of globalization centering in the United Kingdom and United States, the extensive impact of the British Empire during the colonial period, and the dominance of the American economy, culture, technology, and politics in the contemporary

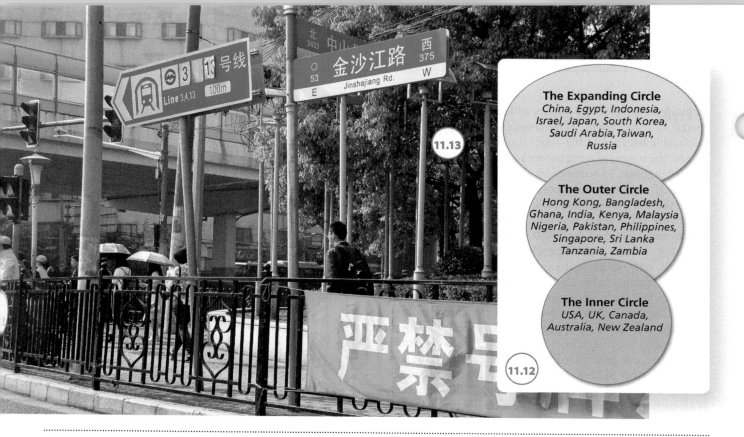

The Expanding Circle
China, Egypt, Indonesia, Israel, Japan, South Korea, Saudi Arabia, Taiwan, Russia

The Outer Circle
Hong Kong, Bangladesh, Ghana, India, Kenya, Malaysia, Nigeria, Pakistan, Philippines, Singapore, Sri Lanka, Tanzania, Zambia

The Inner Circle
USA, UK, Canada, Australia, New Zealand

11.12 Circles of English use. **11.13** Street signs in Shanghai.

world have all led to the dominance of English. It is now the language of global interaction. Most scientific, technological, and academic information in the world is expressed in English, and over 80 percent of all information stored in electronic retrieval systems is in English. It has become the language of international entertainment, popular music, movies, and advertising.

A core-periphery model of global Englishes is suggested by Braj Kachru: an inner ring of native speakers, an outer ring of institutionalized local varieties—often a function of extended periods under British colonization—and an expanding circle of countries that use English as part of international communication (Figure 11.12). The adoption of English has varied. In Jamaica, English became part of a creole language, while in Singapore, English joined Chinese and Malay, and later Hindi, as part of a multilingual society.

Over seventy-five countries officially recognize English as a primary or secondary language. They include Botswana and Cameroon, India and Liberia, Rwanda and Sudan. The number of people who speak English as a first or second language is estimated at 573 million. A further 670 million may have native-like fluency in the language. Even nonspeakers are exposed to the language on a daily basis through advertising, government functions, or interpersonal communication. David Crystal describes them as members of an expanding circle who have been exposed, via the dynamic processes of globalization, to the usage of English on many informal levels. Almost 1.6 billion people, over a fifth of the world's population, exhibit some reasonable competence in the language.

Growing global English competency has accelerated the rate of **cultural globalization** by facilitating the movement of ideas and information. The expanding use of English around the world, however, is not simply the result of diffusion outward from a dominant center; for many people worldwide its adoption is conscious. People in non-English-speaking countries are trying to learn English to participate more fully in international activities. English is required in order to be competitive in global markets. Many countries around the world have adopted English as a second language and emphasize it as an important subject in their schools. In the early days of the People's Republic of China (PRC), English was labeled as an "imperialist language" and suppressed. After the break with the Soviet Union, teaching Russian was abandoned and teaching English was revived and relabeled as "the instrument for struggles in the international stage." In the contemporary PRC, use of English has greatly increased, fueled by the increasing number of students educated in English-speaking countries, especially the United States, and rising demand for English-language skills among professionals and workers in the booming international trade sector. English-language teaching is now an important industry. Even street signs now use English as well as Chinese (Figure 11.13).

For individuals in many countries, English skills are an invaluable asset in the job market. English is a form of cultural capital for national governments seeking to produce a globally competitive workforce and individuals wishing to achieve more employment opportunities in a globalizing world.

Not only has English expanded to become the dominant worldwide medium of communicative practice; it has also undergone considerable reinvention by non-native-speaking communities, many of whom are speaking English in their

11.1 The Linguistic Landscape

Our environment is filled with language signs. A number of different types of **linguistic landscapes** can be found. First, there is the monopolistic landscape. The first figure in this box shows signs on the side of building in Seoul. It is almost entirely in the local language. By contrast, Figure 11.13 shows a streetscape in Shanghai. Notice how the linguistic landscape is now bilingual with signs in Chinese as well as English. The metro system of the city is entirely bilingual, with all signs in English and the automatic voiceover also in English as well as Chinese. In towns and cities across the world, global English words and phrases are now part of the local linguistic landscape.

Then there are trilingual signs such as Figure 11.9, a photo taken in Vancouver's airport. The city-state of Singapore has a bilingual policy. All students are taught English at school but can also learn a second language: Chinese, Malay, or Tamil. The four languages are commonly used in Singapore, and only two of them use Roman script. The second figure in this box, of a warning sign next to a government building, depicts all four languages to make sure everyone gets the message to stay out.

Streetscape in Seoul

Sign in Singapore

TABLE 11.4 Linguistic Landscapes in Israel			
	NEIGHBORHOODS		
LANGUAGE OF SIGNS	**JEWISH**	**ISRAELI PALESTINIAN**	**NON–ISRAELI PALESTINIAN**
Hebrew only	49.6	24.1	—
Arabic only	0.1	5.0	20.9
Hebrew-English	44.6	6.2	—
Hebrew-Arabic	0.9	39.4	5.8
Arabic-English	—	1.2	55.8
Hebrew-Arabic-English	4.9	24.1	17.4

Local language usage can vary from the official policies. In Israel the two official languages are Hebrew and Arabic, with English widely spoken and understood. In 2014 there were demands from some Israeli politicians to make Hebrew the sole official language, in effect hardening the notion of Israel as a Jewish state. A study of the language of signs in three types of neighborhoods in Israel highlighted marked differences. In predominantly Jewish neighborhoods, signs were either in Hebrew or Hebrew-English. In non-Israeli Palestinian areas, dominated by Arab speakers who were not citizens of Israel, most signs were either Arabic-English or Arabic. In Israeli Palestinian neighborhoods, where most people were Arab citizens of Israel, the highest percentage of signs were in Hebrew-Arabic.

REFERENCE

Ben-Rafael, E., Shohamy, E., Hasan Amara, M., & Trumper-Hecht, N. (2006) Linguistic Landscape as Symbolic Construction of the Public Space: The Case of Israel. *International Journal of Multilingualism* 3: 7–30.

own ways. Like many other cultural forms, English is being both **deterritorialized**, as it spreads from its cultural hearth, and **reterritorialized**, as non-native speakers reshape the language to best suit their purposes. Instead of adopting a standard form of English, the language is being reterritorialized as a global medium of communication. And standard English is being changed in the process. The subtitle of David Crystal's survey of global English is "One Language, Many Voices."

Robert McCrum writes of "Globish" as the new spoken international language, primarily used in economic transactions. It joins the list of other hybrid languages, such as Spanglish (Spanish-English). One colleague from Copenhagen assures me that she speaks fluent Danglish (Danish-English).

The complex history of English as a language of both the colonizer and the colonized is summarized in the recent history of India, which was an important part of the British Empire. To govern this vast and populous land, the colonial authorities created an English-speaking Indian intellectual elite to help run the empire efficiently. The English-speaking elite became the leaders of the independence movement. In 1950 the constitution of the newly independent India was written in English. And the languages of the colonized were incorporated into English. Words like *nirvana*, *bungalow*, and *jungle* are now part of Standard English.

The Written Language

English is both spoken and written. While spoken English is a much more flexible form of communication—it is legitimate to speak of global Englishes, including Spanglish—written English, in contrast, is more rule-bound, programmatic, and slow to change. While we can think of a reterritorialization of spoken English, a reterritorialization of written English is less apparent. The use of English as a written global form of communication is particularly visible in the communities of knowledge. English has become the language of global intellectual discourse and the dominant language of intellectual communities involved in the production, reproduction, and circulation of knowledge. English dominates the **global epistemic community**.

Cited References

Crystal, D. (2003) *English as a Global Language*. 2nd ed. Cambridge: Cambridge University Press.

Crystal, D. (2010) *Evolving English: One Language, Many Voices*. London: British Library.

Gorenflo, L. J., Romaine, S., Mittermeier, R. A., and Walker-Painemilla, K. (2012) Co-occurrence of Linguistic and Biological Diversity in Biodiversity Hotspots and High Biodiversity Wilderness Areas. *Proceedings of the National Academy of Sciences* 109: 8032–8037.

Harrison, K. D. (2007) *When Languages Die: The Extinction of the World's Languages and the Erosion of Human Knowledge*. New York: Oxford University Press.

Jameson. F. (1991) *Postmodernism, or, The Cultural Logic of Late Capitalism*. Durham, NC: Duke University Press.

Kachru, B. (1992) *The Other Tongue: English Across Cultures*. Urbana-Champaign: University of Illinois Press.

Krupnik, I. (2011) How Many Eskimo Words for Ice? Collecting Inuit Sea Ice Terminologies in the International Polar Year 2007–2008. *The Canadian Geographer* 55: 56–68.

Kulick, D. (1992) *Language Shift and Cultural Reproduction: Socialization, Self and Syncretism in a Papua New Guinean Village*. Cambridge: Cambridge University Press.

Labov, W. (1972) *Sociolinguistic Patterns*. Philadelphia: University of Pennsylvania Press.

McCrum, R. (2010) *Globish*. New York: Norton.

Nettle, D. and Romaine, S. (2000) *Vanishing Voices: The Extinction of the World's Languages*. New York: Oxford University Press.

Richardson, A. and Galloway, B. (2011) *Nooksack Place Names: Geography, Culture, and Language*. Vancouver and Toronto: UBC Press.

Robb, G. (2007) *The Discovery of France: A Historical Geography from the Revolution to the First World War*. New York: W. W. Norton.

Select Guide to Further Reading

Basso, K. (1996) *Wisdom Sits in Places: Landscape and Language among the Western Apache*. Albuquerque: University of New Mexico Press.

Eira, I. M. G., Jaedicke, C., Magga, O. H., Maynard, N. G., Vikhamar-Schuler, D., and Mathiesen, S. D. (2013) Traditional Sámi Snow Terminology and Physical Snow Classification—Two Ways of Knowing. *Cold Regions Science and Technology* 85: 117–130.

Robson, D. (2012) Chilly Words: How Eskimos Really Say "Snow." *New Scientist* 216: 2896–2897.

Short, J. R., Boniche, A., Kim, Y., and Li, P. (2001) Cultural Globalization, Global English and Geography Journals. *Professional Geographer* 53: 1–11.

Thurman J. (2015) A Loss for Words: Can a Dying Language Be Saved? *The New Yorker*, March 30: 32–39.

Selected Websites

The Washington Post, "The World's Languages, in 7 Maps and Charts"
https://www.washingtonpost.com/news/worldviews/wp/2015/04/23/the-worlds-languages-in-7-maps-and-charts/

Geographic influences on linguistic sounds
http://journals.plos.org/plosone/article?id=10.1371/journal.pone.0065275

Interactive Map Websites

Regional dialects in UK
https://www.youtube.com/watch?v=-8mzWkuOxz8

The second most spoken languages spoken around the world
http://graduate.olivet.edu/news-events/news/second-most-spoken-languages-around-world

Distribution of language
http://www.statsilk.com/maps/world-stats-open-data?l=distribution%20of%20languages%20official%20or%20first%20language

Learning Outcomes

- Language is an important part of culture.
- Local indigenous languages have a sophisticated description of local environmental conditions.
- Historically connected to specific places, language has diffused through the spread of empires and an ever-flattening world.
- The three trends in the geography of language include (1) the emergence of distinct linguistic geographies; (2) the interaction and sometimes replacement of local variants with outside languages because of population movement, social change, and the operation of political power; and (3) the globalization of certain languages.
- Although over 6,000 languages are spoken around the world, half of the world's population speaks only 10 languages. Dominant languages are Chinese, Hindi, English, Spanish, and Arabic.
- We give meaning to our world by describing it with language. Colonial powers expressed their power by naming the landscape, and postcolonial societies embody their independence by renaming it.
- The contact between languages can produces pidgins and creoles.
- Language varies by regional dialects. Socioeconomic differences map onto the difference between standard and regional dialects.
- Language use varies by where people are and who they are. Language varies by context, referred to as language domains.

- Language use is flexible. In code switching, people use different language depending on context. Language shifts occur as people move to different language regions.
- Language us bound up with national identity as national languages are crated and maintained. Language policies of minority languages vary from exclusion to encouragement.
- While the spoken word varies, the written word is more standardized. National written languages emerge with national education systems and shared print cultures of books and newspapers that create a national readership.
- We are losing many of the smaller languages, although there are attempts at language revival. The creation of monolingual geographies has increased global communication, but has decreased linguistic diversity, which is strongly correlated to biodiversity and can lead to an increasingly impoverished world.
- Growing global English competency has accelerated the rate of cultural globalization by facilitating the movement of ideas and information. It has also led to the modification or creolization and ensuing reterritorialization of the language as local communities adapt it to local needs.
- English is the most globally pervasive language, and its worldwide use fosters cultural and economic globalization. It is the language of a global epistemic community.

PART 5

The Global Organization of Space

In this section, we will look at the emergence of a global society through worldwide flows of economic transactions and cultural interactions. Chapter 12 considers the rise of the global economy, explores the idea of space-time convergence, and considers the pace and geography of economic globalization. Chapter 13 examines the formation of global and local cultures in the transmission of language, architecture, and religion.

12

Creating a Global Economy

In this chapter, we focus on the creation of a global economy. In many of the current discussions of globalization, one common assumption is that it is a recent phenomenon. In this reading, globalization goes hand in hand with the construction of the modern world. However, while its pace has increased, globalization is long established as a continuing process of the human occupancy of the globe. Our focus in this chapter is the growing economic interaction between peoples and societies spread around the globe. We will consider this interaction through three related topics: space-time convergence, waves of globalization, and the question, "How flat is the world?"

Space-Time Convergence

Space-time convergence is a collapse in the time it takes to cover distance. This convergence, both a cause and effect of widening economic transactions, reduces the cost of transporting goods and people and pulls places closer to one another. A series of space-time convergences caused by new developments in transportation effectively brought the world closer together. The very first was the invention of the wheel, which allowed more goods to be effectively transported. The wheel was invented around five thousand years ago, appearing simultaneously at a number of sites in the Middle East and Europe. The earliest wheeled vehicles had to move over rough surfaces that made movement slow and difficult. The full flowering of the wheel's advantages had to await the creation of smoother transport surfaces.

A Widening World

For centuries, the easiest way to move people and goods long distance was across water. Natural wind could power sailing ships. Initially, most seaborne trade was limited by navigational knowledge. The development of the compass around 1280 enabled sailors to plot their direction more accurately, thus reducing the time and therefore the cost of long-distance sea travel. Equipped with a compass, sailors could now venture out more confidently from the enclosed seas of the Baltic and the Mediterranean to travel across the wider oceans. This had geopolitical consequences. The Roman Empire, for example, was essentially restricted to the extended basin of the Mediterranean because it had limited ability to sail across the wider seas. Later traders, however, could use compasses to venture across the Atlantic and around the Horn of Africa. Maritime empires such as the Spanish, Portuguese, and British were established and developed because the compass made it easier to move across distant seas and expansive oceans.

Improvements in navigation made it quicker, safer, and hence cheaper to sail across the oceans. Sea travel also avoided the frictions of distance imposed by traveling through different societies, separate countries, and competing empires. The neutrality of the open sea aided long-distance trade and commerce. Because most long-distance trade for the past five hundred years occurred by sea, economic activity is still concentrated in coastal regions and in (current and former) port cities. London and New York, for example, grew as major cities in part because their economies were invigorated by their port status. Amsterdam grew from a small, watery town to a major city as the Dutch commercial empire spread its tentacles across the globe. Port cities also took on a more cosmopolitan feel, filled as they were with exotic goods and foreign peoples. The big commercial ports connected with a world wider than their immediate hinterlands; they had an openness to the wider world. When Peter the Great wanted to modernize Russia, he created the new city of St. Petersburg in 1703, located on the Baltic and therefore more open to foreign influences and international trade. He moved the capital there, from the interior city of Moscow, to create a window on Europe.

The earliest cargo of long-distance maritime global trade was limited to luxury goods and prized items. When transport is costly, it is economically feasible to concentrate on only high-priced items. Early trade between Europe and East Asia, for example, focused on such expensive goods as spices, highly prized for their ability to season food in an era before refrigeration. In the seventeenth century, a pound of nutmeg purchased in the Spice Islands of Southeast Asia cost 60,000 times more when sold in a London market. The spice islands of Banda and Run became vital resources fought over by Portugal, Holland, and England; they also become significant features of European maps of the region.

12.1 The Spice Islands

Figure 12.1 is from a map of Southeast Asia in a 1636 Mercator-Hondius atlas. The Spice Islands are disproportionately large, a reflection of their economic importance at the time.

The Age of Steam

When the age of steam replaced the age of sail in maritime transport, there was an extra tug on the shrinking net of space-time. Around 1890, steamships moved more tonnage than sailing ships. The increasing use of steam allowed bigger ships to carry more goods more quickly and more cheaply. In 1850 it took between eighteen and twenty-four days to sail from Liverpool to Gibraltar. A steamship could make the same journey in between seven and nine days. Not only was steam quicker, it was more reliable, less dependent on the vagaries of the wind. Steam meant faster and more predictable transport times. Highly prized goods and commodities no longer dominated international trade as the declining transport cost per unit "smoothed out" the globe as a transportation surface for ever-cheaper goods and commodities.

Transport developments created new geographies that made the operation of **comparative advantage** more feasible. Comparative advantage refers to the ability of a region or country to produce goods at lower cost than other areas. It works best when the unit costs of transport are low. With a smoother transportation surface and cheaper transport costs, the mechanisms of global comparative advantage generate new economic geographies. It thus became more efficient in the late nineteenth century to grow wheat in the United States than in Britain, because the cost of land in particular, and in some cases labor, was so much less. Steamships made it profitable for grain grown in the

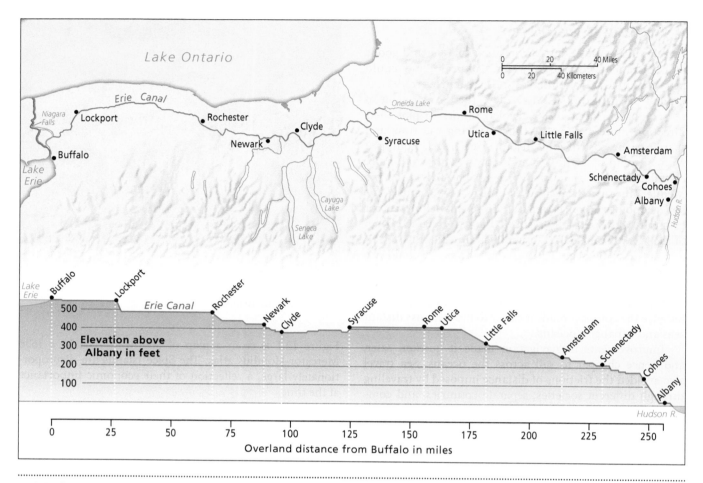

12.2 The Erie Canal

United States and other distant places, such as Australia, to be shipped to meet demand in Europe.

Canals

Before the railways, inland canals were an important transport artery for reducing overland transport costs. Take the case of the Erie Canal, begun in 1817 and completed in 1825: it cut a straight, watery line through the 363 miles between Buffalo and Albany and connected the Great Lakes with New York City (Figure 12.2). The canal compressed space-time. Before the Canal opened, it could take up to twenty days to travel between Buffalo and New York City. After the canal opened, it took less than four days. The cost of moving a ton of freight plummeted from $100 to $5. For twenty miles on either side of the canal, the equivalent of a day's wagon ride, the local economy was transformed. Along the path of the canal, villages turned into towns and towns into cities as local economies were now linked to larger national and global markets. Goods, people, and ideas flowed along the route of the canal.

But just as the canal was creating its greatest impact, a new form of transportation compressed space and time

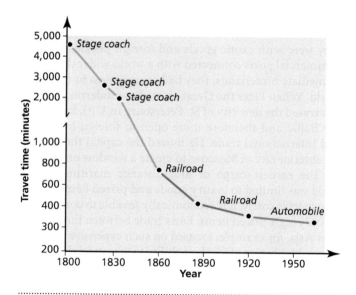

12.3 Space-time convergence between Boston and New York

even more. The coming of the railways further reduced travel time from Buffalo to New York City. Goods moved along the canal at around fifty-five miles a day; the railway covered that distance in an hour. The railway could also enable travel year-round, whereas the canal froze over in wintry upstate New York. By 1850 more freight was being carried on rail than on the canal. The canal soon became obsolete, replaced as a major means of transport by the railroads, and towns along the canal soon withered. Some towns, such as Syracuse, which were located on both the canal and the railway line, grafted new railway-connected growth onto canal-based growth and thus continued to thrive.

Railways

Railways annihilated distance, brought cities closer together, and made interaction easier and cheaper. The geographer Donald Janelle plotted the travel times between

Boston and New York. Figure 12.3 shows the declining travel time between the cities with the introduction of new transport modes and their subsequent improvement. Over time, Boston and New York came closer together, making the movement of people, goods, and ideas between them easier and quicker.

Space-time convergences can bend physical space. London's space-time map is shown in Figure 12.4. Consider the case of the connection between Stranraer and London. Because of the lack of plane connections, the relatively poor railway connections, and the lack of motorways, it can take up to nine hours to travel between the two places. The result is that a space-time map warps the physical distance. London is closer in space-time terms to Paris than it is to Stranraer.

Arnold Horner measured the convergence of rail times between cities in Europe over the period 1964 to 1999. More recently, David Banister has argued that the desire to move farther faster is ultimately unsustainable because of its energy costs. Rather than reducing travel times and increasing distance, he promotes alternative plans that reduce travel distances.

Continuing Convergence

Space-time compression continues apace. Developments in maritime trade include the increasing size of ships and the growing use of containers to pack the goods. The large box containers allow for easier loading on and off the ships; hence the unit cost of transport declines as more goods are shipped more quickly. As the size of the ships increases, new types of port facilities are required. Unlike older, smaller ships, the newer container ships could not sail up narrow, shallow rivers. By the 1970s many port cities worldwide could no longer handle the big container ships. Shipping therefore moved seaward to deepwater ports, and old upstream ports were abandoned. In places as varied as London's Docklands, Baltimore's Inner Harbor, and Sydney's Inner Harbor, these sites—large areas of vacant land at the heart of metropolitan regions—sometimes

12.4 (A) Traffic leaving London. **(B)** Space-time connections of London.

12.5 Space-time convergence: cheap air travel shrinks the globe. Baltimore/Washington International Airport.

became the setting for new postindustrial residential and commercial activity. In this way, each space-time convergence destroys old geographies and creates new ones. Space-time convergence effectively shrinks and re-shapes the globe as economic transactions are brought closer together. The world is flattened and distorted as it is shrunk.

In recent years, the increasing use of air transport has shrunk the globe even more (Figure 12.5). In 1873 Jules Verne depicted rapid travel around the world in the novel *Around the World in Eighty Days*; today, with good air connections, it could now take less than forty hours. Air transport is still more suitable for moving people and high-value goods. Expensive perishable items, such as fresh flowers and fruits, are particularly suitable for transport by air. Flowers grown in Colombia are transported by air to flower markets in North America because the cost of the final item makes the relatively high-cost air freight affordable. There is a distinct seasonality to this trade, as fresh fruit grown in the Southern Hemisphere is shipped to northern markets

in winter. Air freight allows a never-ending supply of seasonal produce. The growing trade in horticultural produce between Kenya and British supermarkets, for example, was possible because produce with a limited life span was shipped through Nairobi's airport. Although air freight allows quick travel along the path of global food chains, unstable fuel prices makes it vulnerable to rapid price changes.

As an economy matures, the transfer and exchange of information becomes as important as the movement of goods. And here we come to the importance of cost rather than time. Once the telephone was invented, messages could be transmitted very quickly. The most important variable became not space-time but **time-cost compression**. In 1919 it cost almost $300 (in 2004 prices) to make a ten-minute call between New York and Los Angeles. Ninety years later the cost is negligible (Figure 12.6). Calling long distance is no longer an expensive proposition. The same process occurs with Internet connections. As the real cost of communication falls, there is an extra tug pulling closer the net of space-time-cost. Each successive wave of space-time-cost convergence creates new markets and new economic geographies.

The waves of globalization have left a legacy. Just as waves sweeping onto a shore affect the topography of the beach, which in turn shapes the flow of subsequent waves, so the tides of globalization have impacted places, and the character of places has in turn affected the subsequent flows of global contact. We can imagine the world as a series of places in a continual process of globalization and reglobalization.

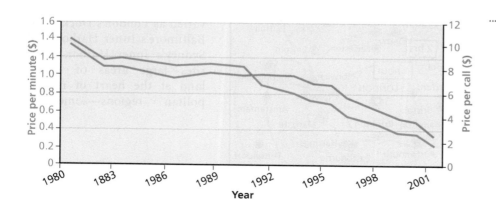

12.6 Decline in telephone costs

Three Waves of Globalization

Space-time convergences do not occur either randomly or regularly. They tend to be bunched together as distinct waves of globalization. We can identify three distinct waves of globalization.

The First Wave

The **first wave of globalization** followed the Columbian Encounter of 1492; it inaugurated a truly global economy. Before, the world was largely a system of regional empires and separate hemispheric ecologies. There was a substantial distinction between the Eastern and Western Hemispheres, what are often called the Old and New Worlds. After 1492, a global system took shape.

A major consequence of the encounter between the Old and New Worlds after 1492 was the demographic holocaust that we discussed in chapter 3. The population estimates of the New World around 1490 range from 54 million to 112 million. By 1650 that population had fallen to just over 5 million. The cause was the introduction of new diseases from the Old World against which the indigenous people had no immunity.

The demographic holocaust was just one consequence of the Columbian Encounter. Others followed. The decline of the local populations led to the need for imported labor to dig in the mines and tend the estates. The slave trade shipped close to 20 million people from Africa to the New World. Then there was the biotic exchange. New World crops such as corn and potatoes enriched the diets of the whole world, leading to population increases in Europe. Turkeys, tomatoes, sunflowers, and peanuts were transplanted from the Western to the Eastern Hemisphere, while horses, hens, cattle, pigs, sugarcane, and wheat moved in the opposite direction. Major environmental changes were wrought as a variety of plants and animals were dispersed around the world. When we pick tomatoes in Europe, or ride horses in America, we are participating in a globalization process that began when Christopher Columbus landed in the New World. The Columbian Encounter bridged the hemispheric divide in a series of transactions and exchanges of people, plants, viruses, and animals that activated a global world.

The Second Wave

A **second wave of globalization** took place from approximately 1865 to 1970. From 1865 to 1914, economic globalization was facilitated by low tariffs, an international labor market, and relatively free capital mobility. Critics of the time called it "British internationalism." British overseas investments were twice those of France and five times those of the United States. It was an economic globalization centered in London. Other countries, such as France, Germany, and the United States, were also involved through their smaller trading empires. The global system was neither fair nor equitable. It was a system dominated by merchants, manufacturers, and investors in the rich countries seeking raw materials, markets, and investment opportunities.

This second wave of globalization was associated with major shifts in space-time convergences triggered by the appearance of the railways, the telegraph, and the steamship. It took less time to travel large distances. People, ideas, and messages could move more quickly. News of events was more quickly transmitted across the globe, in effect shrinking the globe to a more compressed space.

Railways had been introduced before 1865 but were limited to partial networks in only a few countries. After 1865, there is a widening and deepening of the railway network around the world. In 1870 there were only 125,000 miles of track in the world; by 1911 this had increased to 657,000 miles. Far away from the railroad tracks, life may have remained unchanged, but the whistle of the train announced that the world had sped up.

Ocean distances were reduced by the construction and enlargement of canals. The Suez Canal reduced the time it took to sail from Western Europe to East Asia by linking the Mediterranean with the Red Sea. This **"highway to India"** was completed in 1869. The British invasion of Egypt in 1882 was in large part an attempt to control the canal, so vital for maintaining British links with their colony in India. But as ships increased in size and the region became susceptible to political instability, the Suez Canal became a less important transit site. On the other side of the world, the Panama Canal was cut through a narrow isthmus to reduce the distance between the Atlantic and Pacific. After unsuccessful attempts by the French, the United States built the Panama Canal, and it opened to traffic in 1914. In order to control the canal, the United States appropriated property along its route. The Canal Zone became a source of friction between the United States and an increasingly more independent-minded Panama. In 1999 the Canal Zone reverted to Panamanian control. As the container ships got bigger, more traffic was diverted away from the canal. The growing trade between the United States and East Asia, for example, was occurring through the ports on the West Coast of the United States. Almost a third of the world's container ships were too large to pass through the existing canal. The larger locks that can handle the bigger ships opened to traffic in 2016 (see Figure 12.7). The older locks could only cope with ships each carrying 4,800 containers. These new locks allow the passage of much larger ships carrying 12,000 containers with 200,000 tons of freight.

The period from 1865 to 1914 marked a widening and deepening of economic integration around the world. London was the pivot of the international trading system,

12.1 The Flow of Capital

There are distinct phases of global capital mobility. First there was a period of marked upswing from 1880 to 1914, when there were few restrictions on the movement of capital. Much of the capital flowed from London markets to developing economies around the world. The First World War ended this period. A brief resurgence from 1925 to 1930 was followed by the Depression and the collapse of international capital markets. Second, a period of stability emerged after the Second World War. Capital flows were relatively small and regulated. The system ended in 1971 and ushered in the present period of increased capital mobility. Financial deregulation promoted global capital flows. The amount of direct foreign investment has increased in most countries. Four qualifying points need to be made:

1. The level of foreign direct investment is highly variable at the aggregate and national levels. Capital outflows constituted over 20 percent of world GDP in 2007, but this figure rapidly fell in the wake of the 2008 collapse to under 5 percent.

2. The bulk of foreign investment occurs between already rich countries. It is a form of diversified investment rather than development investment. That more investment goes to capital-rich rather than capital-poor countries is termed the Lucas Paradox. The disparity from classic theory, which says that capital should flow to places that need it, is the result of institutional structures, heightened sense of risk, and asymmetric information. Many developing economies are considered too much of a risk. The paradox is that the places that need most foreign investment may be getting less of it.

3. Be careful what you wish for. Increased capital mobility makes it easy to attract investment but comes with the danger that it may lead to rapid and destabilizing capital flows. A smooth surface for capital flows means that a highly mobile capital can undermine national economic planning. Consider the case of Argentina, where the government racked up deficits to reduce unemployment. Investor confidence was shaken, and capital outflows increased. As the currency depreciated, even more capital flowed outward, creating capital shortages and the potential for runs on the bank. The government introduced capital controls and forced companies to repatriate foreign profits.

4. Even when capital flows into developing economies, it comes at a price. Consider the case of an economy with an important commodity that attracts a lot of capital from overseas. Carmen and Vincent Reinhart looked at capital bonanzas in 181 countries from 1980 to 2007. They found that bonanzas have become more frequent because of fewer restrictions on capital mobility and are associated in medium- and lower-income countries with higher incidence of banking, currency, and inflation crises. As they note, "A bonanza is not to be confused with a blessing."

REFERENCES

Lucas, R. (1990) Why Doesn't Capital Flow from Rich to Poor Countries? *American Economic Review* 80: 92–96.

Reinhart, C. M. and Reinhart, V. (2008) Capital Flow Bonanzas: An Encompassing View of the Past and Present. National Bureau of Economic Research Working Paper 14321. http://www.nber.org/papers/w14321.pdf

12.7 Panama Canal: the new locks that opened in 2016

the sun of the trading universe. The formation of a world economy involved the economic and often political incorporation of larger areas of the world. Large parts of Africa and Asia, and even faraway islands in the South Pacific, were annexed to a global economic order. Overseas territories provided cheap raw materials, secure markets for manufactured goods, sources of national prestige, and pawns in the great game of global geopolitics. The 1865–1914 period was also an **age of imperialism**, as a neo-mercantilist doctrine emerged that

12.2 The Flow of Remittances

Remittances are the funds that temporary and permanent migrants send back to their home countries. As both international migration and capital mobility have increased, so has the level of remittances. As it becomes easier, and much cheaper, to send relatively small amounts of money back home, modest remittances contribute to the global flow of money.

By 2010, close to $300 billion in remittances was sent from one country to another. The biggest sender was the United States, with almost $48 billion. The next largest was Saudi Arabia, with $26 billion. The largest receivers include India ($55 billion) and China ($51 billion). However vast these sums are, they constitute less than 5 percent of each country's GDP. Table 12.1 lists a sample of the countries where remittances constitute a relatively large proportion of GDP. What are some of the main characteristics of this subset of countries?

In the case of El Salvador, for example, emigration increased substantially from the late 1970s as bloody civil war wracked the country. Millions fled, and many of them sent money back to their families, villages, and neighborhoods. In 1978 traditional agricultural exports were responsible for most of the hard currency in the country; remittances accounted for only 8 percent. By the first decade of the twenty-first century, remittances made up 70 percent and traditional exports only 12 percent. In many small counties, as well as regions and villages in large countries, remittances are a vital source of much-needed cash and capital, fueling local housing construction and paying for education and the provision of social services.

In some cases, remittances are an integral part of international economic relations. Almost 8 million Mexicans, 15 percent of the entire Mexican-born workforce, are now working in the United States. Many of these migrants send significant parts of their earnings back in the form of remittances. By 2010, the flow of total remittances was close to $25 billion. The flow varies over time. Economic downturns reduce immigration and hence the flow of remittances. The longer migrants stay

TABLE 12.1 Inflows of Remittances, 2010	
COUNTRY	**REMITTANCES AS PERCENTAGE SHARE OF GDP**
Bangladesh	11.8
Bosnia	12.7
El Salvador	15.7
Guyana	17.3
Haiti	15.4
Honduras	19.3
Jamaica	13.8
Jordan	15.6
Lesotho	24.8
Moldova	23.1
Nepal	22.9
Tajikistan	35.1
Tonga	27.7

Source: Based on World Bank and IMF data.

in a country, or if they bring in their own families or start new ones, the more the flow tends to drop off. In countries such as Saudi Arabia, where economic migrants are not encouraged to stay despite being vital to the economic functioning of the country, the level of remittances remains steadily high.

REFERENCE

See a web anthology of migration and resources: http://essays.ssrc.org/remittances_anthology/home/

argued for protection of home industries and the possession of overseas colonies both to ensure cheap raw materials and secure markets and to deny access to competitors.

This second wave of globalization has a major disjuncture, the modern "Thirty Years War" from 1914 to 1945, marked by two world wars and a global economic depression. It was a time of dislocation and economic disintegration, but the experience of world wars led many countries to seek the basis for a renewed integration. The charter of the United Nations was first drawn up in 1945 to provide a forum for international dialogue. The Council of Europe in 1949 was followed by the European Coal and Steel Community in 1952, the European Economic Community

in 1958, and then the European Community in 1967. From the ashes of the war there arose a conscious attempt to create a unified Europe. In the new postwar world, rules and organizations were established to stop the world economy from slipping into the downward spiral caused by the national economic protectionism that occurred during the Great Depression. Delegates from forty-four countries met in 1944 in Bretton Woods, New Hampshire, to establish a new economic order that stressed free trade. They created the World Bank and the International Monetary Fund (IMF) to integrate national economies into a global system. The United Nations, IMF, and World Bank were systems of international regulation that provided the

12.8 Box containers sit on a wharf in Bermuda.

backbone of a globalized world, a world now centered more on the United States. While the UN is headquartered in New York, the IMF and the World Bank are only a few blocks from the White House in Washington, DC.

The Third Wave

A **third wave of globalization** emerged after 1970. It is associated with an even more marked space-time-cost compression, more global production chains, and a deregulated global financial system, creating a "flatter" world with easier and hence quicker worldwide economic links. The icon of the flat world is the **shipping container** (Figure 12.8). First developed in 1956, the container permits the easy and cheap global transportation of trailer-sized loads from ships to trains and trucks without unloading the containers' contents. It is not the shipping container that created economic globalization, but globalization that created the container. The container both flattened the world and emerged from this flattening. A series of decisions between 1968 and 1970 by the International Organization of Standardization created global standards of container size, markings, and corner fittings. The result was a truly global form of cheap transportation. In 2012 close to 20 million containers made over 200 million journeys.

The steep fall in transport costs has made everyday things from around the world readily available. It is the flattening of the world as low transport costs enable relatively cheap products to be made in one country, and then sold in another halfway around the world. Blueberries grown in Chile, T-shirts manufactured in Vietnam, cheap plastic toys made in China, and white wines bottled in New Zealand can all be purchased in the same store in Syracuse, New York. The global trade in goods is an important part of our everyday lives. And global trade now has the power to influence, affect, and even destroy local and national economies. Distance and transport costs are no longer the barrier they once were. A flatter world is a smaller world.

Global Shift

Global shift is used to describe the movement of industrial employment from the industrial regions and cities of the developed world to those of the developing world. Since the 1970s there has been a redistribution of manufacturing employment from Western Europe and North America to Asia, Latin America, and other parts of the world, involving the deindustrialization of the advanced economies and a rapid industrialization of selected cities in a small group of developing countries, including China, Singapore, Taiwan, and South Korea (Figures 12.9 and 12.10).

Smooth global transport networks reduce the cost of transporting goods around the world, and the development of routine manufacturing processes allows manufacturing to become established in areas of unskilled labor. Economic globalization allows companies to search for cheap labor outside their national borders. Economic globalization allows corporations to relocate in order to minimize wage costs and effectively deterritorialize. The old adage that what is good for General Motors is good for America no longer applies, as General Motors now makes cars in China and Europe. Although listed as a US company, Nike's interests do not necessarily parallel US interests. What is good for Nike shareholders is good for Nike shareholders; whether it is good for US workers is beside the point.

Global shift is moving up the manufacturing chain. In the early 1990s, China's economic growth was dominated by traditional labor-intensive manufacturing sectors such as textiles, clothing, and footwear. More recently, however, growth has been more noticeable in the capital-intensive, high-tech sectors such as machinery and electronics.

As manufacturing jobs are routinized and deskilled, they become capable of being offshored. Production is being pried away from pools of skilled and higher-paid labor in traditional manufacturing regions. This shift has two consequences. The first is the rise of a new working class in the newly created industrial regions. A female working class emerges as young women fill many of the new jobs in the maquiladoras along the US–Mexican border or in the industrial cities of coastal China. The second is the decline in the economic and political strength of the organized working class of the old industrial regions of Europe and North America.

The global shift in manufacturing employment was most pronounced from the 1970s to the 2000s. Since 2010 there has been a further redistribution as some manufacturing jobs relocated back to the United States. From 2010 to 2012, 525,000 manufacturing jobs were added in the US, and 50,000 of those arose from overseas firms moving operations

12.3 Transnational Corporations

There are four ideal types of **transnational corporations (TNCs)**:

- *multinational*, with a decentralized organizational structure that operates as a series of semiindependent entities;
- *international*, controlled from one central headquarters with overseas operations as appendages;
- *global*, in which a centralized hub implements parent-company strategies to reach a global market;
- *integrated network* of complex, globally integrated operations.

We can get some idea of the range by considering three TNCs: Nike, Starbucks, and Toyota. Nike is a sportswear and equipment company that exhibits characteristics of the integrated network. In 2011 its revenue was $20.6 billion. Nike does not make any goods directly. There are no Nike factories. Instead, the company buys goods from factory owners and then sells them to consumers. In Nike's early years, these factories were in Japan; by the early 1980s, as labor costs increased in Japan, most Nike shoes were made in South Korea, and the city of Pusan became the capital of Asian shoe manufacturing. Since then Nike has further reduced its costs by having shoes made in China, Indonesia, and Vietnam, where labor costs per worker were only $100 a month. Indonesia is now one of the largest suppliers of Nike shoes: seventeen factories employ 90,000 workers, producing around 7 million pairs of shoes. In southern China, the center of shoe manufacturing is the city of Guangzhou. Just outside the city, one shoe factory, used by Nike, makes 35,000 shoes a day.

Toyota is an automotive company headquartered in Japan but with factories around the world and a global workforce of over 300,000. It strongly exhibits characteristics of the international and integrated network. Toyota emerged in the 1930s and produced its first car in 1935. It made cars in Japan but wanted to sell them abroad. The problem was that many countries, especially those with their own auto industries, put up tariff walls to protect domestic producers. To jump these walls, Toyota established manufacturing plants in selected countries. It became a multinational company headquartered in Japan but with manufacturing plants around the world.

Unlike Nike, it retained direct control of the factories producing its products. Over the years they established a reputation for reliable, well-made cars that ranged from budget to luxury. A successful global brand was established, and in 2008 it became the world's largest automaker and the fifth largest company in the world, with significant market shares in North America, Asia, and Africa and smaller shares in Europe. It makes around 8 million vehicles a year, with manufacturing and assembly plants in twenty-seven different countries. Toyota has five major plants in the United States. Although a global company, Toyota is still anchored in Japan. The company is headquartered there, and the very senior management tends to be Japanese. As late as 2008, almost 5.1 million of the 8.5 million vehicles were produced in Japan. The explicit desire to become the world's number one automaker was an ambitious one. This drive to global dominance is, according to many analysts, the root of its recall problems in 2009–2010, when over 8.3 million vehicles were recalled for either accelerator or braking problems or both. The rapid increase in production perhaps outstripped quality control and the ability of engineers and managers to carefully scrutinize new model developments for design problems and associated safety issues. The problems were widely reported—a global company also gets global publicity when things go wrong.

The first Starbucks opened in 1971 with a small storefront cafe in Seattle. Its globalization is relatively recent. The first store outside the United States opened in Tokyo in 1996. By 2010, Starbucks had 16,858 stores in 52 countries and a global workforce of over 130,000. Starbucks cafes have remarkable similarity around the world. There are few nods to local particularity or national differences. In some cases, the generic nature creates backlash. There was early difficulty breaking into European markets, and Italy is still a holdout—a country with thousands of coffee bars, but no Starbucks.

What Starbucks has done is less to meet a demand than to create a demand. Starbucks now embodies a global coffee culture, sometimes building on existing cultural traditions, in other places creating them anew. It has become a third place between home and workplace where socializing and telecommuting occur as much as coffee drinking—a meeting place as much as a selling place. While in some places it signifies US cultural dominance, in others it is perceived as a symbol of modernity.

there. The reason: the narrowing wage gap between China and United States as wages increase in China and decline in the United States. The gap in wage cost in the respective countries shrank from $20 per hour in 2000 to an estimated $7 per hour in 2015. Rising productivity and increasing innovation are making manufacturing in the United States a more attractive proposition. We may be at the beginning of a global reshifting, especially for the advanced manufacturing sector. The more unskilled routine manufacturing will still continue to shift to areas with lower labor costs.

12.9 Abandoned factory in Syracuse, New York

Big-box retailers reinforce the world's flattening by forcing the global shift in manufacturing. Traditionally, manufacturers made things and then sold them to retailers, who sold them to customers. Retailers were simply in the middle. With the rise of the big-box retailer and a more flexible production system, with many small companies producing small runs of goods relatively quickly, power shifted from the large number of small manufacturers to the small number of larger retailers. The very large retailers have such enormous power that they can force producers to compete with each other to meet strict cost limits. Manufacturers are willing to do this because access to a big retailer ensures a large customer base. The lower price per item is offset by the higher turnover of items.

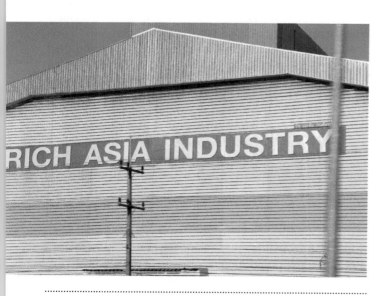

12.10 New factory on outskirts of Bangkok, Thailand

Global Production Chains

Production chains now stretch and snake their way across the globe looking for competitive advantage. The process is driven by competition. At one end, consumers seek greater value, and retailers, competing with one another to provide the best value, force producers into cheaper labor areas and more efficient production zones. At the other end of the chain, producers need to keep finding ways to reduce costs, improve techniques, and hire cheaper labor. As distance shrinks, relocation to ever-cheaper labor areas continues to flatten the world.

The geography of manufacturing is complex, with global supply chains linking iron ore fields in Australia to steel plants in China to fabrication sites throughout Asia, Europe, and North America. Factories are now linked into a dense circuitry of connections as products designed in one country are assembled in another, with parts drawn from suppliers in many other countries.

Textiles are a major sector. Textile manufacturing is a common gateway industry for newly industrializing countries, as the entry costs are low. Textile manufacturing can be established relatively cheaply; it is easier and cheaper to provide sewing machines than a car-making plant, and the skill requirements for workers is relatively modest compared to more complex manufacturing systems. For many nations textile manufacturing is an easier gateway from a primary-based to a manufacturing economy because it requires relatively smaller amounts of capital, lower skills, and less technological sophistication than advanced manufacturing.

As an example of one production chain, consider the manufacturing of clothes for retailers in the United States and Europe. One clothing retailer, Hennes and Mauritz (H&M), a Swedish company with stores in Europe and North America, keeps its prices low by contracting in low-wage areas of the world, such as Bangladesh, where 4 million people work in 5,000 textile factories, most of them women paid around $38 a month to work. H&M contracts with 900 factories worldwide to produce a constantly changing design portfolio. The company has been successful in keeping inventory low; the **just-in-time production** system ensures that goods are made to meet demand. Stores often receive daily supplies. High turnover means that profits can be made through selling many items rather than one; hence the price of individual items can be reduced, which in turn aids turnover. Workers, however, toil for low wages in often poor working conditions. Between 2005 and 2013, more than 1,800 workers died in Bangladesh in factory fires and building collapses in the textile sector. One building collapse in 2013 killed 1,127 workers.

While supply chains are global, there are also regional manufacturing clusters such as the garment factories in Dhaka. The small town of Poole in Dorset, England, is home to the two biggest makers of tiny electric motors that rotate on bearings made of gas molecules. The motors are

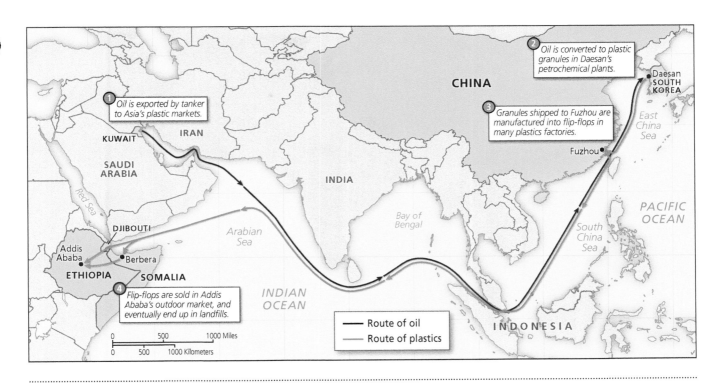

12.11 The flip-flop trail

used to make printed circuit boards. Warsaw, Indiana, is the center of medical implant makers. While global supply chains highlight the wide spread of a globalized economy, these clusters reveal the importance of spatial agglomeration for the regional economy.

On the Trail of the Flip-Flop

We can follow one global chain by looking at the simple flip-flop (Figure 12.11). Caroline Knowles followed the flip-flop trail. It begins in Kuwait where oil is pumped from underground and onto a supertanker that then makes its way to a South Korean petrochemical factory, where it is turned into plastic. The plastic is then shipped to a factory in China where it is transformed into flip-flops. Hundreds of thousands of flip-flops are then shipped to Djibouti and then taken to markets in Addis Ababa, Ethiopia, a major market for China's flip-flops. Even when they are abandoned in a giant landfill site, "scratchers," people eking out a living by recycling the rubbish, salvage them along with other materials. To complete the chain, one of the scratchers is saving the money gained by selling the recycled materials to buy an airline ticket so that she can travel to Dubai and maybe get a job as a maid.

Measuring Risk

Doing business in a global economy can be risky. Countries are now ranked on their vulnerability to crisis and collapse.

The Fund for Peace's Fragile States Index, the Center for Systemic Peace's State Fragility Index, the Economist Intelligence Unit's Political Instability Index, and the PRS Group's International Country Risk Guide (ICRG) all measure risk and state fragility. The ICRG, for example, publishes monthly rankings for 140 countries measured along twelve factors, including socioeconomic conditions, internal conflict, corruption, and ethnic tensions. There is also the Political Instability Task Force's global model, which focuses on political structures and relationships via four independent variables: regime type, infant mortality, conflict-ridden neighborhoods, and state-led discrimination. Figure 12.12 shows the map produced by the Fund for Peace.

International Nongovernment Agencies

Certain **nongovernment agencies** are important actors in the global economy. Three in particular have great influence: the International Monetary Fund (IMF), the World Bank, and the World Trade Organization (WTO). The IMF was initially set up to enhance world trade by providing monetary and currency stability. By 2012 the IMF had

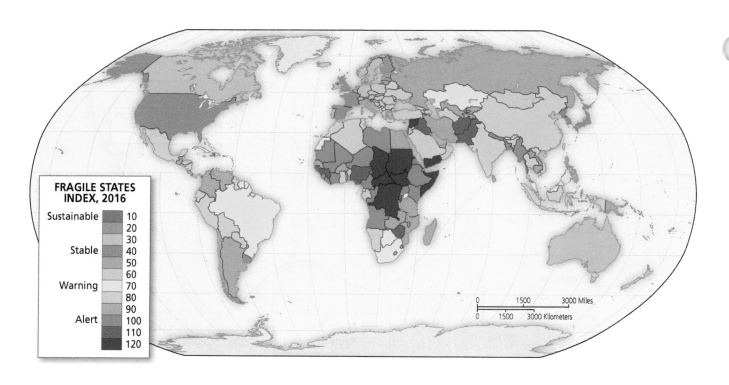

FRAGILE STATES
INDEX, 2016

Sustainable   10
              20
              30
Stable   40
              50
              60
Warning   70
              80
              90
Alert   100
             110
             120

12.12 A risky world

188 members. Each member country contributes a certain sum of money as a credit deposit. Collectively, these deposits provide a pool of money from which members may draw. The IMF acts like an international credit union for states. A country's deposits determine how much it can draw upon, known as "special drawing rights," and its voting power. Rich countries dominate the IMF, as they contribute most of the funds and set the agenda through their voting power. The IMF wields its power through its **surveillance** system and the strings it attaches to lending. Surveillance (the term used by the IMF) involves monitoring of a member's economic policies and evaluation of their economy. One example: South Korea asked for IMF assistance on November 21, 1997, in the wake of depleted foreign reserves and rising short-term loans. The IMF agreed to lend $56 billion and in exchange asked for an economic adjustment package that depressed economic growth, caused unemployment, cut government spending, led to a decline in living standards, and granted foreign financial institutions more access. The IMF imposed market principles and freer international trade on Korean society; its argument was that South Korea's problems were due to lack of transparency in financial dealings, a bloated public sector, and a job market that ensured jobs rather than productivity.

The World Bank was created in 1944. Its initial name, the International Bank for Reconstruction and Development, describes its role (Figure 12.13). It was a fund established to aid the reconstruction of Europe, and its first loan was

$250 million to a war-damaged France. Once Europe got back on its feet, the Bank widened its remit. The first loan to a developing country was $13.5 million given to Chile in 1948 for a hydroelectric scheme.

The World Bank now has almost universal membership. Like the IMF, it was and still is dominated by rich countries. In response to criticism, both internal and external, from both the developing and developed world, from both private investors angry at delays and administrative snafus and aid workers upset at corruption and inefficiency, the World Bank shifted its emphasis in 1989 to eradicating world poverty with more carefully assessed and monitored development schemes.

The WTO is the latest version of what was originally called the General Agreement on Tariffs and Trade (GATT), established in 1948 as a forum for stimulating world trade by reducing customs duties and lowering trade barriers. GATT was a forum for getting rid of protectionism. The first round of multilateral discussions, held between 1948 and 1967, led to some tariff reductions. The long, drawn-out nature of the discussions led many commentators to suggest that the acronym stood for "General Agreement to Talk and Talk." The most decisive round was the Uruguay round, 1986 to 1994, which established new rules for trade in services and intellectual property, new forms of dispute settlement, and trade policy reviews, as well as the creation of the WTO in 1995. The Doha round began in 2001 and is not yet concluded. The WTO now has 159 members that

12.13 Headquarters of World Bank, Washington, DC

Milton Friedman (1912–2006), a Nobel Prize–winning economist, also lauded the virtues of free markets and small, limited governments. His book *Free to Choose*, published in 1980, linked individual freedoms with functioning markets and restrained governments. In this restated neoliberalism, unregulated markets were the solution and governments were the problem.

The neoliberal agenda promotes the privatization of public goods and services, reduced taxes (especially for the wealthy), reduced governmental commitment to the welfare of its ordinary citizens, and an emphasis on enhancing corporate profitability and improving business competitiveness. This agenda's core proposition is that deregulated markets will increase economic growth and raise living standards. Neoliberalism as a political theory has the benefits of being simple, easily understood, and even easier to articulate. As a political practice, however, its greatest difficulty lies in implementation when it comes up against the brute realities of mature markets dominated by powerful sectional interests. These interests promote free trade when it meets their needs but not when it undercuts their power and market share. The tenets of neoliberalism are breached regularly in the face of political realities. Elaborate subsidies and tariff barriers, for example, maintain agricultural producers in the developed world. While efficient Japanese carmakers want free trade, inefficient Japanese rice farmers do not. The US financial services sector wants to penetrate foreign markets, yet the US agricultural lobby funds the erection of import-blocking tariffs.

The neoliberal agenda is promoted by global systems of governance. Both the IMF and the WTO, for example, promote neoliberal economic policies. Around its central ideological core, neoliberalism has subsidiary ideas about accountability, choice, competition, incentives, and performance. It is also a political process that recasts citizens as consumers, reimagines states less as providers of public services and more as promoters of private growth, and shifts governance issues from citizen entitlements to consumer choices. In full-blown neoliberalism, markets trump the state, capital wins out over organized labor, consumers replace citizens, and market choice replaces citizen rights. The private market and the accompanying individual consumerism are enthroned as the means and measure of success.

Yet there are sites of resistance and narratives of dissent. There has been the development of a global discourse

account for 97 percent of the world's trade. China became a full member in 2000. Membership in the WTO is vital to countries seeking access to global markets. But this access comes at a price. For the stronger economies, free trade can mean access to new markets, especially for banking and cultural economies, but it can also mean the decline of traditional industries. For the weaker economies, membership signifies access to the large consumer markets of the rich world, but can mean an inability to control the fate of traditionally protected industries. Membership in WTO implies a willingness to shape national economic policy around the principle of free global trade.

The Promise and Reality of Neoliberalism

Neoliberalism, an economic ideology that promotes deregulation, minimal or small government, low taxation, and free trade, has circulated widely around the world. It has a number of strands, a long history, and, now, pervasive adoption by governments and economic elites all around the world. Classical economists of the eighteenth century, such as Adam Smith, believed in the invisible hand of the market to make things work. More recently, it is part of a broader shift in economic ideologies away from the New Deal Keynesianism that dominated from the 1930s to the late 1970s. Neoliberal theorists included Friedrich Hayek (1899–1992), who, in his 1944 book *The Road to Serfdom*, argued against centralized government planning.

of human rights, environmental protection, and shared projects of poverty reduction. An alternative global vision to neoliberalism is emerging.

A Flat World?

The journalist Thomas Friedman wrote an influential book with the provocative title *The World Is Flat*. But is the world really flat? Is the smoothing of the differences enough to justify using the term "**flat world**"? Two counterarguments can be made. The first is the simple fact that only some of the world is flat, and by "flat" I mean a smoothing of the tyranny of economic distance and a lessening of the friction of national economic differences. Production chains snake their way only into selected countries. Nike athletic shoes are made from components in some countries, including China, Indonesia, and Vietnam, but not in Myanmar, Pakistan, Somalia, or Zimbabwe or most other developing and poor countries. Starbucks has few stores in sub-Saharan Africa. There are limits to economic globalization. The cheapness of labor is not the only issue for a mobile capital. Reliable infrastructure, cheap transport links, the rule of law, and a stable political system are all prerequisites for long-term global shift, and much of the developing world lacks one or many of these elements. The world is not completely flat, but parts of it are, where distance is almost frictionless and economic transactions are easily connected to other parts of the world. In other parts, however, distance is overcome only at great transactional costs. We live in a misshapen world, parts of which are flat, others curved, and yet others pockmarked by disconnected ravines and inaccessible peaks.

The world's topography is more varied than flat. At the international level, there are still many countries only loosely connected to global flows of investment or global shifts in manufacturing. And even within the more connected, flatter countries, there is often a marked demarcation between the more inaccessible rural areas and the metropolitan regions where most of the global economic activity is concentrated. Many of the rural areas of the developing world are far from markets and lack the necessary infrastructure to ensure easy and cheap transportation, the hallmark of a flat world. The flat areas, where much of the world's global economic activities take place, are like landing strips in the middle of rugged terrain, often constituting small parts of larger countries and, in total, only a tiny part of the Earth's economic surface.

The second point is that economic globalization is about creating hierarchies as well as flat surfaces; it generates spatial difference as well as homogeneity. Economic globalization is not just the product of a consumerist culture and a competitive capitalism continually seeking to squeeze costs. Since the 1970s, there has also been a globalization of finance.

Capital now flows in larger and more viscous currents across the world. In this case, the flattening of the world is also tied to the networking of the world.

One particularly dynamic sector of services is the knowledge-based industries, so-called **advanced producer services** such as advertising, banking, and financial services for the larger corporations. Together these sectors constitute the dynamic edge of mature capitalist economy. Since 1980, in the developed world, a region's success rests less on manufacturing employment and more on the extent to which it can generate, retain, and attract knowledge-based employment. This sector is heavily concentrated in the global cities. Flat surfaces are ideal to move goods cheaply and quickly, but in order to do business you need nodes of concentrated activity. A global system of cities, connected by flows of capital, ideas, information, and skilled personnel, "landscapes" the world into concentrated nodes of heightened connectivity and accumulated control. We will discuss this topic more in a later chapter.

In summary, while there has been a flattening of the world, a flat world is not yet a reality.

The differential flattening that does occur has redistributional consequences. Globalization has different effects on capital and labor and on different sectors of labor. While capital is freer to move across the surface of the globe, labor is much more restricted. The low cost of international transport and the growing ease of international trade, crucial requirements of economic globalization, have allowed capital to be more easily disassociated from national interests and local community concerns. Capital is free to roam the space of the world in search of lower wages and higher returns, while labor is more immobile. Retailers can move their production contracts to another factory in another country, but organized labor is more fixed in place. The result is an uneven bargaining arrangement. Globalization has liberated capital but largely restricted workers to national markets.

There are different levels in the flows of globalization that lead to different geographies of globalization. At the top level of the flat world, capital flows with fewer and fewer restrictions. Here are the global rich and the highly skilled, especially in the field of global business. At the bottom are the unskilled and those without capital, whose movements are much more circumscribed; they live in a more rugged topography.

Long ago, distance severely restricted economic interaction. The cost and difficulty of movement meant that most contact was between peoples and areas close together. The world that people inhabited was shaped more by the local than the global. The geographic-historical long view would perceive a world gradually coming together, not as a continuous process but in a series of fits and starts with some areas now pulled together, now pulled apart. World historical geography is in large part the story of global space-time convergence and successive waves of globalization.

Cited References

Banister, D. (2011) The Trilogy of Distance, Speed and Time. *Journal of Transport Geography* 19: 950–959.

Barrett, H., Wilbery, B., Brown, A., and Binns, T. (1999) Globalization and the Changing Networks of Food Supply: The Importation of Fresh Horticultural Produce from Kenya into the UK. *Transactions Institute of British Geographers* 24: 159–174.

Friedman, T. (2006) *The World Is Flat: A Brief History of the Twenty-First Century.* New York: Farrar, Straus & Giroux.

Horner, A. (2000) Changing Rail Travel Times and Time-Space Adjustments in Europe. *Geography* 85: 55–68.

Janelle, D. (1968) Central Place Development in a Space-Time Framework. *Professional Geographer* 20: 5–10.

Knowles, C. (2014) *Flip-Flop: A Journey Through Globalization.* London: Pluto.

Select Guide to Further Reading

Crosby, A. A. (1972) *The Columbian Exchange: Biological and Cultural Consequences of 1492.* Westport, CT: Greenwood.

Dicken, P. (2015) *Global Shift: Mapping the Changing Contours of the World Economy.* 7th ed. New York: Guilford.

Dumenil, G. and Levy, D. (2011) *The Crisis of Neoliberalism.* Cambridge, MA: Harvard University Press.

Engel, C. and Wang, J. (2011) International Trade in Durable Goods: Understanding Volatility, Cyclicality, and Elasticities. *Journal of International Economics* 83: 37–52. http://www.ifw-kiel.de/konfer/staff-seminar/paper/2009/engel.pdf

Goldstone, J. A., Bates, R. H., Epstein, D. L., Gurr, T. R., Lustik, M. B., Marshall, M. G., Ulfelder, J., and Woodward, M. (2010) A Global Model for Forecasting Political Instability. *American Journal of Political Science* 54: 190–208.

Harvey, D. (2007) *A Brief History of Neoliberalism.* New York: Oxford University Press.

Khanna, P. (2016) *Connectography: Mapping the Future of Global Civilizations.* New York: Random House.

Knowles, R. D. (2006) Transport Shaping Space: Differential Collapse in Time-Space. *Journal of Transport Geography* 14: 407–425.

Mann, C. C. (2006) *1491: New Revelations of the Americas Before Columbus.* New York: Knopf.

Mann, C. C. (2011) *1493: Uncovering the New World Columbus Created.* New York: Knopf.

Marsh, P. (2012) *The New Industrial Revolution: Consumers, Globalization and the End of Mass Production.* New Haven and London: Yale University Press.

Sheriff, C. (1997) *The Artificial River: The Erie Canal and the Paradox of Progress.* New York: Hill and Wang.

Short, J. R. (2001) *Global Dimensions: Space, Place and the Contemporary World.* London: Reaktion Books.

Short, J. R. (2004) *Global Metropolitan: Globalizing Cities in a Capitalist World.* London: Routledge.

Sobel, D. (1995) *Longitude: The True Story of a Lone Genius Who Solved the Greatest Scientific Problem of His Time.* New York: Walker.

Steger, M. and Roy, R. K. (2010) *Neoliberalism: A Very Short Introduction.* New York: Oxford University Press.

Websites

Bank for International Settlements' International Financial Statistics
  http://www.bis.org/statistics/about_banking_stats.htm
Center for Systemic Peace State Fragility Index
  http://www.systemicpeace.org/
Connectography
  http://www.paragkhanna.com/connectography/
Economist Intelligence Unit's Political Instability Index
  http://www.eiu.com/
Fund for Peace Failed States Index
  http://www.fundforpeace.org/global/?q=fsi
International Economy
  http://www.esa.doc.gov/about-economic-indicators
  http://www.internationaleconomics.net/data.html
  http://data.worldbank.org
  http://www.imf.org/external/data.htm
  http://libguides.princeton.edu/internationalecon

http://www.newyorkfed.org/research/global_economy/globalindicators.html
International Trade Centre's Trade Map
  http://www.trademap.org
On the trail of the flip-flop
  http://www.flipfloptrail.com
PRS Group's International Country Risk Guide
  http://epub.prsgroup.com/products/international-country-risk-guide-icrg
United Nations Comtrade Yearbook (2015)
  http://comtrade.un.org/pb
World Bank Remittances Data
  http://www.worldbank.org/en/topic/migrationremittancesdiasporaissues
World Bank Trade Indicators
  http://data.worldbank.org/topic/trade

Interactive Map Websites

Global Competitiveness
http://reports.weforum.org/global-competitiveness-report-2015-2016/interactive-heatmap/

Global Economy
http://www.tradingeconomics.com/country-list/maps

World Bank maps
http://econ.worldbank.org/WBSITE/EXTERNAL/DATASTATISTICS/0,,contentMDK:20398804~menuPK:1545601~pagePK:64133150~piPK:64133175~theSitePK:239419,00.html

Learning Outcomes

- While the pace of globalization has increased in modern times, it is long established as a continuing process of the human occupancy of the globe and the interaction among its peoples and societies.
- Space-time convergence has reduced the cost and time it takes to transport people and goods, bringing the world closer together and contributing to globalization.
- A series of space-time convergences caused by new developments in transportation effectively brought the world closer together, including those caused by the invention of the wheel and compass, improved navigation, steam power, new canals and railways, modern ships and containers, and the proliferation of the usage of cars and planes.
- As an economy matures, the transfer and exchange of information become as important as the movement of goods, and the time-cost begins to outweigh space-time compression.
- Space-time convergences, rather than occurring either randomly or regularly, were bunched together in three distinct waves of globalization: the first wave following the Columbian Encounter of 1492; the second wave, spanning 1865 to 1970 and triggered by the appearance of the railways, the telegraph, and the steamship, but also impacted by the two world wars and subsequent worldwide depressions; and the third wave that has emerged since 1970 and created even easier and quicker global economic links.
- The icon of the third wave's "flat world" is the shipping container, which allows easy and cheap global transportation of trailer-sized loads from ships to trains and trucks without breaking bulk. It is not the shipping container that created economic globalization, but globalization that created the container.
- As both international migration and capital mobility have increased, so has the level of remittances. As it becomes easier and much cheaper to send relatively small amounts of money back home, modest remittances have joined the global money flow.
- The global trade in goods is an important part of our everyday lives and has the power to influence, affect, and even destroy local and national economies.
- Since the 1970s a global shift has redistributed manufacturing employment from Western Europe and North America to Asia, Latin America, and other parts of the world.
- The low cost of international transport and the growing ease of international trade, crucial requirements of economic globalization, have allowed capital to be more easily disassociated from national interests and local community concerns.
- Globalization has involved a pronounced global shift of manufacturing employment from the rich industrial countries in North America and Western Europe to newly industrializing regions. Two resulting consequences are the new working class created in the newer regions and the declining economic and political strength of the organized working class in the old.
- Some researchers now point to an increasingly "flat" world in which economic distances and differences are reduced; however, some parts of the world are considered more flat than others.
- Decent, reliable infrastructure, cheap transport links, the rule of law, and a stable political system are all prerequisites for long-term global shift and are inhibitors of economic globalization.

The Global Geography of Culture

13

Culture is what makes us social animals. Cultural forms are rarely fixed; they are best considered as circulating flows of ideas and practices that are always in the process of being embedded in place well as diffusing over space. In this chapter we will elucidate these ideas of fixity and flow with reference to architecture and tourism and show how culture is commodified as well as celebrated.

Culture is a tricky concept. It is used to cover things in the material world as well as systems of belief, ideas, and practices. The word has historically implied the notion of cultivation, of tending to something. This is an important point, because it reminds us that culture is not an a priori set of characteristics; it is something that is produced and cultivated as a response to specific circumstances. Culture is always provisional, elastic, and in a continual process of being made and remade.

Until the late eighteenth century, the word "culture" was often interchangeably with "civilization," referencing a belief in the Enlightenment ideology of human mastery over nature. Yet at the end of the eighteenth century, thinkers such as the German philosopher Johann Gottfried Herder (1744–1803) wrote of the need to conceive of a plurality of cultures. There were entire cultural worlds below the privileged life of the court and beyond the refined manners of the elite. Local peoples also had culture. This view was reinforced in the nineteenth century through a self-conscious creation of "national" and "folk" cultures. Take a single example: in the nineteenth century, the Czech National Revival was a broad-based cultural and political movement in which a Czech identity was crafted in music, arts, and literature. Czech composers such as Bedřich Smetana and Antonín Dvořák drew upon folk motifs in their chamber music and operas. It was an attempt to forge a distinctive national culture at a time when the ruling elite spoke and wrote in German, and globalization, according to some,

LEARNING OBJECTIVES

Understand the concept of the cultural region.

Describe the process of spatial diffusion of culture.

Summarize the sources of cultural production.

Explain how cultural products become deterritorialized or reterritorialized, and cultural practices transplanted or transformed.

Reflect on the global production of culture.

was sweeping away nationally distinct and uniquely local cultures. The defense of culture involves the invention of culture. Wherever and whenever a "culture" is perceived as weakened by contact with either a specific foreign other or a generic global other, local cultures are not only preserved and defended, they are also created and expanded in a complex dance between the local, the national, and the global. Whenever a wave of globalization sweeps around the world, there is a corresponding revival of the local and the national. Resistances to globalization often take the form of more explicit concern and promotion of national and local cultures as a way for groups to forge, redefine, or reestablish a place in the world. Culture is both a resistance to and an accommodation with wider social forces.

Cultural Regions

A **cultural** or **culture region** is defined as an area of similar culture. Different types of cultural region can be identified on the basis of language, religion, ethnicity, or shared history. The historical geographer Don Meinig, for example, identified a Mormon culture region consisting of a core in the Wasatch Mountains, a domain over much of Utah and southeastern Idaho and a wider sphere extending from eastern Oregon to Mexico (see Figure 10.9).

A culture region is different from other areas due to human behaviors and or perceptions. Thus the American South is an area of distinct speech, history, and sense of itself compared to the rest of the USA. The exact boundaries are more fluid than fixed. With cultural regions we need to adopt Meinig's notion of core and surrounding peripheries.

The existence of a cultural region is in part a self-identification by the majority of people in the region as well as by external perceptions. Figure 13.1A shows the level of agreement about the American South by respondents in Michigan and Indiana. Notice how 91 and 96 percent of all respondents agree upon a very narrow demarcation.

The culture in cultural regions is sometimes contested rather than given. Not all in the US South celebrate the traditional definition of the South. Its complex history as a place of segregation and rebellion against federal authority make for complicated cultural celebrations. Cultural regions are not homogeneous. Not everyone in the Bible Belt is Christian or religious. Cultural regions are complex, part created, sometimes imposed, often contested, and never easily defined, especially outside of a very narrow core area.

While traditional cultural geography focused on the identification of cultural regions, much of contemporary cultural geography is concerned with culture as flow. Before we discuss this, however, we need to understand the concept of spatial diffusion.

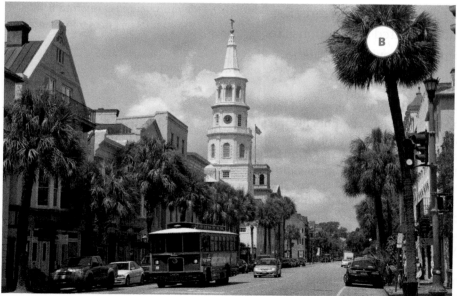

13.1 (A) Levels of agreement about the location of the US South. **(B)** Downtown Charleston, South Carolina.

13.1 The Diffusion of Diseases

The spread of disease is both an example of diffusion as well as of space-time convergence. Space-time convergence, the shrinking of the time taken to cover distance, influences the diffusion of disease. The travel time between India and Fiji during the time of sailing ships was longer than the lifespan of the measles virus. The virus was thus not imported when Indian laborers were shipped to the islands as cheap labor for the plantations. People infected with measles when they left India had enough time to recover from the disease before they docked in Fiji. But travel times during the age of steamships decreased to within the incubation period of measles. Measles contracted in India was then imported into Fiji, with devastating results.

In the case of Iceland, with increasing connection to the outside world, waves of measles outbreaks became more frequent. The coming of the railway in the United States made it easier for cholera epidemics to sweep across the country. In our globalized world, there is a greater possibility of global pandemics. The influenza pandemic of 1918 killed between 3 and 5 percent of the world's population, with total death estimates ranging from 50 to 100 million.

The collapse of distance brought about by recent transport improvements, such as jet travel, has increased the ease and frequency of the transfer of communicable diseases. The swine flu epidemic of 2009, similar to the 1918 pandemic in that it was an H1N1 influenza virus, infected one-fifth of the population and more than half of all children in nineteen countries; it killed more than 200,000 people around the world. Although very contagious, it was not very lethal, killing fewer than 2 in every 10,000 infected. An equally contagious but more lethal virus will generate the kind of global death rate last seen more than a hundred years ago. If a contagious deadly virus does emerge, it will move quickly and inexorably across the globe.

REFERENCES

Cliff, A. and Haggett, P. (2004) Time, Travel and Infection. *British Medical Bulletin* 69: 87–99.

Cliff, A. D., Haggett, P., and Raynor, M. S. (2004) *World Atlas of Epidemic Diseases*. London: Hodder Arnold.

Kerkhove, M. D., Hirve, S., Koukounari, A., and Mounts, A. W. (2013) Estimating Age-Specific Cumulative Incidence for the 2009 Influenza Pandemic: A Meta-Analysis of A(H1N1)pdm09 Serological Studies from 19 Countries. *Influenza and Other Respiratory Viruses* 7: 872–886. http://onlinelibrary.wiley.com/doi/10.1111/irv.12074/abstract

Spatial Diffusion

Spatial diffusion is the movement of new ideas, things, and practices through time and space. The adoption of innovations follows predictable patterns. Let us take for example the adoption of Christianity from its center in the Middle East. It spread along lines of communication, the trade routes that linked the cities around the Mediterranean. From there it spread quickly through towns, into the countryside, and around the world.

Not everyone adopted the new religion. Figure 13.2 shows the typical pattern for a new innovation. Innovators are those people who quickly pick up and adopt new trends and ideas. They are the first person you know with an iPhone and the latest app. Then there are the early majority who come in the wake of the innovators, then the late majority, and finally the laggards. I am most definitely a laggard in the adoption of new technologies. I am always the last person I know with a cellphone or an Uber app.

Diffusions also vary across the urban hierarchy. Figure 13.3 could be used to model the early adoption of Christianity. It occurred first in the towns and then later diffused through the countryside. The slow diffusion into the rural areas leaves us with the linguistic legacy of "pagan," derived from an early Latin word for villager.

Ideas and practices are abandoned as well as adopted. The rejection of a former innovation is referred to as a **paracme**. Figure 13.4 shows three situations. In (a) there is symmetry between a diffusion and a paracme; situation (b) occurs when there is long, slow adoption and quick rejection, as in the case of smoking in the USA, which developed slowly over a century then plummeted with health scares. In (c) there is quick diffusion and a slow paracme, as with the case of the rapid adoption of railways in the nineteenth century, followed by a long, slow decline.

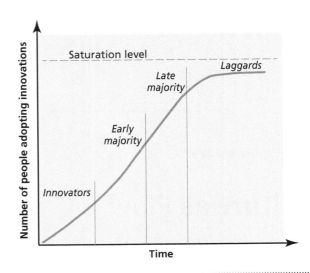

13.2 The spatial diffusion model for a new innovation

13.3 Diffusion of an idea through the urban hierarchy

13.4 Diffusion and paracme

Culture as Flow

Traditional cultural geography highlighted the connection to the local. Emphasis was on vernacular building forms, local ways of doing things, and the construction of specific

landscapes. More recently, the field has been enlivened by an awareness that culture is shaped not only by people's connection to the local, but also by their interaction with the national and the global.

Consider just one example, chocolate. It developed first in the Mayan region of Central America, a region that also saw the first cultivation of corn and tomatoes. The local people made an unsweetened chocolate drink from locally grown cacao seeds. The Aztecs refined the process by pouring the chocolate liquid through blowers. They annexed the region around 1400 and soon levied taxes paid in cacao. Later the Spanish adopted chocolate as a favorite food, adding sugar and cinnamon. The concoction was then imported into Europe where it became popular. The demand grew so high that industrial production levels were needed and plantations were established in West Africa. Today, more than two-thirds of the world's cocoa is produced in West Africa, but the chocolate is manufactured in Belgium, Switzerland, and Hershey, Pennsylvania. So, a food grown from seeds in Mesoamerica is refined by the Aztecs and the Spanish, imported into Europe, and creates huge demand that leads to cocoa plantations in Africa and manufacturing plants around the world. When you bite into a chocolate bar, you are taking part in complex cultural flows that have spanned the globe for at least five hundred years.

There are three dynamically linked sources of production of culture: the connection with the local, the relationship with the national, and the interaction with the global. In this chapter, we will focus on the connections between the global and the local.

Cultures are never static, fixed in time and place. A subtler view of cultures sees them as resulting from ongoing interactions between the local, the global, and the national. The processes that shape local and national cultures are not one-way interactions but rather are dynamic and multifaceted, so that hybrids of the "newly arrived" and the "previously there" are constantly reconfigured and remobilized in and through global flows of people, ideas, and beliefs. Culture is not just passive audiences watching imported television programs or eating fast food from global chains. Culture is an active process; one definition of culture is "tending, husbanding resources," with the same root as "cultivation." A culture is not just consuming things; it is the production of a worldview, an aesthetic sensibility, the active creation of a history and geography, the working out of a place in the world. Culture is the active use of global flows of ideas and practices by local and national groups adapting, adopting, or rejecting these flows in their own specific worldviews, which in turn become part of further flows of culture used and adapted, rejected and resisted, by others in turn.

Global flows of culture are continually shaping and reshaping the world. We can identify flows of people, technologies, capital, media images, and the political configurations of such ideas as freedom, welfare, rights, sovereignty, representation, and democracy. Flows circulate globally and become reexpressed through and in local contexts.

The (Re- and De-) Territorialization of Culture

When we think of culture, we often think of distinctive songs, dances, foods—all those things that comprise collective memories and group identities. In much of our understanding, these are all tied to place: Mexican food, Indian music, Italian design, the Armenian language. The term "**territorialization** of culture" signifies this connection between culture and specific places. However, it is also important to consider the process of the **deterritorialization** of culture, in which cultures are lifted from particular places, and the reterritorialization of culture, whereby cultural practices are transplanted and transformed.

Migrant communities throughout the world are important networks of information flows, capital flows, and economic exchanges. Just as chains of migration are routed through families and friends, so business deals are conducted among family members dispersed through the world. Economic globalization is intimately linked to this particular form of global population dispersal. Businesses are culturally embedded. A flattening world is linked to the deepening economic interactions of far-flung communities, with shared family ties, ethnicity, and race also functioning as conduits of international trade and commerce.

International migration shapes the culture of both destination and origin areas. When migrant communities are established, they bring their culture with them. Sometimes it is frozen in the time of departure and lasts longer than it does at home. Yet cultures always have enough plasticity to adapt to the new and the foreign. Both home and away, origin and destination are constantly changing in response to these continual interactions and flows.

Globalization, it has often been argued, undermines local identities. However, an alternative case can be made that globalization has both deterritorialized and strengthened local cultures. In the nineteenth century, there was a massive immigration to the United States from many European countries. Italians, Germans, Poles, and Irish all moved in the millions to the United States. Their connection with home was always limited, separated as they were by rudimentary communication links. Letters could

take weeks. New identities were created. People from different parts of what is now Italy became "Italians" in the United States: a kind of Italian nationality was achieved in the United States even before there was an Italian nation-state. New identities were shaped. Irish Americans took on the mythic elements of Irish nationalism. Folk memories did not die; indeed, in many cases they became both strengthened and stuck in time. Irish Americans, unlike the Irish in Ireland, could scarcely "remember" anything other than the famine. In more recent years, however, immigrants to the United States and other destination countries have been able to stay in closer connection with their families back home. Cheaper travel, ease of transmitting money, and instant communications all allow diasporic communities to remain in contact, to influence their new surroundings, and to be subtly transformed in the process of being diasporic. All those easier communication systems of a globalized world allow groups to be more in touch with their home areas.

The widespread global diffusion of flows of people and ideas has broken the simple, unproblematic connection between culture and place. On closer inspection, the connection between culture and place is always complex and problematic. "Mexican" food, for example, is a mixture of Spanish and indigenous cuisines that evolved in different parts of the world in different ways for almost five hundred years. When Spaniards came to the New World and enslaved Africans were shipped across the ocean, they combined with indigenous people to create a New World cuisine that was creolized and hybridized right from the beginning. There is no unchanging Mexican cuisine; it is and always was a changing hybrid.

Cultures are always in the process of deterritorialization and reterritorialization, particularly marked during waves of globalization and reinforced by rapid space-time

13.5 The Parthenon, Athens. Work began on this building in 447 BCE. It was dedicated to the goddess Athena.

convergence. We can consider the reterritorialization of culture as revealed by architecture.

Architecture

For the great German polymath Johann Wolfgang von Goethe (1749–1832), architecture is "silent music." Two of the dominant melodies of this music are the vernacular and the global. **Vernacular architecture** uses local resources and traditions in the design and construction of buildings. The materials draw on local sources, whether they are wood and clay, ice or rock, and architectural design is a calibrated response to local conditions of weather and topography. Classic examples would be the Inuit igloo, built of local ice, or the Swedish Lutheran church, built of local birch.

13.6 The neoclassical design of the Lincoln Memorial, Washington, DC. Construction commenced in 1914.

There are also **transnational architectural styles** that transcend particular places. Consider the classical architecture of Greece and Rome, so admired that they have been copied and recopied all over the world (Figure 13.5). **Neoclassical architecture**, the style that echoes the classic architecture of Rome and Greece, became popular in the mid-eighteenth century and soon was used in countries and cities worldwide. Here we come to a complex connection between place and culture. Neoclassicism meant different things in different places. In the United States, it was associated with democratic values and republican virtues (Figures 13.6). In Nazi Germany, it was associated with the attempt to build a thousand-year Reich based on racist beliefs, with the use of "timeless" classical imagery reinforcing the political agenda of Nazi ideology.

From the 1950s to the 1980s, architectural designs around the world were associated with modernist designs. **Modernist architecture** was concerned with form following function, truth to material, and a revulsion against bourgeois decoration. It preferred clean-cut lines and soon became serialized into tall, flat-topped buildings constructed in towns and cities all over the world—so specifically meant for global usage that it was designated "The International Style." And it truly was international: Communist headquarters in Eastern Europe, corporate offices in New York, government buildings in Brasilia, and public housing projects in London and Chicago all used it (Figure 13.7). Despite the differences in place, the buildings all looked the same. In the West, international modernist architecture soon reached dominance and then was so ubiquitous as to become boring. The modernist mantra of "less is more" turned into "less is a bore."

13.7 Modernist buildings along Sixth Avenue, New York

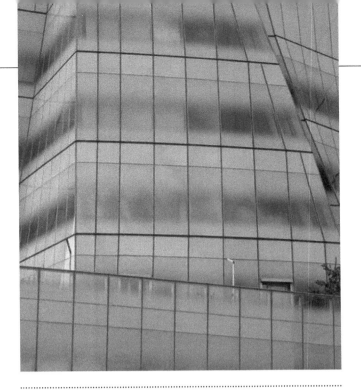

13.8 A postmodern building in New York City, designed by Frank Gehry and completed in 2007.

Since the 1980s, there has been a reaction to modernism, usually called postmodernism. It is less a distinct style and more a conscious distancing from the austere, flat-topped straight line of modernist design; it includes historical referencing, ornamentation, and playfulness with material and design. Office buildings with Georgian facades, museums with Renaissance decoration, and neoclassical houses now became popular. Variants include **deconstructivism**, which involves use of nonrectilinear surfaces (Figure 13.8). In the highly competitive world of global cities, not to have a postmodernist building is to be out of date, too far from the cusp of the future.

This hyper-brief history of architectural styles is just a small indication of the way that buildings and cities reflect transnational forms and styles.

The Global Production of Culture

The global production of culture takes many forms. Here we will consider the production of space, the tourist gaze, and ideas of East and West.

The Production of Global Space

The production of global space has two distinct meanings: the creation of an understanding of the world's geography and the representation of this geography in cartographic forms. A traditional view sees both things simply as a Eurocentric process in which European powers incorporated the world into a global discourse and created new, more modern forms of cartographic representation. It is the story of how Europe explored, discovered, and mapped the world, displacing older, indigenous cartographies with more scientific mappings. Cartographic modernity in this rendering is exported from Europe to the rest of the world. There is now an exciting body of work that points to the creation of new knowledge in the process of interactions and encounters between colonial and colonized, the imperial center and the periphery, the modern and the indigenous. Knowledge was not so much imposed from one region of the world but created and circulated in a complex series of interactions in many different regions of the world. Modernity was created in the space of encounters, continually imported, transformed, and modified, and then reexported in a continuous and ongoing global circulation of ideas and practices.

Kapil Raj describes the integration in the circulation of knowledge between South Asia and Europe from 1650 to 1900. Raj undermines the traditional view that science was exported from the European core to the colonial periphery. He shows that there was an intercultural knowledge encounter that produced new knowledge. In particular, the case of the geographical exploration of British India in the late eighteenth and early nineteenth centuries provides a good illustration of the way in which British and Indian practitioners and skills met around specific projects, how they were reshaped, and how the modern map and its uses coemerged in India and Britain through the colonial encounter.

The Tourist Gaze

Tourism is an important industry. There are almost one billion international tourist visits each year to a vast array of tourist landscapes: Florida beaches, resorts along Spain's Mediterranean coastline, the small Caribbean islands of Dominica and Barbados, and even remoter islands such as the Maldives and Fiji. It is a significant economic sector in large countries. Countries with the largest tourist arrivals are France, the United States, and China, while the countries with the largest receipts from tourism are the United States, Spain, France, and China.

Tourism used to be the preserve of wealthy English aristocrats taking the Grand Tour, visiting classical Mediterranean countries, often purchasing artwork to add culture and class to their residences back home. With space-time convergence, mass tourism became possible and brought more of the world into easier and quicker tourist access. The tourist market is now finely graded, from cheap mass tourism to exclusive and expensive tourist experiences. The cruise industry is finely graded by price, quality, and destinations. Cheaper cruises tend to

13.9 Luxury cruise ship visits Ålesund, Norway

visit sunny recreational sites in the Caribbean, while more upmarket cruises visit a wider range of destinations in luxury ships (Figure 13.9).

As tourist sites have become more popular, there is constant demand for a more restricted experience, away from the crowds. There are a variety of tourisms, including beach holidays, ecotourism, heritage tourism, culinary tourism, and even slum tourism, in places such as the favelas of Brazil. In the wake of Hurricane Katrina, disaster tourists visited New Orleans to see the aftermath of the storm.

John Urry describes what he terms the "**tourist gaze**." When people go on holiday, they look upon different landscapes, settings, and peoples. This gaze is socially organized and systematized. He highlights how the tourist gaze—"gazes" would be more accurate—has changed for different groups and places over the years. There are stock images such as Paris as romantic, Southeast Asia as exotic, English villages as representing an unchanging organic order. The gaze is not an innocent seeing but a perception guided and shaped and freighted with subtle and not-so-subtle messages about history and geography. Cultural geographer Patrick McGreevy explores the changing meaning of visiting Niagara Falls. He shows how the perception of the place is bound up with themes of remoteness, death, nature, and the future. The experience itself has changed over the years from an almost sacred pilgrimage, part of a romanticism of nature, into a more commodified tourist package.

Orientalism, Ornamentalism, and Occidentalism

There is a hierarchy to the production of geographical knowledge. The world is always described by certain people and from certain places. It is a legacy of the British Empire that to this day Iran and Iraq are described as the Middle East. They are only the Middle East if you are describing them from London. Cultural critic Edward Said described **Orientalism** as the writing and description of these places by those in the West. The term is not innocent of political meaning and the operation of military power. To be described and portrayed is to be discursively captured.

Riffing on this idea, historian David Cannadine describes an **Ornamentalism** that he defines as the importance of class and status as much as race and ethnicity, especially for the British Empire. He stresses the construction of affinities between different elites in the encounter between East and West. Cannadine's assertion is that class was as much of a key factor—if not more of one—as race and ethnicity in the social distinctions of the imperial experience.

There is also an **Occidentalism** as well as an Orientalism. While Orientalism is a much-studied phenomenon, Occidentalism receives rather less attention. The West is also understood and imagined by its enemies. Ian Buruma and Avishai Margalit trace the roots of anti-Western ideas. These two authors point to the important role of the sinful Western city in this ideology. Purifying the evil city is a major trope. The United States and especially its cities are often projected as a "rootless, cosmopolitan, superficial, trivial, materialistic, racially mixed, fashion-addicted civilization" (Buruma and Margalit 2004, p. 8).

The Commodification of Culture

Globalization has not disrupted the long connection between culture and place, but it has added some new wrinkles. The commodification of selected cultural forms has transformed some local cultures into globalized forms. Hollywood movies, for example, are now more international, dealing with broader themes than simply US concerns. Indian restaurants serve food that is only ever found in Indian restaurants around the world rather than replicating dishes from India. Sushi restaurants are found not only in Japan but in cities all over the world and have become an important signifier of coolness.

In some cases, cultural globalization is aided by the nature of national cultures. Big-budget Hollywood movies tend to draw upon US themes to provide narrative drive and meaning. But what exactly is this America that is being represented? It is an America that has to appeal to a heterogeneous experience even within the United States. In order for things to sell well in the United States, they already need to have a level of smoothing. Much of the warp and weft of life in specific parts of the United States is filtered through the need to reach a more general national market. A case can be made that American culture has been so successful in achieving global penetration because it has already been partly globalized just to reach a wide US audience. The country is so diverse and heterogeneous that US cultural products have to be smoothed out to succeed, and that primes them for global dispersal.

National and local cultures in turn also influence global cultural forms. As more and more US films are sold abroad, US filmmakers have to make allowances for foreign markets. The particularities of US life are replaced in expensive Hollywood movies by images, plots, and characters that are instantly recognizable by overseas viewers.

The commodification of culture does lead to a certain similarity in the cultural mix available to audiences around the world. But the more astute companies tailor their cultural products for specific audiences. Akio Morita, the legendary head of Sony, first coined the term "**glocalization**," which represented a business strategy to create worldwide operations that were attuned to local markets and conditions. Two waves of Coca-Cola globalization can be identified. The first and earliest involved the marketing and sale of Coca-Cola. The second and more recent involves the sale of Coca-Cola alongside the sale and marketing of "local" brands. In Brazil, soft drinks are often marketed by Coca-Cola as rainforest drinks with "forest" tastes and plant sources.

Culture is not divorced from commerce. Fredric Jameson notes that in the latest stage of capitalism, symbolic meanings and associations increasingly determine the economic value of goods. Imposing symbolism, meanings, values, and emotions onto goods blurs the boundary between the image of those products and their concrete reality. Economics and culture are now intrinsically linked. There is even a range of cultural industries that includes music, dance, theater, literature, the visual arts, crafts, and many newer forms of practice, such as video art, performance art, computer art, and multimedia art. Economic geographer Allen Scott introduces a broader list of "cultural-products sectors," which includes high fashion, furniture design, news media, jewelry, advertising, and architecture.

The Myth of Homogeneity

Debates on cultural globalization often polarize into whether the recent surge of cultural flows and global consciousness has increased or decreased sameness between people and places around the world. One popular argument is that the world is becoming more alike as greater numbers drink Coca-Cola, eat at McDonald's, and watch Disney movies. It is a compelling image that captures some if not all of the complexity. The fact that people across the globe are watching CNN and MTV, McDonald's restaurants are opening throughout the world, and Hollywood films dominate the world film market are taken as indisputable evidence of the homogenization and Americanization of the world (Figure 13.10). The **homogenization thesis** assumes that the same things are consumed in the same ways.

An alternative argument is that, while particular television programs, sport spectacles, network news,

13.10 Fast food in Shanghai

advertisements, and films may rapidly encircle the globe, this does not mean that the responses of those viewing and listening will be uniform. Goods, ideas, and symbols may be diffused globally, but they are consumed within national and local cultures. Ideas, symbols, and goods that circulate around the world are consumed in national contexts and in local circumstances. Similar goods mean different things in different places. There is enough variation to suggest that there is little prospect of a unified global culture; rather, there are a variety of global cultures. Specific cultural backgrounds are not just empty containers for the receipt of global messages; they are critical to how messages are received and consumed. People in the contemporary world have become increasingly familiar with the presence of different cultures, rather than being sucked into a single cultural orbit. To be sure, more people around the world can draw upon a similar range of flows, but there is still a wide variety of cultural formations. In fact, cultural globalization has led to a more explicit concern with more local cultures as certain groups seek to redefine or reestablish their place in the world.

We can replace the popular image of local cultures under assault from a global culture with a more complex picture of local cultures being generated and recreated in response to globalization. The sense of a pervasive globalization has not so much overwhelmed local cultures as helped create them. The search for authenticity in "pure" local cultures (which are in fact mixtures) has helped in the creation of world music, ethnic cuisine, and indigenous culture. Religious fundamentalism is less a return to a pure theocratic ideology and more the self-conscious recreation of religious beliefs in the face of secularization and globalization. Cultural authenticity is less an excavation of the pure and more a contemporary representation of the contrived.

Cited References

Appadurai, A. (1996) *Modernity at Large: Cultural Dimensions of Globalization*. Minneapolis: University of Minnesota Press.

Buruma, I. and Margalit, A. (2004) *Occidentalism: The West in the Eyes of Its Enemies*. New York: Penguin.

Cannadine, D. (2001) *Ornamentalism: How the British Saw Their Empire*. Harmondsworth: Allen Lane.

Gregory, D. (2004) *The Colonial Present: Afghanistan, Palestine, and Iraq*. Malden, MA: Blackwell.

Jameson, F. (1991) *Postmodernism, or, The Cultural Logic of Late Capitalism*. Durham, NC: Duke University Press.

McGreevy, P. (1994) *Imagining Niagara: The Meaning and Making of Niagara*. Amherst: University of Massachusetts Press.

Meinig, D. W. (1965) The Mormon Culture Region: Strategies and Patterns in Geography of the American West, 1847–1954. *Annals of the Association of American Geographers* 55: 191–220.

Raj, K. (2007) *Relocating Modern Science: Circulation and the Construction of Knowledge in South Asia and Europe, 1650–1900*. New York: Palgrave.

Said, E. (1979) *Orientalism*. New York: Vintage.

Urry, J. (2002) *The Tourist Gaze*. 2nd ed. London: Sage.

Select Guide to Further Reading

Knox, P. (2011) *Cities and Design*. New York and London: Routledge.

Lees, L. (2001) Towards a Critical Geography of Architecture: The Case of an Ersatz Colosseum. *Cultural Geographies* 8: 51–86.

Short, J. R. (2012) *Korea: A Cartographic History*. Chicago: University of Chicago Press.

Zelinsky, W. (1992) *The Cultural Geography of the United States*. Englewood Cliffs, NJ: Prentice Hall.

Websites

Culture Regions
http://www.geographynotes.com/human-geography/major-and-minor-cultural-realms-of-the-world-with-maps/1008

Global Tourism
http://www.statista.com/topics/962/global-tourism/

Interactive Map Websites

Culture Regions of the World
   http://eduplace.com/kids/socsci/books/applications/imaps/
   maps/g6_u1/

World Bank Tourism Data
   http://data.worldbank.org/indicator/ST.INT.ARVL?view=map
World Architecture Map
   http://www.worldarchitecturemap.org

Learning Outcomes

- A cultural region shares a similar cultural feature such as language, ideas, religion, or cultural practices. It is often divided into a core and a periphery.
- Spatial diffusion is the movement across space and time of an innovation, idea, or practice.
- There are three dynamically linked sources of production of culture: the connection with the local, the relationship with the national, and the interaction with the global.
- Whenever a wave of globalization sweeps the world, there is a corresponding revival of concern and promotion of national and local cultures as a way for groups to forge, redefine, or reestablish a place in the world.
- It is important to consider the process of the deterritorialization of culture, in which cultures are lifted from particular places, and the reterritorialization of culture, whereby cultural practices are transplanted and transformed.

- Architecture offers examples of the de- and reterritorialization of culture.
- Two forms of architecture that affect human geography are vernacular and global. Vernacular architecture uses local resources and traditions in the design and construction of buildings, while global architecture can incorporate elements of design from around the world, or have elements that eventually transcend national boundaries and impact cultures and geography around the world.
- The world is described from the viewpoints of the West and the East, often referred to as Orientalism and Occidentalism.
- The commodification of selected cultural forms has transformed some local cultures into globalized forms, while allowing other local cultures to influence global cultural forms.

PART 6

The Political Organization of Space

There are many relationships between space and politics. There are the various organizations of political space, such as empires, states, and administrative units. Politics also revolves around issues of space and place, such as when neighboring states have boundary disputes or when a city government tries to locate a new incinerator. Finally, many political processes take place in space, as in the geography of elections. In this part we will look at some of these themes. Chapter 14 will focus on the political geography of world order, with particular emphasis on the rise and fall of empires and identifying the elements of a new world order. Chapter 15 will consider the nation-state.

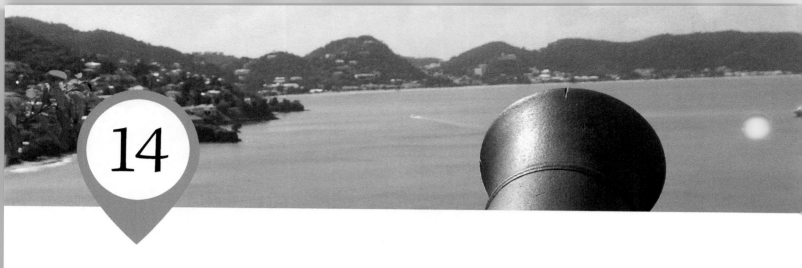

World Orders

LEARNING OBJECTIVES

Summarize the iterations of human societies from their initial small-group forms and into more diverse, complex political forms.

Discuss the rise and fall of global empires and their impact on the current political geography of the world.

Explain how empires promoted global integration as they spread spatially and incorporated a variety of peoples.

Outline examples of the limits to empire, imperial overstretch, and imperial disintegration.

Describe the trends associated with the emergence of a new world geopolitical order.

The very earliest form of human society was small groups of extended families that, over time, merged into larger social units, eventually creating such diverse socio-spatial entities as empires, city-states, and nation-states. In the first part of this chapter we will look at empires as a spatial unit of political organization, and the rise and fall of **imperialism**, which played an important role in shaping global human geography; then we will focus on today's global political geography.

Early Empires

Some of the earliest empires were city-states in Mesopotamia. The city of Uruk emerged around six thousand years ago in the southern part of Mesopotamia. Settled agriculture, sophisticated irrigation, and a marked division of labor propelled Uruk's growth. The city broadened its influence, referred to as the "**Uruk expansion**," through a wider net of control in the upper reaches of Mesopotamia, between the Euphrates and Tigris Rivers. Surrounding peoples were brought into the cultural and economic control of Uruk. The city's elites developed more expensive tastes, and the search for more exotic goods and materials to meet this demand drove the expansion. Eventually, expansion reached its peak and then turned into decline. In a process that is repeated over and over again, the empire collapsed, local traditions reemerged in regions formerly controlled by Uruk, and in the main city, invaders razed the monumental buildings of the temple.

The Geography of Empire

There are a number of themes in this empire's brief story that figure in later imperial experiences: empires emerge, extend their reach, and then collapse from external pressures and internal conflicts. Across time and over space, the process is repeated. What is different is the increasing territorial

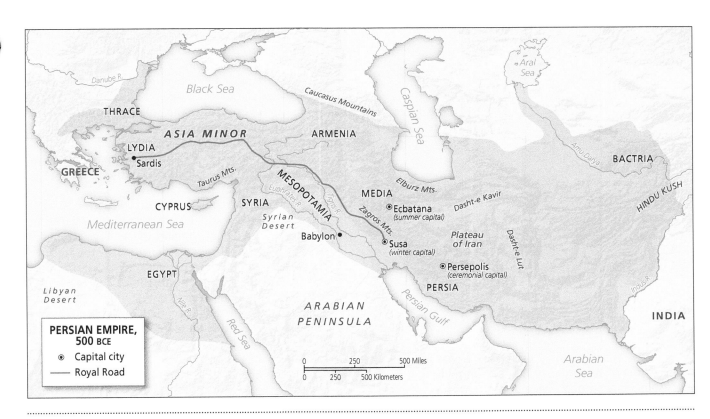

14.1 The Persian Empire

extent of the empires and the enduring consequences of **imperial incorporation**. Subsequent imperial history is of powers becoming larger and their territorial reach more extensive as space-convergence improvements extend the effective imperial range. In the Middle East, after Uruk came the Hittites, the Assyrians, and the Persians. The Persian Empire peaked from around 559 BCE to 331 BCE (Figure 14.1), its spatial expansion aided by the greater use of more mobile cavalry and better roads. At its greatest extent it covered a vast area, from the Indus Valley in the east to Greece in the west and from Central Asia in the north to southern Egypt on its southern reaches.

We can make some generalizations about the geography of empire. Through time, empires grow in size as they are more able to overcome the tyranny of distance. While the Uruk expansion was limited to the Tigris and Euphrates river basins, the Persian Empire stretched from the Mediterranean Sea to the Indus River. Better transport and improved road networks enable empires to extend their spatial reach. Yet empires, no matter how vast, have territorial limits to their power. As they expand further from the core, their ability to wield effective and overpowering force tends to weaken. There is a **distance decay** effect in their military power and political reach. Far from the empire's center, borders are vulnerable. To defend them, early empires relied on force, whose effectiveness declined markedly with distance from the imperial center.

The further an early empire stretched, the more difficult it was to control the ever-more-distant periphery. There were limits on early imperial expansion because of the marked distance decay effect of military reach and political control.

Global Integration

The historical geography of the world is filled with the rise and fall of empires and the spatial spread of imperial influences. Alexander the Great reversed the flow of Persian influence by extending Greek influence throughout Eurasia, taking over much of the remnants of the Persian Empire. The Roman Empire turned the Mediterranean into a Roman lake. These and other empires, such as the Mali in West Africa and the Mayan, Aztec, and Inca in the Americas, created a more homogenous political surface that aided cultural mixing, economic ties, and religious dispersals which diffused creeds and beliefs across space. Empires pulled peoples and region together in shared experiences of (often forced and brutal) spatial convergence. Empires promoted global integration.

The more recent the empire, the greater its enduring consequences. English is spoken around the world because of the wide sphere of influence of the British Empire.

14.2 The Dutch Empire in the New World. Willemstad, Curaçao, retains its Dutch architectural heritage. The Dutch controlled the island from 1634. It was a center of the slave trade.

The experience of empire is one of the most important global phenomena of the past five hundred years. I will discuss three spatial elements: global integration, **imperial overstretch**, and imperial disintegration.

The global integration of space is the defining feature of early **modernity**. In the period 1400 to 1800, empire building across the globe (including the Chinese, Ottomans, Mughals, and Safavids as well as the Spanish and British) created international markets and global exchange networks. These global empires encouraged, forced, and facilitated the movement of people, the spread of new technologies, the diffusion of cultures, and the transmission of religion and scientific practices. The result was a tighter integration of global space. Increasing contact between different cultures led to cross-cultural borrowings. Portugal's trade with Africa, Asia, and South America in the sixteenth and seventeenth centuries resulted in a complex aesthetic blending and technological borrowing: West African wooden and ivory sculpture that incorporated Portuguese figures, precious objects produced in Goa using Christian iconography for the Portuguese market, Japanese paintings that depicted Portuguese merchants, and the introduction of European firearms and maps into Japan.

Whether in intellectual discourses and practices, such as cartography and mapmaking, or in sports, language, religion, and forms of government, modern empires have integrated much of the world: Spanish-speaking South America, cricket playing in India and rugby in Fiji, Catholicism in Goa, and British-inspired representative government in Australia and Canada are just some of the many consequences of the globally integrating effects of modern empires.

Empire was not undertaken just for the opportunity of cross-cultural integration; there was a hard material core.

Empires provided access to resources. The push to empire in the modern world was largely driven by economics. Spain and Portugal laid claims to vast territories in the New World, Africa, and Asia to provide gold, silver, spices, and raw materials for the home countries. The British-based East India Company, and subsequently the British Crown, created a commercial empire on the Indian subcontinent because it provided a rich source of raw materials for British industry as well as a captive market for goods produced in Britain. The Dutch East India Company, and subsequently the Dutch state, annexed land in South America and islands in the Caribbean (Figure 14.2).

It is possible to identify a core-periphery spatial structure to the global political economy that emerged from this imperial relationship. The core consisted of the imperial centers in Europe, with a vast periphery of regions throughout the world under their control. The success of the core was predicated upon the exploitation of the periphery.

From the sixteenth to the twentieth century, much of the world was incorporated by these core powers. Incorporation was achieved through a variety of means, including colonization, informal control, and direct imperial annexation. The British Empire, for example, colonized North America and Australia, had informal control in South America, and annexed much of the Indian subcontinent. Empire could be costly—much better to have economic control without political responsibilities. British economic interests were vigorously pursued in South America without formal annexation. Latin America was the real success story of British commercial empire, because there was market penetration without enacting the responsibilities of formal empire. Direct imperial annexation was, in a sense, a sign of failure to achieve economic ends without direct political intervention. Formal annexation became more common in the late nineteenth century because of increasing imperial rivalries between core countries. Formal imperialism also resulted when the ruling elites of peripheral countries refused to go along with business arrangements.

Colonization involved a rewriting of space to show who was in control and who was controlled. Timothy Mitchell identifies three broad strategies for the framing of colonial states: producing a plan of urban segmentation that includes racial segregation, creating a fixed distinction between inside and outside, and constructing central spaces of observation to keep an eye on things and to show the presence of colonial power. Under British rule in Accra, now the capital of Ghana, new areas of European settlement were created that were sharply demarcated from the "native" areas. Laws barred "natives" from living in areas reserved for whites. Further north, Tunis came under French rule in 1881. The French built neoclassical buildings and created straight boulevards and symmetrical squares as a direct response to the tight and convoluted street pattern of the old Arab town: Gallic "rationalism" counterpoised to Islamic "chaos."

14.1 The Caribbean as Imperial Shatter Zone

The shifting balance of imperial victory and defeat is evident in the changing ownership of islands in the Caribbean. From 1492, the Spanish expanded their imperial control to the large islands of the Caribbean and most of the mainland in the search for gold, precious minerals, and land. It was difficult to retain effective control over the scatter of islands in the Caribbean; they were too many and too widely scattered to police effectively. From 1600 to 1800, other European powers moved into the Caribbean to establish sugar plantations. They were able to annex various groups of islands and the more remote slices of the mainland. The Dutch, for example, gained control of slivers of land in South America as well as of the smaller islands of Aruba, Bonaire, and Curaçao. The French gained control of various islands, some of which remain part of France to this day. In 1657 the Danish West India Company established the town of St. Thomas, in what is now the US Virgin Islands, as the center of a sugar plantation economy based on slave labor.

The need for labor to work the mines and plantations and the decline of the indigenous population led to the slave trade, in which almost 20 million people were captured in Africa and shipped to the New World. The Caribbean was transformed in the process as millions of African slaves became a vital part of the economic geography of the region, an important cultural element, and a nascent political force. The sugar colonies were the scenes of constant eruptions as slaves continually revolted against the appalling living and working conditions. Effective control was always difficult to maintain. These islands were far from the core countries. And after emancipation, the former slaves sought both economic freedom and political independence.

The competition and shifting alliances between European powers were often embodied in changes of territorial control. Consider the island of Dominica, which was under French control from 1632 to 1761, then British control, then French control again from 1778 to 1783, and then British again until it achieved independence in 1978. Most of the small island nations of the Caribbean have similarly complex histories. Trinidad was under Spanish control from 1532 to 1797, when it became a British territory until achieving independence in 1962. St. Lucia was held nine different times by France and six different times by the British, including the last colonial period of 1803 to 1979, when it finally became independent. The French legacy can be heard in the language of the locals.

The imperial legacy lives on. France still controls St. Bart's, Martinique, and Guadeloupe, the latter two being departments of France (equivalent to states in the United States). Montserrat remains a British territory, as do the British Virgin Islands. The Dutch have effective control over Aruba, Curaçao, and Bonaire and share control of the small island of St. Maarten/St. Martin with the French. In 1898 the United States invaded Puerto Rico, and in 1904 it acquired land along what was to be the Panama Canal. In 1917 the United States bought the Virgin Islands from Denmark, less an economic move (as the sugar economy had slumped) and more part of a geopolitical strategy of maintaining a more visible military position in the region and protecting its interests in the Panama Canal.

Local Elites

A key concept is that of the ruling elites, the groups that have power in a country or region. Collaborative elites reproduce the imperial center. This was the case of Anglo-Americans in North America before the American Revolution and the British colonial settlers in Australia and New Zealand. Collaborative elites can also encourage imperial incorporation. Throughout much of Africa and India, for example, the British authorities used indirect rule, working through local rulers who were rewarded for their efforts. Local leaders and existing hierarchies were employed to secure colonial rule. If and when the collaborative elite lost legitimacy and their grip on power, then the colonial powers often had to intervene.

Core-Periphery Spatial Structure

The spatial structure of core-periphery relation, in which the periphery is used as a source of raw materials and as a market for goods produced in the core, laid the basis for subsequent **economic development**. The core industrialized while the periphery remained a cheap resource base and a secure market for manufactured goods. The Industrial Revolution in core countries was built on the foundation of this **unequal exchange** between the core and the periphery. The poverty of much of the developing world is a direct result of the core-periphery structure.

Age of Imperialism

The late nineteenth century also saw the "**age of imperialism**." From around 1880 to 1918, there was a scramble for territory as old and new European powers struggled to incorporate more of the periphery. A map of Africa in 1880 shows the limited range of European influence on the continent. But by 1914, most of the continent had been annexed by European powers, large and small (Figures 14.3 and 14.4). The age of imperialism resulted from growing competition between core powers, a belief

14.3 Africa, 1880

that national economic development was made easier by possessing colonial territory, and the feeling that each significant power had to do something to avoid being left out of the "spoils." Even tiny Belgium laid claim to a vast territory in the middle of the African continent.

By the second half of the twentieth century, the struggle for global supremacy was being fought out between the United States and the USSR. Each had long histories of territorial annexation. The United States expanded from a narrow sliver along the Eastern Seaboard through treaties, purchase, and conquest to become a continental power. In the late nineteenth century, it also began an overseas

expansion, purchasing Alaska from the Russians in 1867 and picking up territories from a declining Spanish Empire, such as Puerto Rico and the Philippines. Hawaii was annexed in 1898. We have an interesting record of this imperial expansion in a popular children's book of the time, *The Navy Alphabet*, first printed in 1900 and written by Frank Baum, who also wrote *The Wizard of Oz*. An illustration that accompanies the text clearly shows the imperial nature of the United States, with explicit use of the term "colonies" (Figure 14.5).

The Russian Empire grew from a tiny area centered on Moscow to become a vast continental power stretching

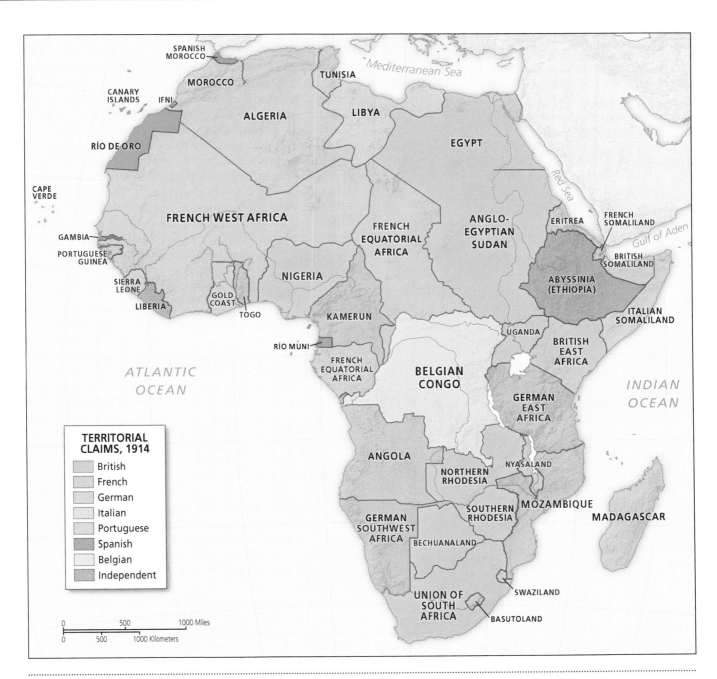

14.4 Africa, 1914

from the Baltic to the Pacific and from the Arctic to the sub-tropics. In 1462 Muscovy was a small isolated region. From this tiny enclave a Russian empire was created through annexation of territory. There was an early move east, drawn in part by the lure of the fur trade. By 1639 the Russians claimed land all the way to the Pacific. The empire was also extended to the west and south. In the north and west, territory was obtained in war with Sweden and Poland. In 1703 St. Petersburg was built by Peter the Great as a major port and as a "window on the west." In the south, the Russians annexed territory from the Tatars and other Turkic peoples, and in 1794 Odessa became the principal Russian port for trade through the Black Sea and Mediterranean. A small landlocked territory had expanded to become a major continental power. When the Bolsheviks grabbed power in 1917, the Russian Empire became the Soviet Empire.

The Cold War

The struggle between the United States and the USSR, called the Cold War, went through several phases. The chilliest period was from 1947 to 1964, as the USSR laid effective claim to most of European Europe by installing and maintaining communist regimes. In 1947 the Truman

Across the OCEAN vast and blue
Our warships plow the billows through
To guard our Colonies afar
And carry aid in time of war
They sail to fair Manila from
The sunny little Isle of Guam,
And then from Porto Rican blooms
To where Hawaii grandly looms.

14.5 Teaching the US Empire, 1900

Doctrine announced that the United States would adopt a global posture against the USSR. Both countries established bases all over the world, formed alliances, and built up their military capabilities. Direct conflict did not occur, and a period of détente (less overt conflict) from 1964 to 1979 (when the USSR invaded Afghanistan) marked a warming of the interactions between the two superpowers. The military buildups and geopolitical maneuverings around the world continued, however, like two chess players with the world as their board, moving pieces to gain advantages. Local elites remained in power with imperial backing if they tied their existence to their backers' imperatives. Throughout much of the Cold War, for example, local elites in countries around the world, such as the Philippines under Ferdinand Marcos, maintained their grip on power by reminding the United States of their anticommunist credentials or, in the case of Cuba, informing the Soviet Union of their anti-capitalist policies.

A new Cold War emerged in 1980 with a significant rise in military buildup. As each side responded to the other's military increases, there was steady escalation, and the dialogue of détente was replaced by the rhetoric of confrontation. The gradual decline of the USSR and its eventual collapse in 1991 meant the end of the bipolar structure of global geopolitics. The United States remains the world's only superpower with a global reach.

The US Empire

The United States is not an empire in the old sense of colonial possessions. However, the US Empire is now a sprawling archipelago of military bases around the world, a vast military-industrial-security complex backing a pervasive worldview dominated by visions of global **hegemony**. There are now at least 700 bases with half a million US personnel spread around the world in at least 130 countries. The United States is the most powerful military state, with the ability to assert global reach. The contemporary US Empire is not like previous empires that annexed territory and peoples into a singular political orbit. The earlier fragments of US Empire, such as Hawaii and Puerto Rico, have long since been incorporated. The US Empire is less about annexing territory and subjugating peoples and more about orchestrating a global order to extend and maintain global economic connectivity.

Imperial Overstretch

The rise and fall of empires is an enduring metaphor for the cyclical nature of success and failure as well as a constant reminder of the spatial limits of territorial expansion. What lies behind the fall of global empires? Paul Kennedy identifies "imperial overstretch," the tendency for empires to expand beyond their ability to maintain economic dominance and military power. The tendency to initiate and expand military commitments undercuts the economic ability to pay for them. In other words, the hubris of empire tends to lead to entanglements that gnaw away at its economic foundations. The collapse of the Soviet Union, for example, was in part due to the inability of the state to maintain its expensive military posture as its economy was disintegrating. The United States is currently facing a fiscal crisis as it tries to pay for expensive social programs as well as the escalating cost of empire.

There are numerous limits to empire. The first is information overload. The bigger the empire, the more the information that needs to be analyzed and turned into convincing narratives. The second and most important is the fiscal limit, as imperial overstretch bumps up against the ability to pay for all the commitments. The possibilities for involvement are infinite, while the ability to pay for them is finite, especially at times of economic distress and declining relative economic power. Third, there are the strategic limits. As strategies and forces are continually restructured according to the demands of the last conflict, there is always the possibility that these will be unable to deal with new conflicts. Thus, in 2001, the US military was very well able to fight against another superpower, but unable to conduct an asymmetric war against small bands of transnational terrorists or undertake the nation building necessary after invading Afghanistan and Iraq. Big military powers have long reaction times that in a quick-changing world make them vulnerable to outdated assumptions, failed strategies, and incorrect tactics. Fourth, there is also the possibility of a "**legitimation crisis**," which, in this particular context, can be defined as a major loss of public support for military involvements. These are more pronounced in more democratic societies, as the public can more easily voice its disapproval. Quick and successful troop deployments are easy to justify. The stunning success of the Gulf War in 1991 was, for many in the United States, a source of national pride, this single campaign erasing many of the wounds of Vietnam as the US military was applauded and honored rather than shunned and vilified. Fast-forward twelve years: as the invasion of Iraq soon turned into a quagmire, with the mounting carnage shown nightly on television, public enthusiasm for the war faltered and waned. When empire involves seemingly ceaseless, costly campaigns, public opinion comes into play. The visible costs of empire can provoke a crisis of legitimation if wars and military engagements are too onerous for too long.

Imperial Disintegration

Empires come and go, but their going is rarely easy.

Decolonization

After the end of the Second World War, there was a wave of decolonization as imperial powers either left or were effectively kicked out of their colonial possessions. From 1945 to 1954, the wave was felt most intensely in Asia. The Dutch fought off independence struggles in Indonesia, but the persistent resistance finally led to the country becoming independent in 1949. The French had control over what was called French Indochina, consisting of what is today Cambodia, Laos, and Vietnam, but here the independence struggle against French colonialism was interpreted by the United States as communist insurgency. An anticolonial nationalist struggle was entangled by a superpower's geopolitical strategy as the Vietnamese fight for independence became the Vietnam War.

The decolonization of Africa came in the second wave, with independence for Sudan (1956), Ghana (1957), and Kenya (1963). In some cases, the struggle was intense. British authority in Kenya was contested especially by the members of the Kikuyu tribe, who mounted a guerilla campaign in pursuit of their goal of independence. The British responded by not only waging a counterinsurgency but also setting up a system of detention camps where up to 1.5 million people were rounded up and imprisoned. In a detailed study, historian Caroline Elkins came to the conclusion that "there was in late colonial Kenya a murderous campaign to eliminate Kikuyu people, a campaign that left tens of thousands, perhaps hundreds of thousands, dead" (Elkins 2005, p. xvi).

With few "national" borders before colonization, many of the independence movements in Africa existed within the boundaries drawn up in arbitrary divisions of spoils by imperial cartographers. The result was that the national boundaries of post-independence Africa emerged from this cartographic legacy. These national boundaries, drawn initially for colonial convenience and compromise, cut across centuries-old tribal and ethnic differences that in some cases reemerged in the wake of decolonization. Compare Figure 14.4 with the contemporary political map of Africa in Figure 14.6 to see the enduring legacy of colonial borders and boundaries on present-day African nation-states. A major problem for the newly independent states was how to create national unity and national identity from such a variegated base. In some cases, the new nations are continually split apart by the resurfacing of tribal and ethnic differences.

Collapse of the Soviet Empire

The fall of the Soviet Union in 1991 is just the latest in a long line of imperial collapses. When popular uprisings took hold in East Germany in 1989, the Soviet leadership did not send in the troops—something they had done in Hungary in 1956 and in Czechoslovakia in 1968, where they snuffed out the popular uprising known as the Prague Spring. The system was economically and politically bankrupt, with its more astute leaders able to see the limits of the Soviet Empire. The unpopularity of the Soviet-dominated Communist system in Eastern Europe fueled public discontent. The people, when given an opportunity, showed a profound lack of support for the regime and a strong desire to be rid of the old leadership and scrap the Communist system. Events moved quickly. In June 1989 the popular political movement in Poland, Solidarity, won the first free election held for two generations. In September the Hungarians set a date for free elections. On November 9, the Berlin Wall ceased to be a meaningful barrier as East Germans scrambled freely over it, and Berliners, from both East and West, delighted in demolishing the symbol of partition and Communist control. Later that year the Communist leadership in Bulgaria, Czechoslovakia, East Germany, and Romania was overthrown by popular uprisings. The corrupt system collapsed as more and more people voiced their disapproval. It was one of the rare examples of imperial decline brought about by popular uprising and bottom-up resistance rather than top-down regime change.

New countries emerged from the break-up of the Soviet Empire. Some of them had mature identities, such as the Baltic republics of Estonia, Latvia, and Lithuania, which had been swallowed whole by Soviet expansion. Others were broken off from larger national units, such as

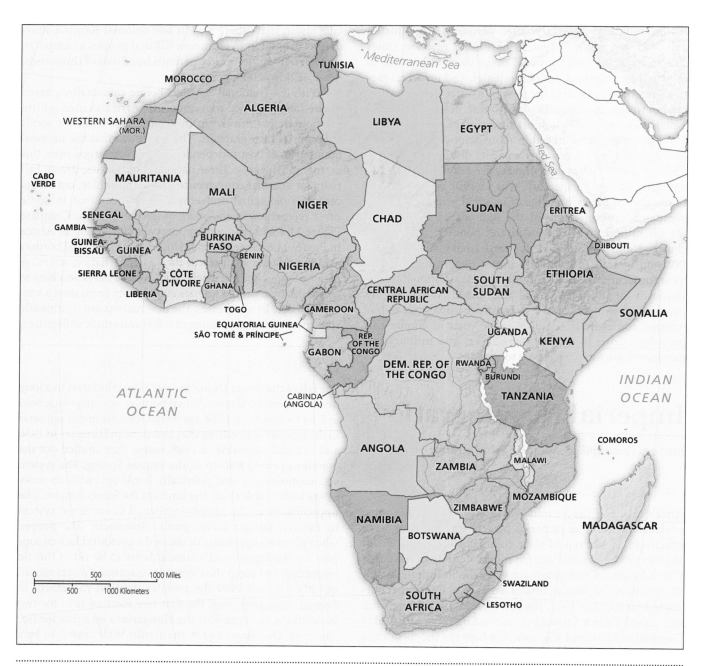

14.6 Political map of contemporary Africa

East Germany, a break that was healed with reunification with West Germany. Some were multiethnic agglomerations, such as Yugoslavia, which quickly began to break up over regional and ethnic tensions after 1989. In yet others, new states emerged in the post-imperial context. Political geographer Nick Megoran tells the story of Kyrgyzstan and Uzbekistan. Before 1924, these countries did not exist. The whole area was a complex, multiethnic place. It was annexed by the expanding Russian Empire in 1876, and in 1924 the Soviet Empire created the administrative structure of separate Kyrgyz and Uzbek units. This division helped forge separate identities, although not separate economic units. The collapse of the Soviet system in 1989 gave space for national emergence (see Figure 14.7). In 1991 the two former regions of the USSR became separate states. Figure 14.8 outlines the borders of the new state of Kyrgyzstan. An uncertain, arbitrary, and hazy distinction now became codified into two separate states sharing an international boundary. Tensions between the two states erupted in 1998 and resulted in the closing of the border. The invented borders of the Soviet era moved from acts of imagination to lines of national conflict.

14.7 The Soviet breakup

14.8 Kyrgyzstan: Soviet imperial disintegration leads to new international boundaries in Central Asia.

The Clash of Civilizations?

In 1993 the political scientist Samuel Huntington wrote an essay with the provocative title "The Clash of Civilizations." He argued that people were divided along cultural lines drawn by religion, history, and geography. He identified nine different types of cultures, including Western, Islamic, Orthodox, Latin American, and Confucian. It is an odd mixture of the religious, such as Islamic and Confucian, and the purely geographical, such as Latin American and African (Figure 14.9). The most contentious conflict was between the West, characterized as societies founded on principles of pluralism, democracy, and individualism, and the Islamic, characterized as less tolerant of difference and authoritarian, with a primary commitment to religion and not to the nation-state.

Several criticisms are made of this argument. First, the categories are very large, homogeneous spatial blocs, whereas the real world is more fractured, with religious and cultural groups more widely dispersed and the spatial units more often heterogeneous than homogeneous. Flows and heterogeneity rather than closed boundaries and homogeneity mark a globalizing world. Second, the historical record of contact between Islam and the West reveals shared human values, commercial trading, and cultural exchange, interaction, and penetration. Third, as the **Arab Spring** of 2011 revealed, the characteristics that Huntington asserted for the Islamic world were not eternal verities but the product of specific regimes and particular times. The demand for democratic inclusiveness

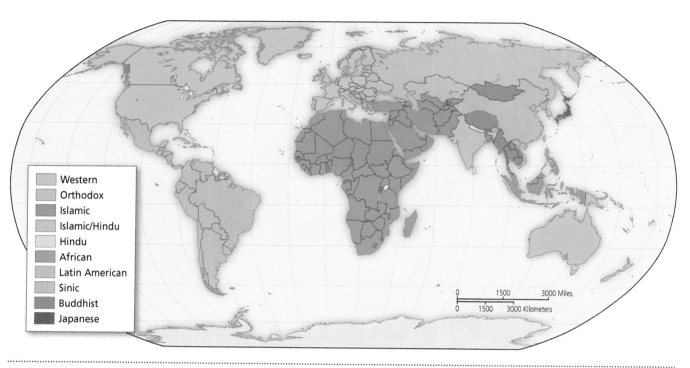

14.9 Huntington's clash of civilizations

is also a very important strain in parts of the Islamic world. Huntington's ideas were given a wider play in the wake of 9/11. The idea of a clash of civilizations between the West and the Islamic world is also central to the ideologies of Islamic terrorist groups such as Al Qaeda. But despite the ongoing terrorist threat, the contemporary world is a more complex place than the broad cultural categories proposed by Huntington can encompass, and are more marked by civilization exchange and dialogue than outright clashes.

Elements of a New World Order

In the past thirty years at least four significant trends can be noted. The first is that the USA remains the world's only global superpower. It has the largest economy, the most powerful military, and the global reach, with bases around the world, that enables troops to be placed anywhere in the world inside 72 hours. The US military hegemony comes at a cost. It is dragged into wars and conflicts around the world, even when its immediate material interests are not at stake. It is blamed for many of the world's ills, leading to a strange asymmetry. While many people in the USA see the country as a force of good, many in other countries see it as a regressive force that props up undemocratic regimes.

A second trend is the rise of security groupings and mutual defense organizations such as the United Nations peacekeeping forces, the North Atlantic Treaty Organization (NATO), and the African Union that consists of all fifty-four countries on the continent except Morocco. The African Union provides troops in certain instances. African Union troops were involved in recent military-policing interventions in Darfur (7,000 troops) and Somalia (8,000 troops) and more recently served in Burundi, Liberia, Côte d'Ivoire, and the Democratic Republic of Congo.

Another trend is the rise—or in some cases a reemergence—of alternative power sources to the USA in the world. And here we can make a distinction between these and organizations such as the EU, which is an alternative economic center of power to the USA that still shares similar global strategic goals. For critics in the US, some countries and organization such as Japan and the EU get to achieve the economic success of a peaceful world underwritten by the military expenditure of the USA. Then there are countries that are both economic and military rivals. The two most obvious cases are Russia and China. Russia emerged as the rump of the former USSR. After a very rough transition from a communist system to something closer to a market system, Russia began to benefit from the sale of its vast oil and natural gas reserves. As world prices of both increased in the 2000s, government coffers swelled, which allowed an increase in military spending and national assertiveness. Especially under President Putin, Russia has reimagined itself as a world power eager to play a larger role in the world order, from annexing Crimea to

14.10 (A) Countries with ISIS provinces. (B) ISIS militants parade through Northern Syria in 2014.

getting involved in the military campaign in Syria to cyber-hacking adventures around the world. But the fall in the rise of oil and gas is already creating a fiscal crisis for the Russian state, which may undercut its future projection of power.

The Rise of China

The most significant entry onto the global stage of the past thirty years is China, which now has the second largest world economy. Before the trade liberalization of the early 1990s, China was a poor country and a bystander in the everyday operation of global society. China's rise is marked by a greater engagement in global organizations such as the World Trade Organization in 2001, a more active role in international affairs that parallels its growing economic connections with countries around the world, and a steady buildup in military capabilities. China's huge demand for resources from across the world has led to significant foreign investments in South America and Africa, including land purchases and building roads and railways to secure access to commodities. Africa is such a large recipient of Chinese money, trade, and migrants that some writers refer to it as China's second continent.

A recent case of China's growing role in the global economy, beyond simply importing vast amounts of raw materials and exporting manufactured goods, is the emergence of the Asian Infrastructure Investment Bank (AIIB). When the US Congress refused to support an increased role for China in the International Monetary Fund, China established the AIIB that, despite US protestations, has now attracted a large number of partners,

14.2 The Pentagon's View of the World

After the ending of the **Cold War** and the disintegration of the **Soviet bloc**, the United States became the world's sole superpower. With bases all over the world, a military budget that dwarfed those of all other countries, and the strategic capability of global reach, it was the most important military power on the planet. How does it see the world? It depends, of course, on the events, but Thomas Barnett provided a possible map of the Pentagon's general worldview. The world is divided into a **functioning core** marked by relative stability, rule of law, and integration into the global economy. Then there is a **non-integrating gap** marked by political instability, social unrest, and lack of integration into the global economy. Here is where the United States is involved in direct combat, shows of force, peacekeeping, anti-insurgency, and drone strikes.

While very revealing, the map hides as much as it reveals. The non-integrating gap is shown as a solid surface, whereas in reality it contains islands of stability and places of connectivity. And in the core-functioning areas there are also areas of instability and insurgency. The map depicts solid homogenous surfaces, whereas the reality is a much more varied mosaic of difference. The map is an ideological rendering of a world divided into safe and unsafe areas by Pentagon strategists, rather than an accurate geographical rendering of global difference.

REFERENCE

Barnett, T. (2004) *The Pentagon's New Map*. New York: Berkeley.

- - - - Boundary of the
      Non-Integrated Gap

0   1500   3000 Miles
0   1500   3000 Kilometers

The Pentagon's world map

including traditional US allies such as Australia, Brazil, Saudi Arabia, and the UK. The AIIB has pledged to provide $40 billion worth of funds for infrastructure improvement, much of it targeted at a new Silk Road of improved communication and transport across Central Asia to link China with Europe. For China it is a way of utilizing its vast foreign exchange, currently valued at a staggering $4 trillion, and of signifying its growing international prestige. The AIIB signals China's emergence as an active stakeholder in global affairs.

Significant brakes on the rise of China include the reliance on high and constant levels of economic growth, which may be difficult to achieve as the economy transitions from a low-wage, low-value to a high-wage, high-value economy; low GDP per person; and potential social unrest as internal criticisms are perceived as existential threats to established political power. Its military buildup has been rapid and impressive but its reach is, as yet, regional rather than global. We will examine the issue of the South China seas in the next chapter. Moreover, other countries are reacting and responding to China's rise. The territorial disputes that are occurring between China and her neighbors are in part a function of China's rise but also of the resistance of its neighbors to territorial claims. The American "pivot to the Pacific" announced in 2010 was in large part a response to China's growing regional assertiveness.

Non-State Actors

Another significant element in the contemporary political geography of the world is the rise of non-state actors. These include social movements, some of them temporary, such as the Arab Spring, which swept across the Middle East in 2011 as young people in particular took to the streets to protest corruption, repression, inequality, and limited economic opportunities. There are also transnational terrorist groups, the most prominent being Al Qaeda and more recently ISIS.

Al Qaeda is a global terrorist organization with significant offshoots including Al Qaeda in Islamic Maghreb, Al Qaeda in Syria, and Al Qaeda in the Arabian Peninsula. It seeks to promote an Islamic fundamentalism that sanctions war against non-Muslims and non–Sunni Muslims. It is marked by violent terrorist acts. ISIS (Islamic State of Iraq), also known as Daesh, is a Sunni Muslim movement that emerged in post-invasion Iraq from Sunni terrorist groups affiliated with Al Qaeda involved in the Sunni–Shia civil war. It became a military force when it linked up with former Baathist military personnel displaced by

the post-invasion Shiite dominance in Iraq. By 2013, it had become a terrorist group motivated by Sunni Islamic fundamentalism, including violence against Shia Muslims and Christians, with a muscular military staffed by former Iraq Army personnel. It became self-funded by capturing oil depots, taxing residents and businesses under its control, extortion, and smuggling. In the failed states of Syria, undergoing civil war, and Iraq, beset by religious tensions, ISIS flourished, and by 2014 it had captured territory across both. Countries with active affiliates, also sometimes described as provinces of ISIS, are shown in Figure 14.10.

ISIS gained some measure of support from alienated and bored Muslim youth who felt marginalized in Europe, North America, and Australia. Of its 31,000 fighters in 2015 in Iraq and Syria, almost 12,000 were foreign nationals from 81 countries. Fundamentalism appeals to a youthful population without jobs, hope, or any sense of belonging. ISIS also has sophisticated marketing using social messaging and the Internet. Its images of pornographic violence also attract the unglued and the violent. ISIS has extended its reach beyond Syria and Iraq though the declaration of separate provinces in the Afghanistan–Pakistan border region, Libya, Egypt, Yemen, Saudi Arabia, Russia (North Caucasus), and Nigeria. In some cases this involves encroaching on Al Qaeda, as in the Afghanistan–Pakistan border area, Russia, and Yemen. In other cases it involves the incorporation of existing insurgent terrorist groups such as Boko Haram in West Africa.

Cited References

Elkins, C. (2005) *Imperial Reckoning: The Untold Story of Britain's Gulag in Kenya.* New York: Henry Holt.

Huntingdon, S. P. (1993) The Clash of Civilizations. *Foreign Affairs* 72: 22–49.

Kennedy, P. (1987) *The Rise and Fall of the Great Powers.* New York: Random House.

Megoran, N. (2012) Rethinking the Study of International Boundaries: A Biography of the Kyrgyzstan–Uzbekistan Boundary. *Annals of Association of American Geographers* 102: 464–481.

Mitchell, T. (1988) *Colonizing Egypt.* Berkeley and Los Angeles: University of California Press.

Parker, C. (2010) *Global Interactions in the Modern Age (1400–1800).* Cambridge: Cambridge University Press.

Select Guide to Further Reading

Alcock, S. E. (2001) *Empires.* Cambridge: Cambridge University Press.

Applebaum, A. (2012) *Iron Curtain: The Crushing of Eastern Europe, 1944–1956.* New York: Doubleday.

Brzezinski, Z. (2012) Balancing the East, Upgrading the West: U.S. Grand Strategy in an Age of Upheaval. *Foreign Affairs* 91: 97–104.

Burbank, J. and Cooper, F. (2010) *Empires in World History: Power and the Politics of Difference.* Princeton: Princeton University Press.

Butlin, R. (2009) *Geographies of Empire: European Empires and Colonies c. 1880–1960.* Cambridge: Cambridge University Press.

Darwin, J. (2010) *After Tamerlane: The Rise and Fall of Global Empires, 1400–2000*. London: Bloomsbury Press.

French, H. (2015) *China's Second Continent: How a Million Migrants Are Building a New Empire in Africa*. New York: Vintage.

Levenson, J. A. (ed) (2007) *Encompassing the Globe: Portugal and the World in the 16th and 17th Centuries*. Washington, DC: Sackler Gallery, Smithsonian Institution.

Pagden, A. (2008) *Worlds at War: The 2,500-Year Struggle Between East and West*. New York: Random House.

Parker, G. (1998) *The Grand Strategy of Philip II*. New Haven: Yale University Press.

Parsons, T. H. (2010) *The Rule of Empires: Those Who Built Them, Those Who Endured Them, and Why They Always Fail*. Oxford: Oxford University Press.

Short, J. R. (2013) *Stress Testing the USA*. New York: Palgrave Macmillan.

Tolan, J. V., Veinstein, G., and Laurens, H. (2012) *Europe and the Islamic World*. Princeton: Princeton University Press.

Websites

A fascinating exhibition on Portugal and the world in the sixteenth and seventeenth centuries
http://www.asia.si.edu/encompassingtheglobe/
Empires through history
http://empires.findthedata.org

Rise and fall of empires
https://www.youtube.com/watch?v=_lTduTwqtjY
https://www.youtube.com/watch?v=hXiLMb-OFgY
https://www.youtube.com/watch?v=abJQD_RTZEU

Interactive Map Websites

Empires in world history
http://www.atlasofworldhistory.com

The British Empire
http://www.britishempire.co.uk/maproom/maproom.htm

Learning Outcomes

- Although the earliest forms of human activity were but small groups of extended families, they eventually merged into larger social units and even highly diverse and widespread empires.
- Across time and over space, the same process repeats itself: empires emerge, extend their reach, and then collapse from external pressures and internal conflicts.
- Better transport and improved road networks enable empires to extend their spatial reach. Yet empires, no matter how vast, have territorial limits to their power.
- Recent empires have greater global influence.
- The recent history of empires has three distinct spatial elements: global integration, imperial overreach, and imperial disintegration.
- Some researchers have identified three broad strategies for the framing of colonial states: producing a plan of urban segmentation that includes racial segregation, creating a fixed distinction between inside and outside, and constructing central spaces of observation to keep an eye on things and to show the presence of colonial power.
- Whether in intellectual discourses and practices, such as cartography and mapmaking, or in sports, language,

religion, and forms of government, modern empires have integrated much of the world.
- The push to empire in the modern world was largely driven by economics.
- There is a tendency for empires to expand beyond their ability to maintain economic dominance and military power, deemed "imperial overreach."
- Among the numerous limits to maintaining and expanding an empire are information overload, fiscal limitations, and strategic limits.
- Although empires may develop into vast economic and geographic worldwide forces, throughout history empires have eventually disintegrated. However, rarely has the process come easily and without dramatic impact on the people and locations left in their wake.
- In the emerging world order, important trends include the continuing hegemony of the USA, the growth of international military authorities, the reemergence of Russia, the rise of China, and the danger of non-state terrorist actors.

The Nation-State

The **state** is an important building block of the political organization of space. In this chapter we will consider the range of different types of states and introduce the notion of the **nation-state** as a spatial entity.

The Range of States

The land surface of the world is divided up into states, separate units of political authority (Figure 15.1). There has been a steady rise in the number of states. In 1900 there were 57 independent countries; there were 70 by 1930, and 160 by 1990. There are now 193 member states in the United Nations. The increase in the past hundred years is due to decolonization, whereby former colonies such as Kenya, India, and Vietnam achieved independence; and breakups, as formerly large empires and states fractured into separate national units. The breakup of the Austro-Hungarian Empire after the First World War led to the creation of a large number of smaller units, including all or parts of Austria, Bosnia, Croatia, the Czech Republic, Hungary, Italy, Montenegro, Romania, Serbia, Slovakia, and Ukraine. Some of these states were reassembled as part of larger units such as the Soviet Union or Yugoslavia. The fall of the Soviet Union in 1991 in turn led to the creation of fifteen new states. The breakup of Yugoslavia after 1989 eventually resulted in six new independent countries. The fracturing continues: in 2008 Kosovo seceded unilaterally from Serbia, and in 2011 South Sudan seceded from Sudan.

At a time of increasing global integration, it is a seeming paradox that the creation of new states continues apace. But global integration allows smaller states to break off from difficult "marriages" as they find it easier to negotiate their place in the world, even as a small, diminished country.

The Power of States

Not all states are equal. Along the dimension of power, we can distinguish between superpowers, major powers, and minor powers. Superpowers

LEARNING OBJECTIVES

Summarize the role of states in the political organization of space and the trends associated with global integration.

Describe the diversity of states.

Recognize the distinction between a state and a nation as differing political entities.

Consider how the state wields power within its territory.

Explain reasons for and patterns of state borders, and how they relate to contested boundaries.

Understand the geographies of voting and elections.

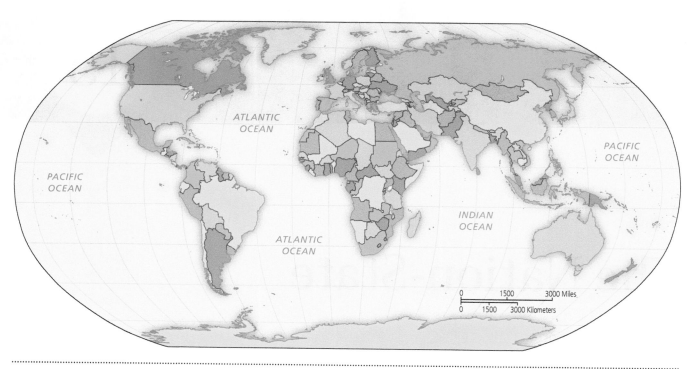

15.1 Political map of the world

have the ability to influence events around the world. Successive superpowers have risen and fallen. In the late sixteenth and early seventeenth centuries, Spain was the first global power, with possessions spread across the world. By the nineteenth century, Britain emerged after years of struggle with France to emerge as the world's superpower, its influence embodied not only by its territorial annexation but also by its naval power, which enabled it to influence events all over the world. Britain's success encouraged emulation. Today the United States remains as the single largest superpower with the ability to achieve global military reach and project its power and influence around the world. This ability is due to the large distribution of military bases spread around the world, a large military force, and commitment to high levels of government military spending. Table 15.1 lists the ten largest military spenders. The United States is clearly dominant, responsible for almost 40 percent of the world's total military spending, more than the next sixteen biggest spenders combined. Military spending is a function of wealth and size. The richer countries, such as France and the United Kingdom, are able to devote some of their considerable national wealth to the military, while the larger countries, such as China, Russia, and India, are able to mobilize their vast resources. Saudi Arabia devotes the largest proportion of GDP to military spending. These ten countries are responsible for more than three-quarters of all military spending in the world.

Only a few countries have the military ability to project national power. While the United States is the superpower, the remaining nine countries are major powers able to exert some national influence beyond their boundaries. The Chinese, for example, are building up their navy enough to back up substantial territorial claims in the South China Sea. Not all states are equal, with the more powerful able to exert an outsize role in world affairs. There is marked asymmetry in global affairs, with only very few countries having the military strength to impose their will across space.

A distinction can be made between soft and hard power. **Hard power** is the ability to coerce others through the superiority of military forces. **Soft power** is the ability to co-opt others to your point of view without the use of force. The United States has hard power through its military superiority, but also elements of soft power through wide dispersal of its cultural forms. Political scientist Joseph Nye, who drew the distinction, argues for the use of smart power, which combines hard and soft power in successful strategies.

The Wealth of Nations

States also vary along the dimension of economic wealth. There are a number of different methods used to classify national economies. The World Bank, for example, uses **gross national income** per capita to produce a fourfold classification of high, upper middle, lower middle, and low

TABLE 15.1	Military Spending, 2015			
	COUNTRY	SPENDING ($ BN.)	% OF GDP	WORLD SHARE (%)
Rank	World Total	1,670	2.3	100
1	United States	596	3.9	38.1
2	China	215	1.9	12.8
3	Saudi Arabia	87	13.7	5.2
4	Russia	66	5.4	3.9
5	United Kingdom	55	2.0	3.2
6	India	51	2.3	3.0
7	France	50	2.1	2.9
8	Japan	40	2.5	2.3
9	Germany	39	1.2	2.3
10	South Korea	35.4	2.6	1.5

Source: Adapted from Stockholm International Peace Research Institute data.

(Table 15.2). There are variations within the groups. Even in the high-income category, for example, the GNI per capita of Switzerland at $84,720 easily exceeds the $54,230 of the USA.

Some countries shift categories as their economies grow or decline. Even in just the past five years, Bangladesh, Kenya, and Myanmar have moved from low income to lower middle, while Mongolia and Peru have climbed from lower middle to upper middle. Argentina and Hungary have made it from the upper middle to high income. Only poor, war-wracked South Sudan shifted downward, from lower middle to low.

There are substantial variations across the world, with the average gross national income of high-income Switzerland 134 times that of low-income countries. We live in an unequal world with national concentrations of wealth and vast areas of poverty. These income figures are more than just abstract statistics; they are a measure of the quality of life, as higher national incomes translate into better jobs, housing, and health care. There is a chance element to the lives we lead. Born into poverty in a poor country, your life chances are more limited and circumscribed than if you are born the

son or daughter of affluent Swiss parents where private affluence is reinforced by public bounty. While we may imagine the character of our lives is self-constructed, it is in large part based on where we were born. Geography in this particular case is destiny. The Economist Intelligence Unit estimates an index for eighty countries based on the number and quality of opportunities for citizens to lead a healthy and prosperous life. The top five countries are Switzerland, Australia, Norway, Sweden, and Denmark. The United States ranks sixteenth and the United Kingdom twenty-seventh. The bottom five are Angola, Bangladesh, Ukraine, Kenya, and Nigeria. You are luckier if you are born in Australia or Switzerland than Angola or Nigeria.

Assessing and comparing national wealth is an exercise fraught with many dangers. A 2012 UN report took up this difficult task. The report's authors measured three kinds of assets: physical capital (buildings, infrastructure, etc.), human capital (including the educational skill levels of the population), and natural capital (land, forest resources, etc.). The United States was the richest country, with over $117.8 trillion of wealth, compared to Japan ($55.1 trillion), China ($20.0 trillion), and Germany ($19.5 trillion). When wealth per person was measured, the rankings changed to Japan, the United States, Canada, Norway, and Australia. These are rough and ready measures, but they begin the process of identifying national wealth and widening the definition of national assets.

Unequal States

The data in Table 15.2 refer only to a national average; there are wider variations within countries, with some regions wealthier than others. In China, for example,

TABLE 15.2	Income Categories, 2015	
INCOME CATEGORY	COUNTRY EXAMPLES	GNI PER CAPITA
Low	Afghanistan, Cambodia, Mali	Below $1,045
Lower middle	India, Ukraine, Vietnam	$1,046–$4,125
Upper middle	China, Peru	$4,126–$12,735
High	Switzerland, USA	Above $12,736

Source: Adapted from World Bank data.

15.1 Depicting Countries in Relative Space

You are perhaps familiar with the standard atlas that shows countries of the world. Most depict nation-states in absolute space. All map projections are distortions, so, depending on the map projection, some areas are exaggerated while others are minimized. In the **Mercator projection**, for example, the size of countries in more northerly and southerly locations is exaggerated, while the size of countries in the tropical zone is minimized.

It is also useful to compare the relative size and comparative size of countries in relative space. At one website (http://www.worldmapper.org/), the size of a country is shown not in relation to land area but in relation to the data under consideration. These maps provide revealing pictures of world geography.

The maps are more accurately termed cartograms. In the world cartogram of total population, China and India "expand" in size because of their large populations. In the map of gross domestic product, by contrast, the United States, European countries, and Japan enlarge dramatically. In the world map of poverty, shown in this box, the size of a country is a function of the proportion of the world's poor living in that country. Notice the small sizes of North America, Europe, and Japan compared to the inflated sizes of Africa and countries in the Indian subcontinent.

This cartogram is both accessible and strange. We can immediately make out the shapes of South America and Africa, so we know it depicts our world. But it is a weird world: North America is thin and spindly, Western Europe has shrunk, and Australia has all but disappeared from view while other regions have expanded. National territory is depicted as a proportion of the world's poor living in that territory. Poverty is measured for most countries with a human poverty index based on life expectancy, literacy, access to clean water, and hunger. Because these indicators make sense in the poor world but not in the rich, thirty of the richest countries use measures of life expectancy, literacy, income, and unemployment. Our understanding of the geography of the world depends upon the measures we use. In this depiction, India, much of South and Southeast Asia, and tropical Africa balloon into major significance as the diagram allows us to identify quickly the geography of global poverty. Effective antipoverty measures in these parts of the world will have a global impact.

REFERENCES

http://www.worldmapper.org/

Dorling, D., Newman, M., and Barford, A. (2010) *The Atlas of the Real World*. Rev. ed. London: Thames and Hudson.

The world's poor

there are substantial differences between the poor rural interior and the wealthier, urban, industrial coastal cities. The average income figure also masks substantial variation between households within one country. The level of income inequality is an important indicator of how income is spread across the national population. The CIA produces a ranking of inequality using the **Gini index**, which measures the spread of wealth across the population. The values range from the most unequal, Lesotho, at 63 to the least unequal, Slovenia at 23. The most unequal states are low- and middle-income countries such as Haiti and Chile. The most equal societies are small, affluent countries such as Sweden, Finland, Hungary, and Norway. However, a very poor country such as Bangladesh has a more equal income distribution than India, China, and Peru—it is even more equal than the United States. This somewhat reinforces the bell-shaped **Kuznets Curve**, which depicts inequality increasing with rapid urbanization and industrialization. The general pattern is outlined in Figure 15.2. So in the case of China, the inequality is to some extent a measure of recent industrialization and the growing economic differences between the interior rural areas and the faster-growing, industrial, coastal urban regions. But notice how the United States, a country that has long since moved from a mainly agricultural to an industrial and service economy, once more has a relatively high level of inequality. The economist James K. Galbraith has produced an amended Kuznets Curve to suggest that in some higher-income countries, rising inequality is created when high-tech and finance become larger parts of the economy. The large wealth amassed by owners and top earners in these sectors skews the national figure toward increased inequality. The smooth bell shape of the original Kuznets Curve then starts to trend upward (see Figure 15.3). The United States is situated after this inflexion in the curve. Thomas Piketty argues that since the returns to capital are higher than the rate of economic growth and so more wealth is appropriated by the richest, then inequality will continue to persist unless there is a marked redistribution of wealth.

The CIA collects data on national income inequality because it is a useful predictor of political stability. More equal societies tend to be more peaceful, less riven by conflict. Even societies that are unequal but have been so for some time have a measure of stability. But societies where

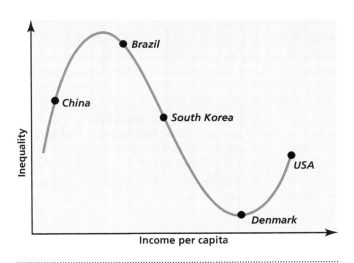

15.3 The Modified Kuznets Curve

the rate of change in inequality is marked and sudden can become unstable as social movements emerge to resist and contest the sudden and marked concentration of wealth in just a few hands. The Arab Spring of 2011, for example, trailed in the wake of sudden increases in inequality due to globalization, privatization, and the adoption of neoliberal policies. Rapid growth that brings a quick increase in inequality can provide the basis for social stress and social upheaval. Even for more stable and affluent societies, the level of inequality is an important indicator of well-being. Richard Wilkinson and Kate Pickett found that in certain states of the United States and in the twenty richest countries, where there was a big gap between the incomes of rich and poor, there was more mental illness, drug and alcohol abuse, obesity, teenage pregnancy, and homicide. The more unequal a society, the shorter is the life expectancy and the poorer is children's educational performance. Scandinavian countries and Japan consistently do well on a range of such social indicators; they have the smallest differences between higher and lower incomes and the best levels of overall physical and mental health. Obesity is twice as common in the United Kingdom as in the more equal societies of Sweden and Norway, and six times more common in the United States than in Japan. It is not just that the lives of the poor are worse; the lives of everyone in the society are made worse. Life expectancy even among the rich is lower in more unequal societies. The United States is wealthier and spends more on health care than any other country, yet a baby born in Greece (at least before the current economic crisis), where average income levels are about half that of the United States, had a lower risk of infant mortality and longer life expectancy than an American baby. It is not only the wealth of society that structures health outcomes and social performances; so does the level of inequality.

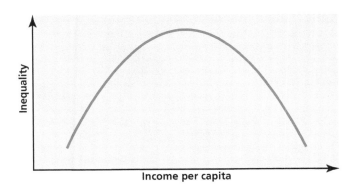

15.2 The Kuznets Curve

Even within one country there are profound differences. National averages mask racial and ethnic differences. The infant mortality rate for the United States, for example, varies by race. For non-Hispanic blacks it is 13.3 per 1,000 live births; for non-Hispanic whites it is 5.6. In other words, black babies born in the United States are 2.3 times more likely to die than white babies. The black infant mortality rate in the United States is higher than the national average in Costa Rica (9.8), Sri Lanka (12.4), or Thailand (12.4) and closer to the national rates of poor countries such as Albania (13.6), Dominica (12.3), and Jamaica (13.9).

Dimensions of Difference

States vary along a host of different dimensions, some easier than others to quantify. There are the hard statistics that measure military power, income levels, and inequality. There are also fuzzier measures that give an indication of political regime and quality of life. In terms of political organization, three main regimes can be identified: *totalitarian*, in which the government has control over wide and deep swathes of social, political, and economic life; *authoritarian*, where power is concentrated but not so deeply entrenched as in totalitarian regimes; and *democracies*, where political power arises from the majority will of the people. There is some variation within these broad categories. Democracies, for example, include monarchies, such as the United Kingdom and Sweden, as well as republics, such as the United States and France. While power may ultimately reside with the people, even in full democracies its expression and implementation are deeply influenced by the distribution of wealth and power in the society.

The Economist Intelligence Unit uses a **democracy index** to gauge the level of democracy using sixty indicators related to electoral process, civil liberties, functioning of government, political participation, and political culture. The index, produced each year, ranges from 10, the most democratic, to 0, the least democratic (Figure 15.4). The most democratic countries include Scandinavian democracies such as Norway, Iceland, Denmark, and Sweden, while the least democratic are North Korea, Syria, and Chad. The United Kingdom stands sixteenth and the United States twentieth.

The democracy index is used to classify countries into different regimes: full democracies, flawed democracies, hybrid regimes, and authoritarian regimes. According to this classification, more than a third of the world's population lives under the duress of authoritarianism (see Table 15.3). The democracy index is a function of the variables employed. Change the mix of variables, and different results may emerge at the margins. But whatever variables are used, some countries, such as North Korea, will always come out as deeply undemocratic. At this extreme end, there are such depressing categories as **kleptocracies**, states geared to the enrichment of the tiny political elite. Some of the greediest include former

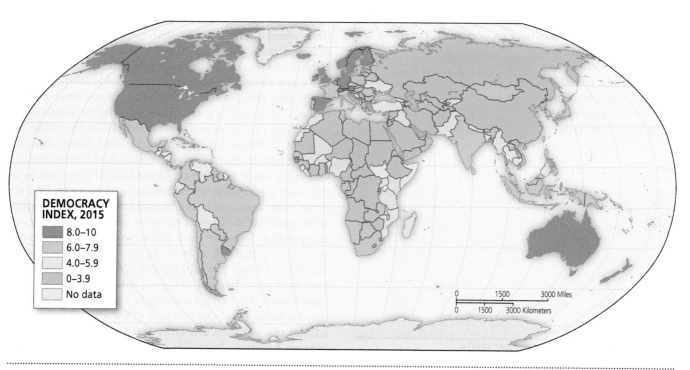

DEMOCRACY INDEX, 2015

- 8.0–10
- 6.0–7.9
- 4.0–5.9
- 0–3.9
- No data

15.4 The democracy index

TABLE 15.3 Political Regimes and Democracy Scores

CATEGORY	SCORE	% OF WORLD POP.	EXAMPLES
Full democracies	8.01–10	8.9	Canada, United States, United Kingdom
Flawed democracies	6.01–8	39.5	Argentina, Serbia, Philippines
Hybrid regimes	4.01–6	17.5	Armenia, Bolivia, Pakistan
Authoritarian	0–4	34.1	China, Russia, Saudi Arabia

Source: Based on Economist Intelligence Unit data.

Indonesian president Suharto, who reputedly stole between $15 billion and $35 billion; former president of the Philippines Ferdinand Marcos ($5 billion–$10 billion); and Mobutu, former president of Zaire, now the Democratic Republic of the Congo ($5 billion), who plundered the country and pauperized the population. Current kleptocracies include Equatorial Guinea, where Teodoro Obiang Nguema, president since 1979, has amassed a personal fortune estimated at around $600 million. He has personal control over the nation's treasury and works well with foreign oil companies eager to drill for oil. Despite the fact that Equatorial Guinea is one of the world's poorest countries, the president lives a life of unrivaled opulence.

Classifications are snapshots in time. Popular movements and internal regime changes can alter them. Marcos was overthrown by a popular uprising, and the Philippines are now classified as a flawed democracy—not ideal, but more democratic than in the Marcos era. Indonesia is also now in the same category. Myanmar, after years of authoritarian military control, seems to be moving toward something approaching greater democracy. Congo remains in the authoritarian category.

States vary by quality of life as well as political system. A standard economic argument is that rising income levels create a greater a sense of well-being. More income, so the argument goes, creates happier people. The **Easterlin Paradox**, based on the work of the economist Richard Easterlin, suggests that once basic needs are met, however, there is little difference in reported happiness with increasing incomes. In other words, the paradox is that once a certain base level of income is reached, increasing income further may not necessarily increase the level of happiness in the country. The debate continues. Including variables such as life expectancy, hours worked, and unemployment confirms the paradox. More recent papers highlight the finding that people in rich countries with more progressive taxation tend to report higher levels of happiness than those with lower and more regressive tax policies. Perhaps we need to measure **gross national happiness** as well as gross national income. What variables would you include? One of the few countries to measure happiness is Bhutan. However,

their index does not capture the fears and anxieties of the ethnic Nepali minority in the south of the country, many of whom have fled to neighboring Nepal since the 1990s.

The Rise and Fall of States

States vary enormously along a variety of dimensions, including military might, economic wealth, and political freedom. What is also worthy of note is the rapid rate of change, with poor countries sometimes becoming richer and authoritarian systems becoming more democratic. To be sure, there are opposing trends, as some countries become poorer and governments slide into authoritarianism. The system of nation-states is a dynamic jigsaw puzzle, constantly changing, sometimes slowly and at other times with dramatic speed. Perhaps we need maps of changing status to get a firmer sense of this dynamism.

The openness of a state is defined as the extent and ease to which people, ideas, information, goods, and services flow across its borders. Very closed societies such as North Korea are very stable. But once such a closed society becomes more open, there is a risk of political instability. That is why the rulers of closed states strive so hard to keep the lid on; it keeps them in power. The position of some individual states is difficult to assess. China, for example, encourages the free movement of capital and goods but keeps a firmer control over ideas that come over the Internet. It is open to the globalization of goods for its consumers, but less open to the globalization of ideas and information for its citizens.

The political geography of states is complex and dynamic. States rise and fall, expand and contract. The United States, for example, started off as thirteen colonial states pinned up against the Eastern Seaboard, but through annexation, invasion, purchase, and treaties it extended to continental proportions. Historian Norman Davies relates the fascinating story of states and kingdoms in Europe that no longer exist. He recounts the rise and fall of, among others,

Byzantium (330–1453), Rosenau (1826–1918), Tsernagora (1910–1918), and the Soviet Union (1924–1989).

The Crises of the State

States, despite their power and seeming permanence, are, in the long-term, fragile organisms.

Their fragility results from the variety of crises that can beset them. Following on and developing the work of the German sociologist Jürgen Habermas, we can identify three contemporary and interlinked crises; fiscal, legitimation, and rationality. A **fiscal crisis** is when a state has more expenditure than revenue. Military and social spending, either singly or together, can outmatch the revenue from taxes and tariffs. In the short term the state can borrow money, and the biggest economies can borrow more, but ultimately a fiscal crisis turns into a political crisis as debts have to be repaid and unpopular cuts in spending have to be made. Fiscal crisis is an endemic feature of poor countries but is increasingly a problem for richer countries as well: either they spend a lot on social welfare and the military (the United States) or just on social welfare (Western European countries), or they have a low tax base and/or rampant tax evasion (e.g., Bulgaria, Italy). The **shadow economy**, in which people pay no taxes, constitutes 32 percent of the Bulgarian GDP, 21 percent of the Italian economy, but only 7 percent of the US economy. Fiscal crises prompt political crises.

A **legitimation crisis** is when the state loses its popular appeal and its ability to govern. When it loses the support of an increasing part of the population, its legitimacy is at risk. During the Arab Spring, for example, the governing elites of Egypt, Libya, and Tunisia no longer had the support of the majority of the population. In Syria, the Assad regime lost the support of the majority of the Sunni population. Declining living standards, a sense of injustice, and a feeling of profound alienation between the government and the governed prompt a legitimation crisis. The crisis can take a shallow form when the party in power is highly unpopular, or can be deeper when the entire system or regime is unpopular and wholesale change is demanded.

A **rationality crisis** occurs when the state makes enough poor decisions that other crises emerge. The US decision to invade Iraq in 2003 was based on faulty assumptions and led to the death of almost 100,000 Iraqis, the displacement of millions more, and the loss of over 4,000 US troops. In another case, in 2008 the Irish government guaranteed the losses of Irish banks. The banks received almost $4.5 billion each. By 2010, government support for the banks constituted almost a third of the country's entire GDP. The ruling party at the time of the bank bailouts, Fianna Fáil, the largest party since 1932, lost the 2011 election in a landslide defeat with only 17 percent of the votes.

These examples show the interlinked nature of fiscal, legitimation, and rationality crises.

The Spatial Nature of the State

A state is a self-governing political unit. It has monopoly control over a territory and boundaries with other states. For it to be legitimate, it has to be recognized by other states and most of its population. There are currently almost 200 states in the world that cover the land surface of the Earth and make claims on the surrounding and adjacent seas and oceans. States are spatial entities.

The Size of the State

States vary in their spatial characteristics. There are large and small states. Size can be measured by territory and by population. Table 15.4 lists the top ten countries by area and population. Some countries, such as Australia and Canada, have a large territory but relatively small population. Some are small in area but large in population, such as Bangladesh. Others are big in both respects. Besides the United States, four other countries make both top tens: Brazil, Russia, India, China, and South Africa. These big five are often grouped together—termed the **BRICs**, after the first letter of each country's name—because they all combine large size with developing economies. Their joint population of 2.8 billion constitutes just over 40 percent of the total world population. Indonesia, with 237 million people, is a possible future member of the BRICS. Although its economy is perhaps poised to become more dynamic, in part fueled by a demographic dividend,

TABLE 15.4 The Top Ten "Big" Countries	
BY AREA	**BY POPULATION**
Russia	China
Canada	India
China	US
US	Indonesia
Brazil	Brazil
Australia	Pakistan
India	Nigeria
Argentina	Bangladesh
Kazakhstan	Russia
Algeria	Japan

Indonesia's problems of poor infrastructure, corruption, underemployment, and poverty may limit its future growth. Despite the hurdles faced by each of these countries, they have the necessary size and rate of growth so that their combined potential represents a possible shift in geopolitical and geo-economic power away from the core of rich countries. Their potential is in part based on their territorial sizes, which include large amounts of natural resources (as in Brazil and Russia) and great potential for growth through the demographic dividend (especially pronounced in Brazil, India, and Indonesia). Size does matter.

One way that smaller countries try to offset their limited size is through organizations with other states. For instance, **economic unions** try to extend the size of the market. The European Union, for example, grew from the perceived need of European countries to avoid further conflict in the wake of two world wars, but also to provide a large enough economic entity to compete with the United States. It began as the European Economic Community of six countries formed by the Treaty of Rome in 1957. It was renamed as the European Union and expanded to include twenty-seven countries with a single market. The recent fiscal crisis, highlighted by the debt problems of Greece but also the very real potential for similar crises in Portugal, Italy, and Spain, reveals that unions do not solve all economic problems.

States also come together for specific economic interests. We already noted the important role that OPEC plays in regulating the global supply and price of oil. There are also looser economic organizations like NAFTA (the North American Free Trade Agreement) between Canada, Mexico, and the United States.

The Location of States

States also vary by location. One significant feature is access to sea transport, which in general tends to be far cheaper than overland or air transport. For landlocked states to access global markets, they have to ship goods overland through another state, which increases costs and tariffs (Figure 15.5). There is no simple relationship between level of economic development and being landlocked. On a variety of different measures, Switzerland is one of the richest countries in the world, with many inhabitants enjoying high per capita income and a full range of quality public services. However, for many poor countries such as Bolivia, Chad, Mali, Mongolia, Niger, and Paraguay, landlocked status makes economic growth all that more difficult.

As spatial units, states can be lucky or unlucky in their territorial endowment. States that have rich oil reserves, for example, can wield effective political power because their resource endowment provides huge revenues. The Saudi government's oil revenues pay for its projection of soft power through the promotion of fundamentalist Islam around the world. In some cases, rich resources provide welcome revenues. In 2011, when the price of a

15.5 Landlocked states

15.6 This picture a of rustic church at Geirangerfjord embodies an idealized picture of the national landscape of Norway.

barrel of oil was $111, revenues from oil and gas accounted for one-half of Russia's federal budget, and raw materials constituted 80 percent of exports. While a generous resource endowment may provide short- and medium-term riches, it can, over the longer term, lead to problems. Easy reliance on nature's bounty may restrict the innovation necessary for long-term economic growth. This is the so-called **Dutch Disease**, based on that country's experience with oil revenues that created a huge trade surplus and large currency reserves, which led to overvalued domestic industries that were uncompetitive in global markets. Resource prices are also very volatile, so that revenues may vary from one year to the next. In high-price years, governments may spend to shore up popular support, but that support may vanish when prices fall and spending has to be curtailed. By 2016 the price of barrel of oil had fallen to $33. The revenues of the Russian government were effectively reduced by 70 percent in just five years. Commodity price volatility is one manifestation of the "**resource curse**," sometimes known as the paradox of plenty. While valuable resources can create short-term booms, they may depress long-term economic growth. In less democratic countries, wealth from resources is easily diverted, stolen, and misspent.

National Environmental Ideologies

Nations occupy territory. National identity is bound up with attachment to territory and its creative representation (Figure 15.6). National environmental ideologies build on myths of wilderness, countryside, and city. These myths often have a bipolar quality. Consider wilderness. Its defeat is part of many national histories. In the United States, for example, the defeat of the wilderness in the nineteenth and early twentieth centuries was an important element in the sense of national identity, and the moving frontier gave a sense of an expansive national trajectory. Slowly an alternative began to emerge, that of preserving the remaining wilderness. As wilderness disappears, its survival can also become a national project. The national park movement in the United States took as its guiding principle the defense of remaining wilderness areas. The wilderness and its point of contact with the civilized world, the frontier, are the background to one of the essential US texts, the Western movie. Countryside can also be idealized, especially in a rapidly industrializing and urbanizing country. An idealized representation of countryside is still at the core of English environmental ideology. The idea of the city can range from Jerusalem, a place of social uplift, the "city on the hill" of the early Puritans in North America, to Babylon, the place of vice and debauchery populated by the foreign Other. In the United States, there is a long anti-urban tradition that extols the virtue of the small town and rural life.

Environmental ideologies are expressed in paintings, novels, and cinema. Australian landscape painting, for example, depicts the changing idea of the outback; many English novels deal with the tension of modernity in an idealized countryside; and the United States Western provides mythic depictions of the frontier between the settled and the wild.

While the general categories of wilderness, countryside, and city are used in varying ways to construct and challenge national environmental ideologies, more specific features are also employed. In a fascinating study, Tricia Cusack tells the story of how five rivers became part of national identity. She tells the story of the Hudson as an icon of American nationalism; the Thames as embodying ideas of monarchy, empire, and commerce in England and Great Britain; the Seine and consumer culture in France following the War of 1870; the Volga and its role in Russian romantic nationalism; and the Shannon and its connection with Celtic Irish cultural identity. Rivers are more than just hydro-geomorphic features. They are texts that are used to generate and sometime unify diverse narratives and competing claims on national identity. Forests, mountains, rivers, coasts, valleys, and other natural features become elements of a national landscape.

15.2 Imagined Communities

The construction of national identity is an important subject of recent writings. In an argument of great subtlety, Benedict Anderson argued that nations are not so much facts of race or ethnicity; rather, they are what he terms "imagined communities." Anderson paid particular attention to the role of print capitalism in creating a national discourse. In a later elaboration, Anderson identified three institutions of power: the *census,* the *map,* and the *museum,* which together allow the state to imagine the people under its dominance, the geographic territory under its control, and the nature of historical legitimacy.

National imaginaries are rarely coherent, consistent, or stable. The dominant national imaginary in Australia has shifted from celebration as an Anglo-Celtic colonial outpost to the more uncertain embrace of a postcolonial, multicultural society. A major element in this shift is the changing role assigned to indigenous peoples.

There are national histories and national geographies, national characteristics and national claims to greatness. "National" events are enacted and reenacted; "national" stories are told and retold. National identities are also created through commemorative activities. Exhibitions, fairs, and sites of historical memory and commemoration are common vehicles for celebrations and claims of national identity. In the centennial celebrations of the United States (1876) and Australia (1888), the indigenous peoples were excluded. Fast-forward a hundred years and ritual and symbol were mobilized once again. Both bicentennial celebrations, of 1976 and 1988, praised political liberties and economic prosperity, yet the fundamental belief in the forward march of progress had weakened over the course of the intervening century. Now cultural diversity was celebrated rather than ignored. The indigenous peoples were given a more central role; in Australia, this emerged mainly as a result of Aboriginal protest at the concept and the practice of the celebrations.

REFERENCES

Anderson, B. (2006) *Imagined Communities: Reflections on the Origin and Spread of Nationalism.* Rev. ed. London: Verso.

Spillman, L. (1997) *Nation and Commemoration.* Cambridge: Cambridge University Press.

National identity in the United States, for example, was shaped by contact with forests. The seemingly endless supply of trees created a throwaway culture in which things were not built to last. Only when the forest cover was threatened was a conservation movement initiated. Trees shaped American identity just as Americans transformed the forest.

A "national" landscape is crafted from and represented by the raw materials of the environment into texts of historic meaning and political significance.

Nation, State, and Minorities

A distinction can be made between a state and **nation**. A state is a political organization that controls a particular territory. A nation is a community of people with a common identity, shared cultural values, and a commitment and attachment to a particular area. In some cases, there is congruence between nation and state. In the case of the small landlocked country of Swaziland, the nation and the state are almost identical, with the great majority of the population ethnically Swazi. Such instances are rare, especially with international migration creating more complex patterns.

Nations and States

The incongruence between nation and state takes two main forms. First, some nations lack a state. For example, the traditional homeland of the Kurds is in the Near East. After the fall of the Ottoman Empire, just after the First World War, their homeland was divided by Western powers that drew boundaries to suit their geopolitical interests rather than to honor national differences. The Kurdish nation was divided among separate states, including Turkey, Syria, Iran, and Iraq, and Kurds became vulnerable minorities within them (see Figure 15.7). In some cases the Kurds have been successful in creating regions of relative autonomy. After the US invasion of Iraq, the Kurdish region in northern Iraq achieved a large measure of autonomy. However, the Kurdish region in eastern Turkey is still a scene of ongoing conflict between the Turkish government and Kurdish separatists.

There are also states with more than one nation. In Belgium, for example, the centuries-old rift between the Dutch-speaking north and the French-speaking south, despite their shared Catholic religion and rising living standards, remains always just below the surface and always capable of breaking out into explicit calls for separation. In some cases, rifts widen to civil war. Sudan was a colonial possession of the United Kingdom. It consisted of two different regions: a northern part centered on Khartoum in which most people are African-Arab and Muslim, and

Black Sea

RUSSIA

KAZAKHSTAN

GEORGIA

Caspian
Sea

ARMENIA   AZERBAIJAN

TURKMENISTAN

TURKEY

KURDISTAN

NAKHCHIVAN
(AZER.)

CYPRUS

IRAN

SYRIA

LEBANON

Mediterranean
Sea

IRAQ

ISRAEL   JORDAN

EGYPT

SAUDI
ARABIA

0   100   200 Miles
0   100   200 Kilometers

75–100% Kurdish
30–74% Kurdish
Less than 30% Kurdish

15.7 (A) Distribution of Kurds in the Middle East. **(B)** Kurds celebrating the festival of Newroz ("New Day") in Van, Turkey.

B

targeted when economic difficulties provoke social unrest. In Indonesia, for example, there is a significant ethnic Chinese population, located in specific neighborhoods of cities and towns. Out of a total population of 240 million, more than 2 million are estimated to be ethnic Chinese. Their experience has not been a happy one. There is a long history of anti-Chinese feeling in Indonesia. In 1740, economic unrest led to demonstrations and mass killings. Ethnic Chinese could travel only with special permits. After independence from the Dutch in 1949, the ruling junta regularly provoked anti-Chinese feeling. In 1998, anti-Chinese rioting in the capital city of Jakarta erupted in violence and damage to Chinese-Indonesian communities. Minorities can often become scapegoats for wider economic issues. When economic conditions decline, the Other is often blamed.

In many states, there is some measure of tolerance for people different from one another. In others, however, the differences remain, often exacerbated and embodied in years of struggle. In Northern Ireland, for example, the latest era of struggles between the Catholic and Protestant communities lasted for almost forty years, from 1970 until a power-sharing administration began in 2007.

The protection of minority rights is a fundamental part of the Charter of the United Nations and is now an important element of the global discourse of international politics. However, the cases of Bosnia, Darfur, and Rwanda indicate that the human rights of minorities can still be violated. When a state is committed to harming minority groups within its borders, the international community has to overcome distance, time, and political inertia in order to stop that state from killing its own citizens. International involvement is slow, costly, and difficult to organize.

a southern one where most people are black Africans and either animist or Christian. When the country achieved independence in 1956, the two parts were joined in one state. It was never a happy marriage. Power was retained in the north, and the south languished. Decades of civil war caused casualties and created famine. More than 2 million people died, and millions more were displaced. The south was able to secede only after years of war and starvation. The state of South Sudan came into existence on July 9, 2011.

The population of a state is rarely homogenous. Differences in religion, ethnicity, and language, if they have distinct spatial expression, can create tensions in the internal coherence of the state. Minority groups are often

The Territory of the State

A distinction can be made between government, state, and national territory. The government is the political expression of power. "State" is a wider term that covers the more embedded power structures, such as the army, the police, and the educational system. In some cases, a "**deep state**" of entrenched power operates despite changes of government. In more authoritarian regimes, government and state are more interconnected, while in more democratic states there is space between the government and the state.

Centripetal Forces

We can identify **centripetal forces** that unify the state's power across space and **centrifugal forces** that disrupt it. Centripetal forces include external aggression, which may stimulate national bonding against a common enemy; federal structures that allow the safe expression of regional and other subnational differences; and national mass media and education, which create a shared culture and language. In some cases a measure of regional autonomy can blunt separatist claims. In Spain, for example, under the Franco regime (1936–1975) power was centered in Madrid and used to suppress both Basque and Catalan autonomy. Support for the soccer team of Barcelona was and still is a vigorous expression of Catalan identity. Since the fall of the Franco dictatorship, there has been a more pronounced commitment to power sharing. In other cases, separatist movements may lie dormant but then erupt. Scotland joined with England in 1707, and for three centuries Scottish nationalism in lowland Scotland was little more than a whisper, although many in the Highlands resisted vigorously in 1715 and 1745; recent years, however, have seen more vocal calls for separation.

Centrifugal Forces

Centrifugal forces include political and economic inequality. One reason behind a resurgence of Scottish nationalism is the large gap in living standards between Scotland and the southeast of England, the center of power and home of the elites. Despite the homogenizing abilities of the state, religious, ethnic, and language differences can persist and survive attempts at their eradication. In Spain, the Basque language is still a principal vehicle and embodiment of Basque identity and difference from Castilian Spain.

The state's responses to separatist movements within its national boundaries can range from repression to some form of accommodation. Many states began with repression. When the British state defeated the Highland rebels at the Battle of Culloden in 1746, it criminalized wearing tartan, playing bagpipes, or speaking Gaelic. The territorial expansion of both the United States and Canada involved the defeat and removal of many indigenous nations. States can also accommodate separatist movement by granting a degree of political autonomy. In the past twenty years the British state has given Scotland its own parliament with greater fiscal powers in order to head off the appeal of Scottish nationalism.

Territorial Integrity

A state has a monopoly of legitimate physical force. But in some instances, the rule of the state does not extend to all of its territory; when it cannot effectively control it all, a territorial crisis occurs. In some cases, the territorial divisions erupt into civil war. The US Civil War (1861–1865) was fought between northern and southern states, with the South declaring itself independent from the United States. It was a bloody war. By 1865 over 620,000 Americans had died as a direct result of the war, more than all the Americans killed in twentieth-century conflicts. On one day, September 17, 1862, at the battle of Antietam in Maryland, more than 4,000 men were killed and 17,000 more wounded.

The territorial control of the state varies along a continuum. At one extreme are states that wield undisputed monopoly power over their territory. At the other are dysfunctional states with limited territorial control. In Somalia, for example, the government's control barely extends beyond the government compound in Mogadishu. In Afghanistan, the president of the country is known as the mayor of Kabul, emphasizing his limited influence. In Colombia, a fifty-year civil war was fought between the central government and an insurgency. In its remote mountain areas during the 1950s, Communist guerillas and members of the Liberal Party started to organize against the army and the Conservative Party in a struggle over control of land and political dominance. More than 200,000 were killed. In 1964, FARC (Revolutionary Armed Forces of Colombia) came into existence and waged a forty-year war against the central government. The rebels were funded through their proceeds from the drug trade, while the government received aid from the United States. Only by 2016 did a signed agreement lead to a possible end to the decades-old conflict.

The internal territorial integrity of the state is more assured where there is a homogenous population. But marked differences, overlaid by economic inequality and profound cultural differences, can exert pressures on the integrity of state control over national space. When states cannot control the national territory, they become dysfunctional. Piracy developed along the coast of Somalia because of the power vacuum created by a dysfunctional state. The pirates were halted only with the intervention of outside powers.

Weak states lacking territorial control are the breeding ground for non-state actors such as FARC, Al Qaeda, and ISIS. Dysfunctional states or states characterized by internal convulsions along ethnic and religious lines and an economic system that offers few opportunities for the majority of people, especially young people, are the necessary oxygen for non-state actors that wish to challenge the legitimacy and territorial integrity of states.

The Example of Brexit

In 2016 the voters of the UK were asked to vote whether to stay in or leave the European Union. It was knows as Brexit, short for "Britain's exit."

A majority in the UK voted to leave the EU. But a look at the geography of the vote provides a good example of the political geography of the United Kingdom (Figure 15.8)

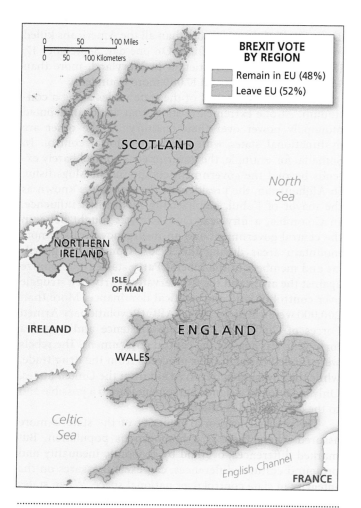

BREXIT VOTE
BY REGION

Remain in EU (48%)
Leave EU (52%)

15.8 The Brexit vote

and the centrifugal and centripetal forces that affect it. A major area of support for remaining in the EU was centered on London. Greater London comprises 7.5 million people, but the greater metropolitan region has a population of approximately 21 million. The reach of the city extends into most of the southeast part of the country and beyond. It is by far the most affluent part of Britain. London has emerged as a global financial center, attracting expertise and investment from across the globe and around Europe—it is hardwired into the financial circuits of the EU and global economy.

For the past thirty years the principal aim of government economic policy was to protect London's position, largely ignoring the rest of the country. In the UK version of trickle-down economics, London is the golden goose whose droppings fertilize the rest of the country. In reality there was not much trickle and the UK, like the US, is becoming a more unequal society. The UK has the most marked regional inequality in Europe.

The rich and wealthy are concentrated in London and the southeast, where household incomes are higher than in the rest of the country. As the UK became a more unequal and divided society, the cleavage between London and the southeast and the rest of the country has become more marked.

The rest of the country in England and Wales, and especially the lower-income groups in those regions, was bypassed by the emphasis on the London money machine. But discontent could not feed into demands for political separation as it could in Scotland—there was no constitutional basis—and so the rest of England and Wales seethed at the rising inequality and neglect. The exit vote was also a vote against the dominance of London and the political establishment that captured popular resentment against the status quo.

All districts in Scotland voted to remain in the EU—a complete reversal of the 1975 referendum when Scotland voted against joining the EU while much of the rest of the country embraced the European project. What happened? The EU provided benefits to Scotland that the London-biased UK government did not. Scotland's political culture, it turns out, is closer to the EU than that of London-based Tories. And years of punishing Thatcher rule soured many in Scotland against relying on the UK political system.

The Brexit vote highlights the problem of sharing a political space with divergent interests. The Scottish Parliament in Edinburgh will now move to draft legislation for another referendum on Scottish independence. The result could be different from the one in 2014 when 55 percent voted to stay in. The Brexit vote may herald the breakup of a Union first created in 1707.

Northern Ireland voted to remain in the EU. Its economic position is similar to Scotland's but in a very different political context. The province is split between those who want to join up with Eire and those who want to stay in the UK. Sinn Féin, the party that seeks union with Ireland, has already proposed a vote on all-Ireland unification.

The Brexit vote revealed the deep divide in the UK between the different regions of England and Wales and especially between the affluent London and the southeast and the rest of the UK. This division is unlikely to heal soon.

Boundaries and Frontiers

States have edges, and these boundaries create both tensions that can lead to conflict and sites for interstate cooperation. Political geographers used to make a distinction between artificial and natural borders. People made artificial borders, while natural borders were coastlines, mountain ranges, rivers, and the like. The limits of the United States are shaped in the east and west by the natural borders of oceanic coastlines but are more artificial in the north and south. More recent work tends to view borders as social constructions, the result of bordering practices rather than the embodiment of "natural" differences.

Viewing borders as socio-spatial constructions focuses attention on their origin and evolution and interprets them as stages for the performance of national identity. Borders arise, shift, and change shape over time, their changing configuration embodying competition for territorial space. Consider the case of Hungary. During the Austro-Hungarian Empire (1867–1918), Hungary was joined with Austria in ruling over a large empire in central Europe. The breakup of its empire after the First World War reduced Austria to a small German-speaking state and started the process of shrinking Hungary's territory. In the Treaty of Trianon (1920), Hungary lost Vojvodina, Croatia, and Slavonia to the new state that became Yugoslavia, and forfeited its coastline. In the north, Hungary lost what is now Slovakia to newly formed Czechoslovakia. In the east, Hungary had to cede Transylvania and the Banat to Romania. To the west, Hungary lost Burgenland to Austria. In all, Hungary forfeited two-thirds of its former area.

The borders of many countries are complex texts that tell of the country's history. Poland, located in central Europe, has a more complicated history than most. Poland's current boundary encompasses territory that once belonged to Russia, Hungary, and Germany, while formerly Polish territory can be found in Belarus, the Czech Republic, Germany, Russia, and Ukraine. One Polish city, Wrocław, was a municipality of the Habsburg Empire, of the Kingdom of Prussia, of the German Empire, and then of Poland. Its name changed in the process from Vratislavia to Breslau to Wrocław. Over the longer term, boundaries are fluid and changeable.

Boundaries are used to define and contain. The Great Wall of China, a centuries-old building project first begun 2,500 years ago, was built to keep out marauders from the steppe. During the Cold War, the boundary between East and West Germany was used to define the limits of effective power but also to limit mobility and movement. Boundaries can be used to constrain the movement of people and goods, but in a globalized, interconnected world, it is increasingly difficult to limit the flow of information. But states try. The newest "border construction" in China is the government's control over Internet traffic.

The Porosity of Borders

A distinction can be made in the porosity of boundaries. Hard, impervious boundaries are created and maintained by the state as a display of power, a performance of national identity. The increasingly fenced boundary between the United States and Mexico, for example, is a form of border theater to display US intentions against illegal immigration and cross-border drug trade. The United States's boundary with Canada is more porous than its boundary with Mexico. We can thus make a distinction between **hard** and **soft borders**, **porous** and **nonporous boundaries**. The boundaries between the United States and its two neighbors provide example of hard, nonporous borders and soft, porous boundaries, although in reality they are both more porous than official pronouncements would suggest.

Along the edge of boundaries, frontier regions may develop. There are the landscapes of hard boundaries: the barbed wire, the border patrols, and the checkpoints. Between North and South Korea, the boundary is made visible on either side by a demilitarized zone of unpopulated territory filled with landmines. Figure 15.9 depicts this hard, nonporous boundary. Frontier zones in such cases may develop as liminal spaces lacking investment and development.

Conflict

Boundaries can be places of conflict, especially where territorial sovereignty is disputed. Around the world, boundaries are often neither settled nor agreed upon. Let us consider one example in Southeast Asia.

French cartographers created the boundary between Thailand and Cambodia in 1907, when France was the

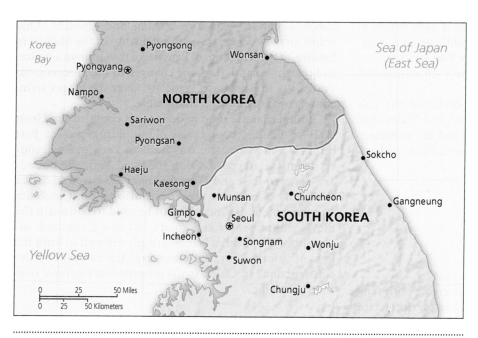

15.9 Boundary between North and South Korea

Exclusive Economic Zones (EEZs), first introduced in the 1982 UN Convention on the Law of the Sea (UNCLOS), are areas where states can claim exclusionary rights over marine resources. They have a limit of 200 nautical miles (370 km) from the coastal baseline, the seaward edge of its territorial sea, but may extend beyond the 200-mile limit into the continental shelf. Where EEZs overlap, it is up to the states to delineate the maritime border. Matters are made more complex when small islands are used to define a nation's EEZ and are claimed by more than one country. A country can establish territorial seas up to 12 nautical miles from a naturally formed island and a 200-nautical-mile EEZ. Part of China's strategy in the waters of the South China Sea is to lay claim over islands, natural and created, to claim the surrounding waters.

China's claim impinges on the EEZs of Vietnam, the Philippines, Malaysia, and Brunei (Figure 15.10). There are disputes over the various islands in the sea. The Paracel Islands are claimed by both China and Vietnam. When China moved a giant oil rig to the region in 2013, it was seen as a direct threat by the Vietnamese and undermined China–Vietnam relations. The Chinese have built a surface-to-air missile defense system on Woody Island, one of the largest of the Paracel Islands, and recently added a helicopter base to Duncan Island.

The Scarborough Shoal, a 60-square-mile chain of rocks and reefs, is claimed by both China and the Philippines. In 2012 China took effective control by reconstructing seven islands. Control over the islands in the Shoal allows China to monitor US military activity by triangulating with Chinese bases and listening posts on the Paracel and Spratly Islands. The Philippines responded by taking the issue to the UN rather then pursuing bilateral discussion. The Philippines, unlike other states in the region, have fewer economic ties with China and close military ties with the USA.

Six countries—China, Vietnam, Taiwan, Malaysia, Brunei, and the Philippines—make claim to the Spratly Islands, more than 740 reefs, islets, atolls, and islands. All but Brunei have built or claimed outposts among them. However, China's activity has ramped up since 2013 when it started to build on existing rock outcrops and reefs to create 2,900 acres of new land. Vietnam and the Philippines have undertaken similar ventures, but China has outpaced every

colonial power in the region. An area around the ancient Khmer temple in Preah Vihear was given to Cambodia. Thailand occupied this area when the French withdrew in 1953, but the International Court of Justice confirmed Cambodian sovereignty in 1962. In 2008 Cambodia submitted an application to have the temple listed as a World Heritage site. Nationalist parties in both countries fanned the controversy to inflame public opinion for their own political agendas. In 2008 troops from both countries exchanged fire. There were sporadic exchanges until April–May 2011, when more sustained conflict occurred. In a ten-day skirmish, Cambodia claimed that Thai forces fired 50,000 shells along the border. A ceasefire was brokered on May 4, 2011. The area around the temple remains in dispute.

Maritime Boundaries: The Case of the South China Sea

International boundary disputes also involve islands and control over the sea. Maritime sovereignty disputes are likely to grow as the search for valuable minerals and resources extends into the sea.

China claims a wide swath of territory in the South China Sea, first claimed by the Nationalist Government in Taiwan at the end of the Second World War. It is an area of rich fishing grounds, vast oil and gas reserves, and vital sea lanes. It is of major economic and geopolitical significance.

15.10 The South China Sea claims

military presence in Philippines and Australia, and a fleet of super-carriers able to project power and influence. The US has a significant military and naval presence in the region that the Chinese see as a threat to their regional power status.

Cooperation

Boundaries can also be sites of cooperation. The opening of formerly closed boundaries and the easing of movement across once nonporous boundaries are often highly symbolic markers of regime change and interstate cooperation. The fall of the Berlin Wall in 1989 marked the end of the Cold War in Europe. When Berliners chipped away at the Wall, they were demolishing not only a concrete wall but also a political regime. A hard, nonporous boundary was replaced by the free flow of national unification.

New Perspectives

The political geography of borders is now a lively area of geographical scholarship. The current emphasis is on the social construction of borders, practices of bordering, and the narration of borders as important parts of national identity. Reece Jones recounts the bordering between India and Bangladesh. He begins his interesting paper with a description of Moushumi, a domestic worker in Bangladesh who crosses the border into India to visit her son. Prior to 1947, the trip would not have been a border crossing, as both areas were part of the same space of colonial British control. After 1947, however, the partition of the continent created an international border. It took a while for the border to become bordered. A map was drawn in 1947, but the border was not properly surveyed until 1952, and only in the 1960s were border security forces deployed. The Indian government has now fenced more than three-quarters of the 4,000-kilometer border. This hard border is now "a key site for the state to establish the binaries of power that frame the world as citizen–alien, national–foreign, here–there, and we–they" (Jones 2012, p. 691). Despite this

other country combined. Sand is pumped up from the sea floor and spread across islands, and even the live coral of submerged reefs, which are then concreted over to become bases capable of providing docks for ships and runways for planes. Military runways have been built on Mischief Reef and Subi Reef. The Chinese have militarized the disputed islands.

On July 12, 2016, the Permanent Court of Arbitration announced that China's claim has no basis in international maritime law. China will ignore the ruling as it seeks to reach bilateral agreement with countries such as the Philippines that took the case to the Court in 2013. China's claim is best interpreted as a claiming of world power status and importance. The Chinese look around the region and see the projection of US power: US bases in South Korea and Japan, a nuclear submarine fleet, a hardening, cross-boundary flows still occur. The paper is interesting because it tells the story of the border as a place where the performances of state power are played out in conspicuous acts of security theater, but also highlights the limits to and often ragged nature of state territorial control. Borders are transgressed by smugglers and illegal immigrants and in unrecorded economic transactions. Borders are sites of performance of territorial control as well as examples of resistance, refusal, and transgression.

Geography of Elections

States maintain their popular legitimacy through the performances of elections. There is a geography to these elections. Three distinct geographies can be identified: of voting, of representation, and of electoral systems.

The Geography of Voting

This geography examines the distribution in votes cast for different parties and candidates. Consider the voting for presidential candidates in Nigeria in 2011 (see Figure 15.11). The three candidates appealed to different parts of the country. One candidate won most of the northern region, another won the south, while a third candidate's support was very localized in one state. The differences reflect tribal and ethnic differences. The northern states are predominantly made up of Muslim and Hausa-Fulani tribes, while the south is Christian and dominated by Yoruba and Igbo tribal members.

There are national and regional cleavages in voting. In the 1968 US presidential election, there were three candidates. While the Republican, Richard Nixon, won most of the country, the Democrat, Hubert Humphrey, won most of the northeastern states as well as his home state of Minnesota. The third-party candidate, George Wallace,

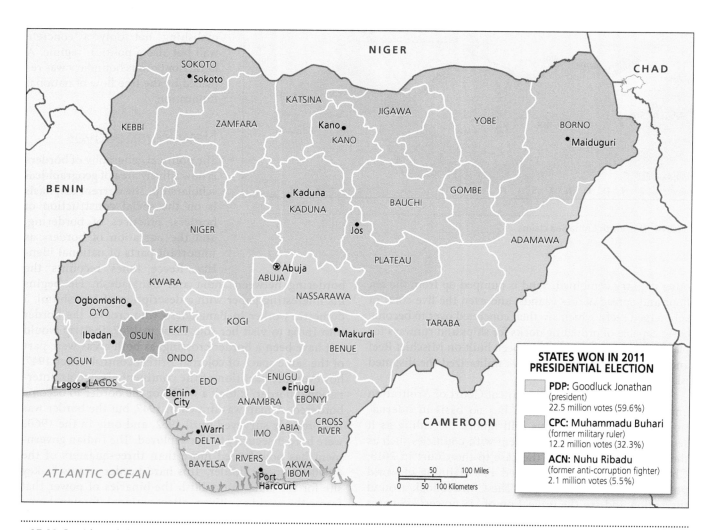

15.11 Presidential election voting in Nigeria

who pledged to maintain racial segregation, won only states in the Deep South.

There is also an even more local effect, known as the **neighborhood effect**. This refers to the fact that neighborhood can trump traditional allegiance. Democratic-leaning voters in a predominantly Republican neighborhood will vote for Republicans, and vice versa. We have detailed empirical demonstration for this neighborhood effect in voting outcomes. The precise causal factors are difficult to disentangle. Does the neighborhood cause the shift, or do people predisposed to the shift mark their intentions by moving to the neighborhood in the first place? The geography of voting is influenced by distribution of income and racial and ethnic differences, but also by local factors such as neighborhood and the effect of friends and neighbors.

A marked spatial polarization in voting has been noted in the USA. Since 1992 across the different spatial scales of 3,077 counties, 49 states, and 9 census divisions, there was an increase in the polarization of votes for one of the two main political parties in presidential elections. The vote distribution shifted from areas with a relatively even split to areas more heavily skewed toward either Democrat or Republican. This may be the result of selective relocation as the population sifts into communities of similar political persuasion. And these communities, in turn, may influence those who subsequently move in. Whatever the exact combination of factors, the end result is that more people live in areas where neighbors amplify and reinforce rather than contest and question their political beliefs and prejudices. This spatial polarization may be one of the causes behind the increasing social polarization along party lines in the USA.

The 2016 US presidential election provides an interesting example of the geography of voting. A map of Trump's America, places where he picked up the majority of votes, consists of a vast landmass of the interior of the country. Clinton's America, on the other hand, consists of the large cities along the coasts, and large inland cities and places with college towns and substantial black and Hispanic populations (see Figure 15.12). The swing

15.12 Clinton's America

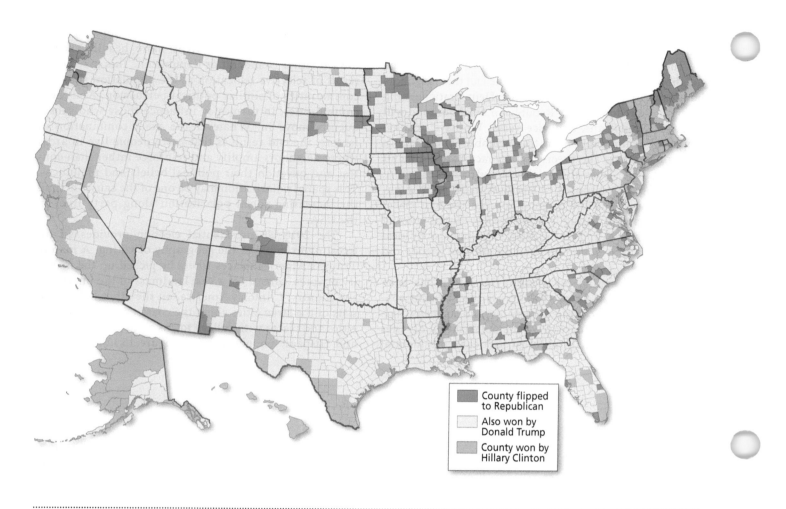

15.13 The swing to Trump

to Trump, as indicated by those counties that voted for Obama in 2012 but Trump in 2016, highlights his appeal in the old industrial heartland of the upper Midwest and Northeast (Figure 15.13). Trump's wins in Michigan, Pennsylvania, and Wisconsin extended Republican support beyond its usual base and ensured his victory in the Electoral College.

The Geography of Representation

National territories are divided into political constituencies. The geography of this representation has an impact on the results. **Malapportionment** refers to the imbalance between the number of voters and the population in each constituency. In some cases, numerical malapportionment is accepted, indeed codified into the legislative framework. In the United States, for example, each state receives two seats in the Senate, yet in 2016,

while California and North Dakota each had two senators, California had a population of 39.1 million versus North Dakota's 756,927. This disparity was accepted by the Founding Fathers as the price to be paid to create a union of fiercely independent states. Malapportionment is a function of political power. Those who gain from the existing system rarely want to change the system.

There is also **gerrymandering**, which refers to the manipulation of voting boundaries to engineer specific political outcomes. The term originated in the activities of Elbridge Gerry, who in 1810 as governor of Massachusetts signed a bill that demarcated boundaries that favored his party (Democratic-Republican). The bizarrely shaped new district was depicted in a political cartoon as a salamander, which opposition politicians quickly described as the "gerrymander" The system was so "gerrymandered" that while the Democratic-Republicans won only 50,164 of the votes compared to the 51,766 gained by the Federalists, they won twenty-nine of the forty seats. Gerrymandering is a recurring feature of political boundary making and an ongoing reality of congressional boundaries in US states.

The US Constitution requires each state to establish new congressional districts every ten years to reflect the population changes measured by the census. This congressional redistricting is freighted with partisan political interests. In Utah, the results of the 2010 census revealed enough population increase to justify another congressional district, increasing the number from three to four. The political geography of Utah consists of the more Democratic-leaning metro area of Salt Lake City and the Republican-dominated rural areas. The Republicans control the state legislature and thus the redistricting. If the metro area were one congressional district, it could possibly ensure a Democratic victory. To avoid this possibility the Republican-dominated state legislature produced a redistricting plan that split the metro area across three separate districts. The Democratic vote was thus spread and diluted within more rural districts. In Democratic-controlled Maryland, the post-2010 redistricting proposals ensured that Republican support was diffused to favor Democratic voters. A new District 6 was drawn up so that the more Republican voters in rural counties were lumped together but swamped by the more populous Democratic-voting suburban areas. District 3, the least compact congressional seat in the country, was the result of drawing boundaries to ensure a Democratic victory (Figure 15.14). The redistricting plan was placed on the November 2012 general election ballot; it passed.

The Geography of Electoral Systems

States vary in the geography of their electoral systems. The biggest distinction is between the "**first past the post**" method, one of what are called plurality voting systems, that ensures that the politician with the majority of votes in a district wins that district. This dominates the electoral scene in Canada, the United Kingdom, and the United States. The system is simple and easy to understand, but it tends to solidify the political representation of large nationwide parties and marginalize third and fourth parties with wide, but not necessarily nationwide, depth of support. Table 15.5 provides the results of the

15.14 Congressional boundary redistricting in Maryland

15.3 Geopolitics

Geopolitics is the study of the power relations between empires and states. Early proponents include Rear Admiral A. T. Mahan of the US Navy, who wrote in 1890 about the influence of sea power, and the Englishman Sir Halford Mackinder, who provided a geopolitical strategy for the British Empire. Mackinder's 1919 book *Democratic Ideals and Reality* is concerned with geopolitical strategies in the immediate aftermath of the First World War.

More recent contributions include George Kennan, writing during the Cold War of the need for a US policy of containing the USSR. Even more recently, popular writers such as Robert Kaplan have argued that a nation's position on the world map is a primary determinant of conflict. The world is viewed as a series of states competing over space, with emphasis on the notion of geographic pivots.

Since the 1990s, a more critical form of geopolitics has emerged that looks at the social construction of political spaces, exposes the material interests involved in the narratives used to explain these spaces, and explores the spatial construction of social identity. It also interrogates the ideological underpinning of standard geopolitics. Thus Gerry Kearns reexamines the imperial context of Mackinder's life, works, and geopolitical ideas. There is also a popular geopolitics that looks at the role of popular culture and mass media in structuring national identities and popular geographical understandings of the world, as well as a reworking of the spatial nature of war and conflict. Derek Gregory, for example, looks at the geographical dimension of contemporary war through an examination of three global borderlands: Afghanistan–Pakistan, US–Mexico, and cyberspace.

Geopolitics—traditional, popular, and critical—explores the meaning, contestation, construction, and maintenance of the political organization of space.

REFERENCES

Dittmer, J. (2010) *Popular Culture, Geopolitics, and Identity*. Lanham, MD: Rowman & Littlefield.

Dodds, K., Kuus, M., and Sharp, J. (eds.) (2013) *The Ashgate Research Companion to Critical Geopolitics*. Farnham, UK: Ashgate.

Gregory, D. (2011) The Everywhere War. *Geographical Journal* 177: 238–250.

Kaplan, R. (2012) *The Revenge of Geography: What the Map Tells Us About Coming Conflicts and the Battle Against Fate*. New York: Random House.

Kearns, G. (2009) *Geopolitics and Empire: The Legacy of Halford Mackinder*. Oxford: Oxford University Press.

Mackinder, H. J. (1919) *Democratic Ideals and Reality: A Study in the Politics of Reconstruction*. London: Constable.

TABLE 15.5 General Election Results in the United Kingdom, 2010		
PARTY	**% OF VOTES**	**% OF SEATS**
Conservative	36.1	47.0
Labour	29.0	39.6
Liberal Democrat	23.0	8.7

2010 general election in the United Kingdom for the three main parties. Notice how the Liberal Democrat Party won almost a quarter of all votes but managed to gain less than 10 percent of the seats. Smaller parties with wide support, but not deep enough to win particular seats, are penalized by the first-past-the-post system. When the Liberal Democrats joined with the Conservatives in a coalition government, their price was the promise of electoral reform.

A large number of countries, Ireland for example, now have a system of **proportional representation** that translates votes into seats in a more equitable manner. This can be done through giving seats in proportion to the votes cast. Some countries have a mix of seats allocated by first-past-the-post and then some by proportional representation methods. Under a single transferable vote system, electors rank candidates in order of preference. If their preferred candidate does not gain enough votes, the votes are given to the second ranked candidate. There are also preferential systems. In an alternative vote system, voters have an opportunity to express their second choice, as happens in Australia; or, as in French presidential elections, in a second round of voting if no candidate has an absolute majority in the first round. Each system has a mix of advantages and disadvantages. The first-past-the-post system ensures one party has a majority, but small and minority parties are underrepresented. In the proportional and preferential system, the legislature reflects voting preferences more closely, but the large number of parties can make for fragile and unstable coalition governments.

Political outcomes in part depend on geography. The geography of voting, representation, and electoral systems all play an important part in turning votes into representation. Consider again the 2016 US presidential election. One candidate, Donald Trump, gained 46.1 percent of the popular vote, a total of 62.9 million votes, while the other main candidate, Hillary Clinton, won 48.2 percent of all votes cast, 65.8 million votes. But because the US electoral system, in the form of the Electoral College, rewards those who win the most votes in each state, the winning candidate had fewer popular votes than his defeated rival.

Cited References

Cusak, T. (2010) *Riverscapes and National Identity.* Syracuse, NY: Syracuse University Press.

Davies, N. (2012) *Vanished Kingdoms: The Rise and Fall of States and Nations.* New York: Viking.

Easterlin, R. (1974) Does Economic Growth Improve the Human Lot? Some Empirical Evidence. In Paul A. David and Melvin W. Reder (eds), *Nations and Households in Economic Growth: Essays in Honor of Moses Abramovitz.* New York: Academic Press, 89–125.

Galbraith, J. K. (2011) Inequality and Economic and Political Change: A Comparative Perspective. *Cambridge Journal of Regions, Economy and Society* 4: 13–27.

Habermas, J. (1975) *Legitimation Crisis.* Boston: Beacon Press.

Jones, R. (2012) Spaces of Refusal: Rethinking Sovereign Power and Resistance at the Border. *Annals of the Association of American Geographers* 102: 685–699.

Johnston, R. J., Jones, K., and Manley, D. J. (2016) Spatial Polarization of Presidential Voting in the United States 1992–2012: The Big Sort Revisited. *Annals of the Association of American Geographers* 106: 1047–1062.

Nye, J. (2004) *Soft Power: The Means to Success in World Politics.* New York: Public Affairs.

Piketty, T. (2014) *Capital in the Twenty-First Century.* Cambridge, MA: Harvard University Press.

Short, J. R. (2005) *Imagined Country.* (Reprint with new introduction; first published 1991.) Syracuse, NY: Syracuse University Press.

Wilkinson, R. and Pickett, K. (2009) *The Spirit Level: Why More Equal Societies Almost Always Do Better.* London: Allen Lane.

Select Guide to Further Reading

Bremmer, I. (2006) *The J Curve: A New Way to Understand Why Nations Rise and Fall.* New York: Simon & Schuster.

Brooks, K. (2011) Is Indonesia Bound for the BRICs? *Foreign Affairs,* November–December, pp. 109–117.

Davies, N. and Moorhouse, R. (2003) *Microcosm: Portrait of a Central European City.* London: Pimlico.

Diener, A. C. and Hagen, J. (2012) *Borders: A Very Short Introduction.* New York: Oxford University Press.

Dittmer, J. and Sharp, J. (eds) *Geopolitics: An Introduction.* London: Routledge.

Elden, S. (2013) *The Birth of Territory.* Chicago: University of Chicago Press.

Filkins, D. (2012) The Deep State. *The New Yorker,* March 12. http://www.newyorker.com/magazine/2012/03/12/the-deep-state.

Kolossov, V. and O'Loughlin, J. (1998) New Borders for New World Orders: Territorialities at the Fin-de-siecle. *GeoJournal* 44: 259–273.

Megoran, N. (2012) Rethinking the Study of International Boundaries: A Biography of the Kyrgyzstan-Uzbekistan Boundary. *Annals of the Association of American Geographers* 102: 464–481.

Popescu, G. (2012) *Bordering and Ordering the Twenty-First Century.* Lanham, MD: Rowman & Littlefield.

Richardson, T. (ed) (2013) Special Issue on Borders and Mobilities. *Mobilities* 8: 1–165.

Scuzzarello, S. and Kinnvall, C. (2013) Rebordering France and Denmark: Narratives and Practices of Border Construction in Two European Countries. *Mobilities* 8: 90–106.

Short, J. R. (2014) The Supreme Court, the Voting Rights Act and Competing National Imaginaries of the USA. *Territory, Politics, Governance* 2: 94–108.

Stevenson, B. and Wolfers, J. (2008) *Economic Growth and Subjective Well-Being: Reassessing the Easterlin Paradox.* NBER Working Paper No. 14282. http://www.nber.org/papers/w14282

Webster, G. (2013) Reflections on Current Criteria to Evaluate Redistricting Plans. *Political Geography* 32: 3–14.

Wrong, M. (2001) *In the Footsteps of Mr. Kurtz: Living on the Brink of Disaster in Mobuto's Congo.* New York: HarperCollins.

Websites

CIA world and regional maps
  https://www.cia.gov/library/publications/the-world-fact-book/docs/refmaps.html
UN report on national wealth
  http://www.ihdp.unu.edu/docs/Publications/Secretariat/Reports/SDMs/IWR%20SDM%20Low%20Resolution.pdf
Military spending as a percentage of GDP in 2014
  https://upload.wikimedia.org/wikipedia/commons/5/5e/Countries_by_Military_expenditures_%28%25_of_GDP%29_in_2014.svg
National comparative data
  http://www.MyCountryRankings.org/
  https://www.cia.gov/library/publications/the-world-factbook/

http://data.worldbank.org/
http://en.wikipedia.org/wiki/Democracy_Index#2011_rankings
https://www.prsgroup.com/about-us/our-two-methodologies/icrg
http://www.worldmapper.org/
Stockholm International Peace Research Institute
  https://www.sipri.org
Gross National Happiness
  http://www.grossnationalhappiness.com/
Freedom House
  http://www.freedomhouse.org/

Interactive Mapping Websites

World Bank
  http://data.worldbank.org/indicator/NY.GNP.PCAP.CD/countries?display=map

CIA World Factbook
  https://www.cia.gov/library/publications/resources/the-world-factbook/

Learning Outcomes

- States are separate units of political authority that are important building blocks of the organization of space.
- States have increased in number, especially since the beginning of the twentieth century.
- Not all states are equal. The more powerful are able to use their military strength to exert an outsized role in the world and impose their will on others. Some also have stronger economies that afford different levels of world influence and power.
- Income inequality is a useful predictor of political stability. More equal societies tend to be more peaceful and less riven by conflict.
- It is not only the wealth of society that structures health outcomes and social performances; it is also the level of inequality.
- In terms of political organization, three main regimes can be identified: totalitarian systems, in which the government has control over wide and deep swathes of social, political, and economic life; authoritarian regimes, where power is concentrated but not so deeply entrenched as in the totalitarian regimes; and democracies, where political power arises from the majority will of the people.
- States vary enormously along a variety of dimensions, including military might, economic wealth, and political freedom.
- We can identify three contemporary and interlinked types of crises that nations face: fiscal, legitimation, and rationality.
- A state is a self-governing political unit that must be recognized by other states and the majority of its own population in order to be legitimate.

- One way that smaller countries try to offset their limited size is through organizations with other states, such as in economic unions (e.g., the European Union) that try to extend the size of the market.
- As spatial units, states can be lucky or unlucky in their territorial endowment. For example, some have vast oil reserves, while others are landlocked with few available natural resources.
- A distinction can be made between a state and a nation. A state is a political organization that controls a particular territory. A nation is a community of people with a common identity, shared cultural values, and a commitment and attachment to a particular area.
- There are nations without states, and most states have more than one nation.
- The population of a state is rarely homogenous. Differences in religion, ethnicity, and language, if they have distinct spatial expression, can create tensions in the internal coherence of the state.
- A distinction can also be made between government and state. The government is the political expression of power. "State" is a wider term that covers the more embedded power structures, such as the army, the police, and the educational system.
- We can identify centripetal forces that unify the state's power across space and centrifugal forces that disrupt it. Centripetal forces include external aggression, which may stimulate national bonding against a common enemy; federal structures that allow the safe expression of regional and other subnational differences; and national mass media

and education, which create a shared culture and language. Centrifugal forces include political and economic inequality.

- A state has a monopoly of legitimate physical force, but in some instances, the rule of the state does not extend to all of its territory, which may lead to a territorial crisis.
- The internal territorial integrity of the state is more assured where there is a homogenous population. When marked differences are overlaid by economic inequality and profound cultural differences, there are pressures on the integrity of state control over national space.

- Political geographers used to make a distinction between artificial and natural borders, but more recent work tends to view borders as social constructions, the result of bordering practices rather than the embodiment of "natural" differences.
- Viewing borders as socio-spatial constructions focuses attention on their origin and evolution and interprets them as stages for the performance of national identity.
- Three distinct elements can be identified in the geography of elections: those of voting, representation, and electoral systems.

16

The Urban Transformation

A majority of the world's population now lives in urban areas. This shift constitutes a major change, because for most of humanity's occupancy of the Earth, most people lived in rural areas. In this chapter we will consider the rise of cities and the nature of this urban transformation. We can identify three distinct urban revolutions.

The First Urban Revolution

The very first cities emerged between 11,000 and 5,000 years ago. They developed alongside settled agriculture, the domestication of animals, and the creation of complex irrigation systems. A traditional argument claims that the **First Urban Revolution** was predicated upon the agricultural revolution: it was agricultural surplus that created cities. Most commentators now reverse the direction of the causal arrow. It was cities, or at least permanent settlements, that created agriculture. An 11,000-year-old complex of megaliths in Turkey, called Göbekli Tepe, provides some insights. In an apparent act of religious devotion, the monuments were built by hunter-gatherers. The need for large numbers of people to stay in one place and cooperate to build these monuments created a demand for a stable and secure food supply; hence the invention of agriculture. It was a solution that allowed early societies to remain in place to build monuments. Cities caused the development of agriculture because they generated enough demand in one place to stimulate new sources of food production.

Progress or Regress?

The development of agriculture and the creation of cities are often depicted as an upward trajectory of increasing civilization. However, the hunting-gathering societies that preceded this revolution spent less time working than agricultural societies; they are what one commentator terms the "original affluent society." In other words, pre-urban, pre-agricultural societies had more disposable time and more freedom. Even today, the Bushmen of the Kalahari spend only around 1,000 hours a year hunting and gathering food, and that is in a harsh environment; hunter-gatherers who lived in more hospitable climes probably would have spent much less time. The rest of the time is spent entertaining, relaxing, and being with friends and families. Rice farmers in southern China, in contrast, spend around 3,000 hours a year tending to their crops.

The shift from hunting-gathering to settled agriculture involved more work. Mortality increased, and food intake declined to such a narrow range of foods that it led to an increase in anemia and vitamin deficiencies. The supply of food did not keep pace with the increasing population, so more farming land was needed, creating the constant growth that was an essential feature of the early urban empires. Agriculture and cities were a mixed blessing that, in the short term, meant declining living standards for the majority of the population while an elite prospered.

The urban-agricultural revolution, for the vast majority, marked a loss of freedom, greater work discipline, and more time devoted to the drudgery of work and the compulsion of social order. It was a social order that had to be imposed. An example: there are remains of an important urban culture in the desert Southwest of the United States known as the Anasazi. This independent urban civilization was centered on Chaco Canyon, New Mexico. The traditional rendition goes like this: between the tenth and twelfth centuries, the Anasazi culture, based on efficient agriculture, flowered into cities of vast cliff dwellings and major feats of engineering, architecture, and art. Brilliant pottery, sophisticated irrigation systems, and keen solar and astronomical observations round out a picture of an urban civilization that follows the old precept that cities equal civilization. Yet there is another interpretation of the Anasazi, a darker side suggesting that the Anasazi culture developed

from the Toltec Empire, which lasted from the ninth to the twelfth centuries in central Mexico. This was an empire centered on human sacrifice and cannibalism. Thugs from the Toltec Empire moved north into what is now New Mexico and found a pliant population of docile farmers whom they terrorized into a theocratic society. Social control was maintained through acts of cannibalistic terror. The Anasazi culture, so long admired, was one in which the bad and powerful controlled the weak and the vulnerable. The great feats of art and astronomy, road building and city formation, were less sparks of human ingenuity and more the mark of organized social terrorism.

There was a distinct pattern to the spatial layout of the first cities. We can picture a model of the **preindustrial city**. At the city center were the political elite and sites

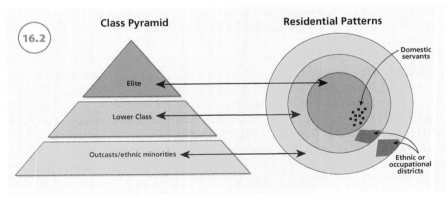

16.1 The Mayan city of Tolum, now in Mexico, was a typical preindustrial city with a prominent central site. As with many early cities around the world, the architecture at the center of the city was monumental and symbolic, connecting the sacred and the profane, the religious and the political. It prospered from the thirteenth to the fifteenth centuries, but the coming of the Spanish brought disease and death. 16.2 Model of the preindustrial city.

of religious devotion, the temples and altars that forged a direct link between the sacred and profane (see Figure 16.1 and 16.2). The city was like a map of the cosmos. Elites were located in the center of the city, with the poor at the periphery. The homology between cosmos and city was a way to legitimize the social hierarchy. The city not only housed people; it was a text that explained and justified the social world and embodied the wider cosmos.

Urban Vulnerability

The early cities were also vulnerable to ecological collapse. Over 4,000 years ago, the cities of Mesopotamia, including Ur, Uruk, and Umma, were brought under the unified control of the Akkadian Empire, centered on the city of Akkad. At least four generations of kings ruled over the sophisticated empire. And then, quite suddenly, the empire collapsed. One text, written a century after the collapse, noted,

> *For the first time since the cities were built and founded*
> *The great agricultural tracts produced no grain.*
> *The inundated tracts produced no fish,*
> *The irrigated orchards produced neither syrup nor wine,*
> *The gathered clouds did not rain.*

Archeological research has revealed that the collapse was due to a prolonged drought. The "gathered clouds did not rain" is not a poetical conceit but a record of severe drought.

Around the world, there is evidence of urban collapses due to ecological change. Beginning in the eighth century, many of the great Mayan cities in the Yucatán Peninsula were abandoned as wars raged and people fled. A 200-year dry spell, starting around 750, caused a significant decline in regional rainfall. By 930, the Mayan heartland had lost 95 percent of its population. The "Mayan collapse" was caused by widespread drought.

Angkor Thom was the capital of the Khmer Empire, which controlled a large area of what is today Cambodia, Thailand, Laos, and Vietnam from the ninth to the fourteenth centuries. The city area encompassed more than 385 square miles. At its peak, it housed up to one million people, with more than 1,000 temples. Water collected from the hills was stored and distributed for a wide variety of purposes, including flood control, agriculture, and ritual bathing. A system of overflows and bypasses carried surplus water to the lake to the south. The network of reservoirs was extensive and supplied tremendous amounts of water to the city. The fall of the Khmer Empire in the fourteenth and fifteenth centuries

16.3 Ruins of Carthage in North Africa. The city was destroyed in the Punic Wars between Rome and Carthage in 146 BCE. It became a Roman city and also a center of early Christianity; it was then destroyed by a Muslim army in 698 CE. The ruins are silent testimony to the rise and fall of cities and urban empires.

was due to the combination of a dramatic decrease in rainfall in the region and the deleterious impact of the built environments on the surrounding environment. The city's growth led to deforestation, which increased flooding.

The earliest cities, like the cities of today, had an environmental vulnerability. The constant demand for food and water and the risk of climate change and ecological disruption means that most urban empires, sooner or later, collapse (Figure 16.3).

Cities and Social Change

Urbanization is often described as the geographical redistribution of population from rural areas to urban areas. However, it is not just the spatial reorganization of society but also the social reorganization of space. Cities incubate social and political change.

The **merchant city**, for example, began to develop in association with the creation of a money economy, the extension of trade, and the emergence of a merchant class. In Europe, early merchant cities developed, such as Amsterdam, Bruges, Florence, and Venice (Figure 16.4). The merchant city was a vital cog in the development of international trade and commerce, but also in the creation of a commercial society where private interest was regulated, collective rules were established, and civic communities were forged. The people of the merchant cities created the notion of the public realm and civic society.

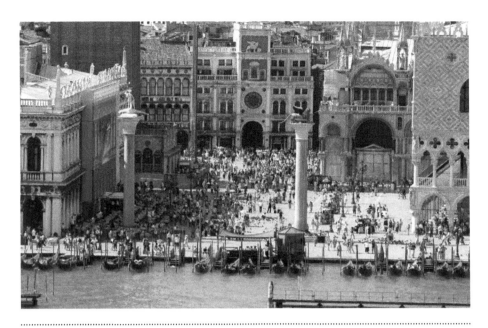

16.4 Venice was an important merchant city with trade links throughout Europe and with the Ottoman and Chinese Empires.

A Second Urban Revolution

A **Second Urban Revolution** began in the late eighteenth century with the creation of the industrial city and unparalleled rates of urban growth. The **industrial city** was the crucible of early industrial capitalism, giving us factory life and class struggles. Factory production replaced household production, and industry replaced agriculture. The factories of Manchester and other cities in Britain in the early nineteenth century were wonders of the modern world—indeed, it was the world made modern; centuries-old agricultural dominance crumbled in a sudden seismic economic shift. In 1801 almost 70 percent of the British population lived in places smaller than 2,500 population, but by 1851 over 40 percent were living in cities with a population greater than 100,000. New cities were built on greenfield sites, and major cities developed from tiny hamlets.

Although the cities of Britain were at the forefront of the industrial revolution, the rest of Europe and North America were not far behind. Cities like Essen, Cologne, Toronto, and Melbourne began to grow rapidly by the late nineteenth century. Toronto's population increased from 30,000 in 1851 to 181,000 in 1891. Swift economic change and population growth also occurred in cities such as Boston, Pittsburgh, Cleveland, Milwaukee, Cincinnati, and Philadelphia. Industrialization generated unprecedented

levels of urbanization. In 1830 Cincinnati's population was 24,800, but by 1850 it had quadrupled to 115,400 and by 1870 it was 216,000. The population of Paterson, New Jersey, increased from 11,334 in 1850 to 125,600 by 1910.

Cities and Class Struggle

The industrial city created the arena for intense struggle between capital and labor. There were social commentators like Karl Marx who saw the industrial city as the harbinger of social revolution. There were also social reformers who used the appalling conditions of the industrial city to demand more environmental regulations, better public health, and a more livable urban environment. The city, in its various emblematic forms, creates the opportunities, contexts, and laboratories for new social developments, innovative political developments, and far-reaching economic changes. Cities are the accelerants of social, economic, and political change. Modernity emerges in the rise of the city.

The Third Urban Revolution

Today we are in the midst of the global phenomenon of a **Third Urban Revolution**. The first was concentrated in the fertile plains of only a few river basins. The second was restricted to cities in countries undergoing rapid industrialization. The third is a truly global phenomenon that picked up pace in the last half of the twentieth century. It is marked by five distinct characteristics.

Rapid Change

The first is the sheer scale and pace of change. A majority of the world's population is now urban, and in many countries this new urban majority appeared in less than a generation. In 1900 only 10 percent of the world's population lived in cities. By 2010 it was more than 50 percent, and by 2050 almost 70 percent—more than two out of every three people on the planet—will live in cities. Table 16.1 shows the urban percentages in 1950 and 2010

16.1 The Metropolitan United States

The US Census Bureau employs the term "**metropolitan statistical area**" (MSA) to refer to urban areas with a core area of 50,000 or more people and economic links to surrounding counties. Using this statistical, rather than political, division of municipal boundaries, it is possible to measure the metropolitanization of the US population. In 1950 the metropolitan population constituted 56.1 percent of the total US population. By 2010, the figure was 83.6 percent. The US population is increasingly and overwhelmingly concentrated in metropolitan areas. More than 90 percent of the country's entire population growth in the last decade occurred within MSAs.

Charlotte MSA is the thirty-fourth largest MSA and the fourth fastest growing in the nation. Notice the rapid growth in the ring of suburbia along the main highways.

The geography of MSAs sometimes differs from the traditional political boundaries, especially in highly urbanized regions of the country. The counties of the state of New Jersey, for example, are part of seven different MSAs. Twelve of the counties are part of MSAs that include counties in New York State; five are part of MSAs that include counties in Pennsylvania. Only four (Atlantic, Cape May, Cumberland, and Mercer) are considered to be specifically New Jersey MSAs. Metropolitan realities are sometimes different from traditional political boundaries.

A further 10 percent of the US population lives in **micropolitan statistical areas**, which contain an urban core of at least 10,000 and, in total, have populations less than 50,000. Only 6.3 percent live outside these two types of urban areas. The United States continues to become a more urban and metropolitan society.

REFERENCE

Short, J. R. (2012) Metropolitan USA: Evidence from the 2010 Census. *International Journal of Population Research* Article ID 207532, doi:10.1155/2012/207532 http://www.hindawi.com/journals/ijpr/2012/207532/

ANNUAL POPULATION GROWTH BY ZIP CODE

- -0.67–1.53
- 1.54–3.21
- 3.22–12.31

Charlotte's MSA

Ocean City — MSA region

County boundary

MSAs within and extending into New Jersey

TABLE 16.1 Percentage Urban, 1950–2050

	1950	2010	2050
World	29.4	51.6	67.2
More developed regions	54.5	77.5	85.9
Less developed regions	17.6	46.0	64.1

Source: Population Division of the Department of Economic and Social Affairs of the United Nations Secretariat, *World Population Prospects: The 2010 Revision* and *World Urbanization Prospects: The 2011 Revision*, http://esa.un.org/unpd/wup/index.html.

and estimates for 2050. The faster rates of urban growth occur in the developing regions of the world. In 2010 3.5 billion people lived in cities, of whom 2.6 billion lived in the developing world. By 2050 these figures will be 6.2 billion and 5.1 billion, respectively. The world's urban population is increasingly living in cities in the developing world.

Big Cities

The second characteristic is the increasing size of individual cities. Throughout the world, cities have continued to grow larger. In 1800 there were only two cities—London and Beijing—that had more than one million inhabitants; by 1900, there were thirteen. Today, several hundred cities exceed one million in population, and over thirty-five have more than 5 million inhabitants. By 2030, it is estimated that there will be around 662 cities with at least one million inhabitants. The average size of cities has grown dramatically. One of the more visible aspects of contemporary urbanization is the rise of megacities, large urban agglomerations with more than 10 million inhabitants. They are a recent addition to the urban scene. In 1900 no metro region in the world had a population greater than 10 million. In 1950 only New York and Tokyo had populations of more than 10 million. Today, there are thirty-nine cities with at least 10 million people. In 1950 the Nigerian city of Lagos had a population of only 320,000. By 1965 it had surpassed one million, and in 2002 it became the first sub-Saharan African **megacity** when it topped 10 million. With an annual population growth rate of 9 percent, it is one of the fastest-growing cities in the world. In 2016 the metro area had a population of 21 million.

Megacities are not just big cities; they are a new, distinctive, spatial form of social organization that radically transforms the city–nature relationship. Their sheer size exerts a large and heavy environmental footprint. Megacities can impose a heavy environmental toll. Continual city growth generates tremendous rural-to-urban land use changes and associated ecosystem transformations. The increasing population also puts extra pressure on the biophysical systems that provide land, air, and water.

Metropolitanization

Third, this revolution exhibits a marked **metropolitanization**. Improvements in transport have allowed dispersal of people and activities away from the tight urban cores of preindustrial and industrial cities. Large metropolitan regions rather than individual cities are the new building blocks of both national and global economies. Three giant urban regions in the Asia Pacific region, Bangkok (15 million population), Seoul (25 million), and Jakarta (38 million), have between 35 and 75 percent of all foreign direct investment into their respective countries. In China, the three city regions of Beijing, Shanghai, and Hong Kong constitute less than 8 percent of the national population, yet they attract 73 percent of foreign investment and produce 75 percent of all exports. China is less a national economy than the aggregate of three large metropolitan economies. In the United States, ten **megapolitan regions**, defined as clustered networks of metropolitan regions that have a population of more than 10 million, constitute only 19.8 percent of the nation's land surface, yet comprise 67.4 percent of the population.

Population and economic activities have spread beyond the municipal boundaries. In the past fifty years, small towns across the world have grown into cities and big cities have sprawled into giant metropolitan regions.

Urban Sprawl

Fourth, a distinctive global urban trend is **urban sprawl**. Big-city regions are now characterized by more dispersed forms of urban development. The steady suburbanization of jobs and residences has extended the urban region farther out from central cities. In the United States in 1950, only 23 percent of the US population was living in suburbs. This figure had increased to 46.8 percent by 2010. More people live in metropolitan areas, and more of these people are living in extending suburbs. The process, while particularly evident in the developed world, is not limited to it. Consider the case of Shanghai. Since 1991, new residential and industrial complexes have developed along the rural–urban fringe, aided and encouraged by local governments,

16.2 Megalopolis

One of the largest contiguous areas of metropolitan counties is what is sometimes called **Megalopolis**, a region spanning 600 miles from north of Richmond in Virginia to just north of Portland in Maine and from the shores of the Northern Atlantic to the Appalachians. The region includes the consolidated metropolitan areas of Washington-Baltimore, Philadelphia, New York, and Boston and covers 52,000 square miles, with a 2010 population of 44.6 million. It contains just over 14 percent of the entire US population.

In 2000 the combined population of Boston, New York, Philadelphia, and Washington was 10.64 million; by 2010 this had increased to 10.88 million. Baltimore, with a relatively large manufacturing base, declined from 651,154 in 2000 to 620,961 in 2010.

Megalopolis is overwhelmingly suburban, with three out of every four persons living in the suburbs. Like other large city regions across the country, it is a place of increasing racial diversity, one of the most diverse in America. In the four combined MSAs that make up this extended region, the percentages of blacks, Asians, and Hispanics are, respectively, 20.2, 8.9, and 18.6, versus 12.6, 4.8, and 16.3 for the United States. In some metro areas, there are more minorities living in the suburbs than in the central city. In the Washington, DC, MSA, for example, there were 857,380 Hispanics living in the suburbs, compared to only 54,749 living in the city. The Hispanic experience in the Washington MSA is predominantly a suburban one. Across the entire region, the suburbs are becoming more racially diverse. Megalopolis is the destination of significant amounts of immigration from overseas. The foreign-born population increased from 10 percent in 1960 to 23.8 percent in 2010, almost double the national average of 12.4 percent. Migrants are found both in central cities and in suburban areas; particular concentrations can be identified as immigrant gateways. One example is Tyson's Corner, VA, an archetypal edge city located off the Washington Beltway. The population of Tyson's Corner is 19,627, with almost 35 percent foreign-born. Immigrant suburbs are now an important part of the metropolitan United States.

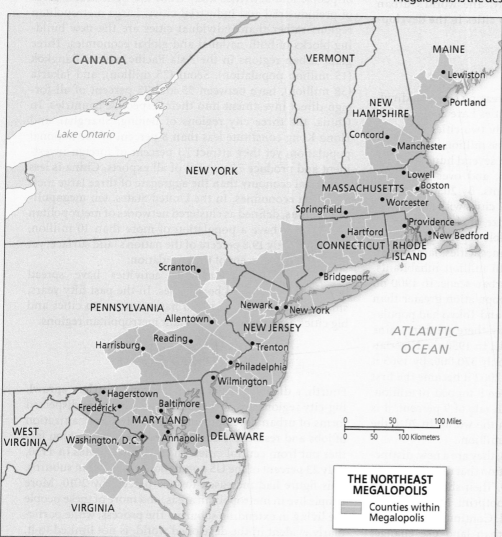

Megalopolis, United States

REFERENCES

Gottmann, J. (1961) *Megalopolis: The Urbanized Northeastern Seaboard of the United States.* New York: Twentieth Century Fund.

Short, J. R. (2011) The Liquid City of Megalopolis. In *The New Blackwell Companion to the City,* ed. G. Bridge and S. Watson. Malden, MA: Blackwell, pp. 26–37.

Short, J. R. (2007) *Liquid City: Megalopolis Revisited.* Washington, DC: Resources for the Future Press/Johns Hopkins University Press.

Electronic Atlas of Megalopolis http://www.umbc.edu/ges/student_projects/digital_atlas/instructions.htm. Accessed October 19, 2012.

Sprawl is a form of development that is very often too diffuse to support public transport or easy walking. The heavy, and in some cases total, reliance on private auto transport in the United States, for example, imposes a great environmental price in terms of air pollution and the increasing dedication of space for roads and parking. The reliance of a built form precariously balanced on one fossil fuel with large and fluctuating costs raises issues of long-term sustainability. There is also an emerging body of literature that points to the negative public health effects of suburban sprawl, including a link between the promotion of a driving lifestyle and both less physical activity and higher obesity.

There is also the fundamental issue of the long-term sustainability of sprawl. Low-density suburban sprawl is possible only with relatively cheap fuel and lack of accountability for its environmental impact. It is unlikely that the cheap gasoline that literally lubricated suburbanization will ever return. Where does that leave low-density suburban sprawl, which is so reliant on large-scale private car usage? The general answer: in a very precarious position. The long-term sustainability of low-density, energy-profligate sprawl with its heavy ecological footprint is now a matter for serious consideration.

Slums

Fifth, there is a distinct feature of urban growth in the rapidly growing cities of the developing world (also referred to as the global south), namely **slums**. The term "slum" (also called "shantytown," "informal housing," and "squatter housing") refers to unplanned, often illegal, informal housing. The United Nations has a working definition of slums as marked by:

1. Inadequate access to safe water
2. Inadequate access to sanitation and infrastructure
3. Poor structural quality of housing
4. Overcrowding
5. Insecure residential status

The insecure residential status refers to the uncertain legal tenure of many slum dwellers, many of whom lack legal title to the land that they occupy. They are therefore subject to government removal, especially if the land becomes valuable and sought by developers and politicians. Slum dwellers often have a precarious hold on their accommodations; the lack of legal ownership makes slum dwellers particularly vulnerable to government clearances. One particularly brutal example of a process repeated throughout the world: on May 19, 2005, the government of Zimbabwe, under the dictatorial control of Robert Mugabe, launched a new urban policy. It was called Operation Murambatsvina, literally "Drive out the

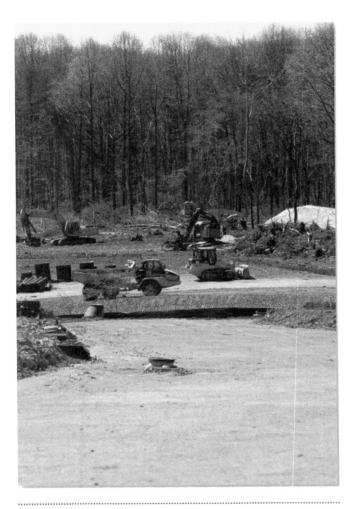

16.5 Urban sprawl outside Baltimore, Maryland

which are key stakeholders in the land development system. The end result is a more widely dispersed urban system with land development often leapfrogging across the landscape. Shanghai is now a huge polycentric city.

Dispersed urban expansion triggers land cover changes, especially the disappearance or fragmentation of croplands and woodlands, the destruction of both low- and high-quality agricultural lands, and the emergence of a hybrid rural–nonfarm landscape (see Figure 16.5).

Sprawl is now associated with a host of negative environmental impacts, including:

- loss of agricultural land in the wake of low-density sprawling development;
- increased impermeable surfaces, which leads to flooding and large discharges of polluted and contaminated water that overwhelm drainage systems and damage ecosystems;
- heavier vehicular traffic, which leads to increased air pollution and global warming.

The city's population in 1949 was around 2.4 million before it embarked on decades of sustained economic growth, fueled by export-led manufacturing and population growth based on rural-to-urban migration within South Korea. In the 1960s and 1970s, the city population was increasing at the rate of a half million every two years. The enforced industrialization and

Seoul, an older part of the central city that is rapidly transforming

urbanization led to major environmental damage and social dislocation, but it did lift the majority of South Koreans out of poverty. Authoritarian governments ensured that breakneck economic development was not halted by democratic discussion. Massive urban renewal involved loss of cheap housing, clearance of squatter settlements, the widening of roads, and the channelization of the Han River with high embankments. Seoul was the center of the country's rapid urbanization. The proportion of the nation's population classified as urban increased from 14.5 percent in 1950 to 88.3 percent in 2000.

By 2015, the city population was just under 10 million, while the wider metro area had a population of 25.6 million. Despite the growth of other cities in South Korea, Seoul retains its national prominence, dominating the country's economy, politics, art, and culture. The city is moving rapidly toward a postindustrial economy. There was a rapid rise and relative fall in manufacturing employment in the city that peaked at 1.3 million in 1990, 30 percent of the labor force. By 2014, manufacturing constituted only 15 percent of total employment. The city's economy is now dominated by the service sector, which employs 3.6 million people, 80 percent of the city's total workforce. It has also spread out as urban growth leapfrogged the greenbelts. A once concentrated city is becoming a polycentric metropolitan region tied together by mass transit and increasing car usage.

Seoul is South Korea's most important global hub. However, the ethnic homogeneity of the city continues to be very obvious. With less than 1 percent foreign-born, the city's ethnic and racial homogeneity is maintained by strong controls on foreign immigration.

REFERENCE

Kim, K.-J. (ed) (2003) *Seoul: Twentieth Century*. Seoul: Seoul Development Institute.

rubbish." As part of a nationwide campaign to beautify the cities, destroy the black market, reduce crime, and undermine the support base of its political opponents, the government authorized the bulldozing of squatter settlements that fringed the city's larger cities. The squatter settlements had long been the main source of opposition to his rule. The campaign destroyed the homes of 700,000 people and wiped out an informal economy that provided a livelihood for 40 percent of the population. Almost 2.4 million

poor people were faced with increased economic hardship. Another example: in July 2012 the Lagos state government in Nigeria moved to evict residents of Makoko, a slum neighborhood of around 200,000 people living in dwellings built on stilts in a lagoon. The residents were given seventy-two hours to vacate before men started chopping down the dwellings. City officials argued that the slums were an environmental nuisance, impeded waterfront development, and undermined the megacity status of Lagos.

They felt that a 2010 BBC television documentary program, *Welcome to Lagos*, which told the story of Makoko's residents, presented the country and the city in a negative light. In fact, the documentary was a celebration of the ingenuity and vibrancy of slum life in the city. While it noted problems of flooding, irregular power supply, and poor environmental conditions, it was lyrical in its invocation of the resiliency and vitality of marginal groups living in slums on the economic margins. Rather than depicting gloom and darkness, it highlighted the slum dwellers' incredible dynamism and energetic entrepreneurship.

The process continues: in the buildup to the 2014 World Cup and 2016 Olympic Games, many of the favelas in Rio de Janeiro came under assault from clearance, forced eviction, and destruction. One plan produced by city authorities envisioned the removal of 13,000 families from 123 different neighborhoods in the city.

Slums arise due to the inability of formal markets and public authorities to provide enough affordable and accessible housing. There are numerous names for these slum settlements. They are called "ranchos" in Venezuela, "pueblos jóvenes" in Peru, "favelas" in Brazil, and "barong-barongs" in the Philippines. A UN survey estimated that around a billion people lived in slums. The figure is estimated to rise to 2 billion by 2030 as people migrate to the cities and urban populations grow. Regions of the world with the largest numbers living in slums are sub-Saharan Africa (196 million), South Asia (191 million). East Asia (190 million), and Latin America (110 million).

Some slums go through a development from temporary accommodation built on the most marginal sites to fully formed urban neighborhoods. In the 1950s, some rural migrants moved to Ciudad Nezahualcóyotl, an area on the outskirts of Mexico City, where they built their own homes on appropriated land using whatever materials they could find. Over time, more permanent buildings were constructed. The area was designated a municipality in 1963, allowing the formal provision of public services such as potable water, pavements, sewerage, and electric lighting. By the 1980s, public buildings such as hospitals and schools were being built. In 1995 the area exceeded one million residents. In this process, temporary shelter for recent migrants became a slum and then an integral part of the city. Peter Ward first studied 300 self-builders in neighborhoods in Bogotá and Mexico City in the 1970s. A follow-up study in 2007 allowed him to identify changes. He found that more than four out of every five households stayed in the houses that they built, but densities rose along with the sizes of the extended families, and the values of the properties increased substantially. "Temporary" self-built housing provided stable family accommodation for over forty years as well as an important family asset.

There are different types of slums. In Abidjan, Côte d'Ivoire, where slum dwellers represent one-fifth of the city's population, there are three types of slums. Zoé Bruno, for example, contains buildings of permanent material and basic

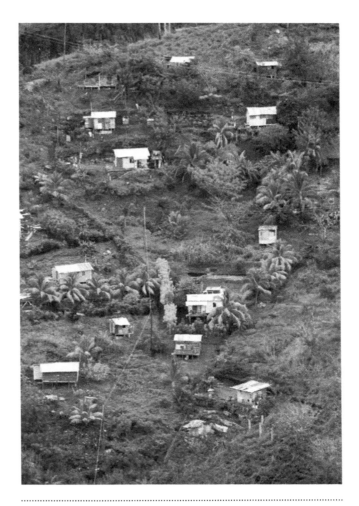

16.6 Informal housing on a steep slope on the Caribbean island of St. Vincent.

infrastructure and is different from the formal areas of the city only by the fact of the illegal land occupation. In Blingue, the buildings are made of nonpermanent materials and the area has low levels of infrastructure. In the worst areas, such as Alliodan, makeshift buildings have no infrastructure.

The urban explosion of the past fifty years has forced expansion onto new spaces on more vulnerable sites such as steep hillsides and flood plains, areas with unstable soil conditions and subject to landslides, and places with other environmental and social hazards. Many slum structures are erected on such marginal lands because standard, legal housing long ago claimed the best, most secure land in the city (see Figure 16.6). The poor are often forced to settle on land subject to higher risks. Slum homes perched precariously on hillsides are often swept away by heavy rains, which also inundate slums located on floodplains. Heavy rain contributes to the landslides and the flooding, but the real causes are the fact that low-income groups could find no land site that was safe and the government failed to ensure a safer site or to take measures to make existing sites safer. The worst slums therefore often have to occupy the most hazardous sites.

16.4 Shanghai

After the 1842 Treaty of Nanjing, when Shanghai was opened up as a port under colonial control, it became one of the largest shipping ports in the world. It exported food and raw material from the Chinese interior to world markets. Tea, silk, and raw materials were shipped through the city, which became China's major industrial center, with mills, factories, chemical plants, and shipyards. In the 1930s, there were 150,000 workers in textile factories, and the city's population approached 3 million, including 100,000 foreigners. At its pre-Communist peak it was the fifth largest city in the world, a cosmopolitan city with at least sixty different nationalities.

Shanghai became a Communist city on May 25, 1949, when troops of the People's Liberation Army marched in and took it over. The earliest economic plans of the Party aimed to build up heavy industry. The city population quickly grew from 4 million in 1950 to almost 6.5 million in 1960, a staggering 50 percent increase.

During the Cultural Revolution, from 1966 to 1976, more than 30 million urban dwellers were forced to move to and work in the countryside. Shanghai was especially targeted. As a place of foreign influences and conspicuous consumption, it was treated as a dangerous hybrid place—Chinese, yet contaminated with anti-Communist tendencies and bourgeois sensibilities. From a peak of 6.4 million in 1961, the city's population declined to 5.4 million in 1978.

In 1984, fourteen coastal "open" cities, including Shanghai, were declared. The reglobalization of Shanghai involved, as with other open cities, the creation of special economic and technological zones to foster concentrations in export-led manufacturing, high-tech industries, and financial services. From 1978 to 2005, the annual growth rate of GDP was close to 10 percent, and average wages grew sixfold. In 1990, a new open economic development zone called Pudong was created. An area of former rice paddies is now home to over 3 million people.

There is a frenzy to the remaking of the city: an expanding network of inner rings roads and outer motorways, new subway lines, tunnels and bridges, airport upgrades, and new high-speed trains. There are the new spaces of consumption: the malls, shopping centers, and gated communities. At the extremes are exclusive gated communities for the wealthy and enclaves of marginalized migrants. The post-Communist city now exhibits marked social and spatial segregation.

REFERENCE

Short, J. R. (2012) *Globalization, Modernity and the City.* New York: Routledge.

Shanghai. A new skyscraper in the Pudong neighborhood.

Across the urban world, then, we can identify a continuum from **slums of hope** at one end to **slums of despair** on the other. Slums of hope provide a platform for rural urban migrants to make their way in the city. In slums of despair, residents are imprisoned in webs of multiple deprivation. Both type of slums are vulnerable to assault and destruction from urban development projects.

Cited References

Batuman, E. (2011) The Sanctuary: The World's Oldest Temple and the Dawn of Civilization. *The New Yorker*, December 19 and 26, pp. 72–83.

Sahlins, M. (1972) *Stone Age Economics*. Chicago: Aldine.

Sjoberg, G. (1960) *The Pre-industrial City*. Chicago: Free Press.

Turner, C. and Turner, J. (1999) *Man Corn: Cannibalism and Violence in the Prehistoric Southwest*. Salt Lake City: University of Utah Press.

UN Human Settlements Program (2003) *The Challenge of Slums*. London: Earthscan.

Ward, P. M. (2012) "A Patrimony for the Children": Low-Income Homeownership and Housing (Im)Mobility in Latin American Cities. *Annals of the Association of American Geographers* 102: 1489–1510.

Select Guide to Further Reading

Benton-Short, L. and Short, J. R. (2013) *Cities and Nature*. London: Routledge.

Davis, M. (2006) *Planet of Slums*. London: Verso.

Duncan, J. S. (1990) *The City as Text: The Politics of Landscape Interpretation in the Kandyan Kingdom*. Cambridge: Cambridge University Press.

Gruebner, O., Khan, A., Lautenbach, S., Muller, D., Kramer, A., Lakes, T., and Hostert, P. (2012) Mental Health in the Slums of Dhaka: A Geoepidemiological Study. *BMC Public Health* 12: 177. doi:10.1186/1471-2458-12-177.

Haug, G., Gunther, D., Peterson, L., Sigman, D., Hughen, K., and Aeschlimann, B. (2003) Climate Change and the Collapse of the Maya Civilization. *Science* 299: 1731–1735.

Kaniewski, D., Campo, E. V., and Weiss, H. (2012) Drought Is a Recurring Challenge in the Middle East. *Proceedings of the National Academy of Science* 109: 3862–3867.

Nelson, A. and Lang, R. (2011) *Megapolitan America: A New Vision for Understanding America's Metropolitan Geography*. Chicago: APA Planners.

Schneider, A. and Woodcock, C. E. (2008) Compact, Dispersed, Fragmented, Extensive? A Comparison of Urban Growth in Twenty-Five Global Cities Using Remotely Sensed Data, Pattern Metrics and Census Information. *Urban Studies* 45: 659–692.

Short, J. R. (2012) *Globalization, Modernity and the City*. London: Routledge.

Short, J. R. (ed) (2017) *A Research Agenda for Cities*. Cheltenham: Edward Elgar.

Websites

Megacities
http://www.megacitiesproject.org

Urban health
http://www.who.int/topics/urban_health/en

Urban population
http://www.unicef.org/sowc2012/urbanmap/

The United Nations Human Settlements Programme
http://www.unhabitat.org/categories.asp?catid=9

The urban transformation
http://www.wilsoncenter.org/program/comparative-urban-studies-project

World Bank urban data
http://data.worldbank.org/indicator/SP.URB.TOTL

World urban areas
http://www.demographia.com/db-worldua.pdf

Interactive Map Websites

History of urbanization
http://metrocosm.com/history-of-cities/

Urban megaregions in USA
http://www.rpa.org/sites/rpa.civicactions.net/files/2050_Map_Megaregions2008_150.png

Urban population growth
http://www.smithsonianmag.com/science-nature/make-cities-explode-size-these-interactive-maps-180952832/?no-ist

World city populations
http://luminocity3d.org/WorldCity/#3/12.00/10.00
http://www.iied.org/cities-interactive-data-visual
http://www.economist.com/node/21642053

Learning Outcomes

- Although people have lived in rural areas for most of history, we can identify three distinct urban revolutions.
- The First Urban Revolution occurred between 11,000 and 5,000 years ago and was marked by the development of the world's first cities alongside settled agriculture, domesticated animals, and complex irrigation systems.
- As cities began to develop, so did the need for advanced agricultural practices to supply larger groups of stationary people.
- The urban-agricultural revolution actually marked a loss of freedom, greater work discipline, and more time devoted to the drudgery of work and the compulsion of social order.
- The distinct pattern of the first cities was marked by a legitimization of social order by locating elites in the center, with poor at the periphery.
- The Second Urban Revolution began in the late eighteenth century with the creation of the industrial city.
- While the first merchant cities developed during the First Urban Revolution, industrial cities dominated the Second Urban Revolution, leading to early capitalism, factory life, class struggle, and unprecedented levels of urbanization.
- The Third Urban Revolution of today is marked by five distinct characteristics: rapid urbanization, increases in the size of individual cities, increased metropolitanization, increased urban sprawl, and the proliferation of slums.
- A majority of the world's population is now urban—expected to include two out of every three people by 2050—and in many countries this new urban majority appeared in less than a generation.
- The world's cities have also substantially grown in size during this Third Revolution, with several hundred exceeding populations of 1 million, more than thirty exceeding 5 million, and the inception of cities with populations greater than 10 million, a new phenomenon known as a megacity.
- Technological improvements in transportation have allowed for greater metropolitanization, the dispersal of people and activities away from the tight urban cores of preindustrial and industrial cities.
- The vast expansions of urban sprawl during the Third Revolution are now associated with a number of negative environmental impacts, such as lost agricultural land; increased impermeable surfaces allowing for more pollution runoff; and heavy, pollution-emitting vehicular traffic.
- The size and number of slums throughout the world have increased due to the inability of formal markets and public authorities to provide enough affordable and accessible housing for low-income populations.

17

Networks of Cities

Cities exist not only as singular places but also as nodes in a variety of networks. In this chapter we will explore regional, national, and global networks.

LEARNING OBJECTIVES

Consider the city as a node in a network.

Understand central place theory.

Discuss the distribution of cities in a national network.

Outline the various methods for defining and identifying the global scale network of cities.

Describe the flows that connect cities in global, national, and regional urban networks.

Regional Networks

One of the earliest forms of urban network is the **periodic market system**. Periodic markets occur in villages and towns when demand density is low and vendors are relatively mobile. At the simplest level they may consist of a seller by the roadside with seasonal wares or produce (see Figure 17.1). There are also the larger weekly, monthly, and annual markets with larger numbers of vendors and customers. There is a geometry to this periodicity, as markets seek to either minimize travel for vendors by bringing them closer together (Figure 17.2A) or maximize demand by spreading markets across space (see Figure 17.2B). A study in northern Nigeria found that markets held on the same day are generally located 10.6 miles apart, whereas markets with meetings two days apart have an average spacing of 3.3 miles.

As demand grows, some traders stay in place, and eventually periodic markets are replaced by **fixed market systems**. Figure 17.3 is a picture of a fixed market in Curaçao that used to be held weekly. However, there are still examples of periodic markets throughout much of the developing world. And even in the rich developed world, farmers' markets continue to attract customers.

Central Place Theory

With rising demand, fixed markets tend to replace periodic markets. In 1915, C. J. Galpin, who was studying rural communities in Wisconsin, suggested that towns with the same number of services tend to locate at regular intervals. People in the surrounding rural areas tend go to the nearest town to shop for goods and services. The ideal pattern of a flat agricultural plain is highlighted in Figure 17.4, where a regular distribution

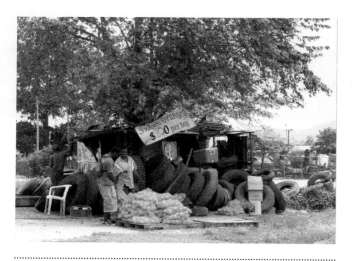

17.1 Roadside sale in Jamaica

17.3 Fixed market in Curaçao

of towns is surrounded by circular complementary regions that overlap. If we assume distance minimization by customers, the resultant complementary region of each town takes the form of a hexagon, because if we draw a straight line through shared areas it forms that

shape. Walter Christaller (1893–1969) looked for a similar pattern in southern Germany in the 1930s. He identified what he termed the **range** and the **threshold** of a good or service. The range of a good or service is the distance consumers are willing to travel. While a buyer of a car may travel long distances to seek out good bargains, the buyer of a carton of milk will only travel much shorter distances. The threshold of a good is the necessary minimum population to support the continued supply of that good. Car dealers need a larger population to support their business than do milk sellers. People buy cars more infrequently than milk. Higher-order goods and services have larger thresholds and larger ranges

A) minimizing distance

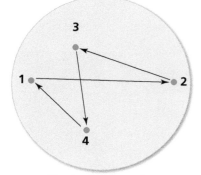

B) maximizing demand

Numbers refer to days when market is open

17.2 Periodic markets

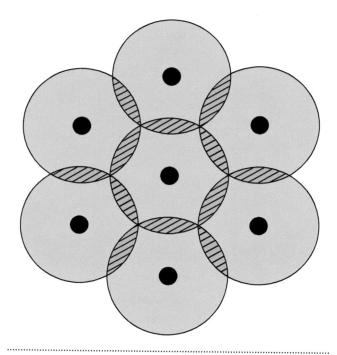

17.4 The Galpin Model

than lower-order goods and services. Market towns are distributed in a hierarchy, with higher-order goods and services located in the larger towns. Christaller visualized a range of hexagonal structures depending on whether the system was geared to optimizing markets, minimizing travel, or optimizing administrative boundaries (Figure 17.5).

Christaller's work, first published in 1933 and resuscitated in the 1960s, was seen as a contribution to economic and urban geography. More recently, it has undergone piercing criticism as the context of his work has been uncovered. Christaller joined the German Nazi Party in 1940. He joined the Reichskommissariat für die Festigung deutschen Volkstums (Reichskommissariat for the Strengthening of German Nationhood), where he worked on plans to occupy the lands of Czechoslovakia, Poland,

and the eastern USSR. His central place theory was employed to organize and settle the occupied lands. The seemingly innocent geometry of central place theory was devised to organize the Nazi occupation of an Eastern Europe emptied of its original inhabitants. One journalist describes him as "Hitler's geographer." The story of Christaller reminds us of the tangled connections between academics and politics, theory and practice, and the contested politics of space.

Central place theory was extended in a range of case studies. Brian Berry looked at the urban system in rural areas of the US Midwest, while Bromley and Bromley uncovered the hierarchy in a region of Ecuador by looking at bus services. However, it has fallen from favor as a major topic of study. The theory is more suitable for stable agricultural regions of the world. It gives little insight into the massive urbanization or rapid metropolitanization of today.

The basics of central place theory, and in particular the notions of complementary regions and of the range and threshold of goods and services, are still important tools for uncovering some of the socio-spatial relations in urban networks. The measurement and definition of complementary regions allows us to identify the effective influence of a city across space. The US Census Bureau, for example, uses commuting patterns to identify city regions.

The changing range and threshold of goods also allow us to understand some of the changes in local urban systems. In many rural areas, small towns are shriveling up and dying. Urban networks are thinning as goods and service provision move up the urban hierarchy to larger towns and cities. Small towns shrink in economic importance when the range and threshold of goods and services both increase. Urban decline is in part a function of changes in these variables.

National Networks

Cities are also part of national networks. There are numerous ways to measure a national urban network. One way is to consider the relationship between a city's population and its rank. In a **rank-size distribution**, also referred to as **Zipf's Law**, there is a simple and regular relationship between a city's rank and its population. This relationship is found mainly in large countries with a large number of big cities, such as China, India, and the United States. Figure 17.6 plots US city population against expected values according to Zipf's Law.

17.5 Market areas in central place systems

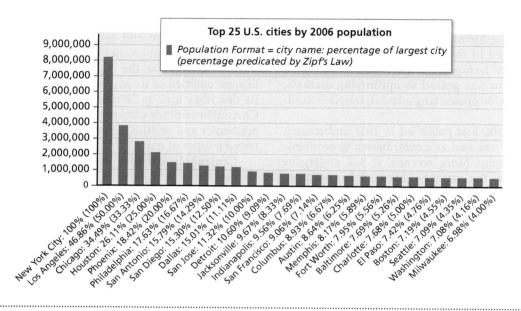

17.6 Rank-size in the United States

Urban Primacy

In contrast, a **primate distribution** occurs when one city dominates the national urban system. Primate cities are a different order of magnitude and significance from all the other cities in a national urban hierarchy. Short and Pinet-Peralta calculate a **primacy ratio** using the ratio of the population of the largest city in a country divided by the combined population of the next two largest cities. A value of 3.0 tells us that the primate city's population is three times the combined population of the next two largest cities. This is a substantial concentration. Table 17.1 notes all the countries with a primacy value above 5.0. Let

us consider the most primate urban system to tease out some of the possible causal connections.

The population of Thailand exhibits the most pronounced degree of urban primacy. Out of a population of almost 65 million, just over 15 million live in the urban region of Bangkok. Since 1782 the city has been the capital of the country. The court was based in the city, and there were few other towns of any size. Bangkok was and remains its unchallenged center of political, intellectual, and religious life. In the past fifty years, this embedded centrality of the city to the wider life of the country has shaped subsequent growth, especially export-led growth. The city became the main transmission hub of economic globalization for the country and the wider region. The centralizing forces of globalization reinforced the national primacy. By 2005, the city was responsible for almost half of all gross domestic product. Thailand is one extended urban region centered in Bangkok. The city is an important global city not only for the country but for the wider region of Southeast Asia.

Primate cities are home to the elites, the population center of gravity, and the economic hub of the national space economy. Primacy was reinforced as economies shifted from primary to secondary and then tertiary economic sectors, rural-to-urban migration increased, and foreign investment connected the local to the global. People, jobs, and investment moved to the major city.

A number of richer countries, including Austria (3.2), Hungary (4.4), and the United Kingdom (4.4) also exhibit primacy. Vienna, Budapest, and London were all centers of far-flung empires rather than just capitals of individual countries. Nineteenth-century Vienna was one of

TABLE 17.1	High Urban Primacy
COUNTRY	**PRIMACY**
Thailand	9.48
Suriname	8.24
Togo	7.92
Uruguay	7.37
Nigeria	5.94
Uganda	5.94
Ethiopia	5.82
Mongolia	5.67
Peru	5.43
Guinea	5.27

17.1 Estimating City Population

An obvious source of urban population data is the United Nations at http://urbandata.unhabitat.org, This source has the benefits of ready accessibility, constant updates, and ease of use. The downside is that, as with all international urban data, it is deeply flawed. Different countries take censuses at different times, use different measures (e.g., some use formal urban political boundaries, while others give the data for functional urban agglomerations), and have wildly varying degrees of accuracy. Take the case of Nigeria and its major city of Lagos. At the UN source, accessed in February 2007, the city's population is available only for 1975 and is listed at just over one million. According to the latest official figures in Nigeria, the population of Lagos in 2006 was almost 9 million, but local officials accuse the federal government of deliberately undercounting this southern Nigerian city and its encompassing region to favor the northern part of the country. More neutral observers estimate the current population of Lagos at between 13 million and 15 million.

Economics and urban studies professor J. Vernon Henderson has also created a documentation site for world cities data over the period 1960 to 2000, from UN data as well as from world gazetteers and national censuses. This is freely available at http://www.econ.brown.edu/faculty/henderson/worldcities.html. This site has the advantage of providing time series analysis, but, as the website informs readers, the data are based on "incomplete documentation and data only partially cleaned." The data for the United Kingdom include the city data for London but the metropolitan data for Birmingham and Manchester.

A more user-friendly source is available at http://www.citypopulation.de/, which lists the population of all principal cities for most countries. The data are drawn primarily from official censuses and estimates, but even the webmaster reminds the user that the data "are all of varying, and some of suspect accuracy."

Another major problem with urban population data is the distinction between population estimates for the city and for the wider metropolitan region. London's population is calculated with reference to the city boundary. Almost 7.5 million people live within this boundary. Yet its commuting range stretches much wider. The wider city region stretches from the Wash to the Isle of Wight. Within this urban agglomeration live 21 million people, or almost 35 percent of the UK's total population.

The "dirty little secret" of global urban research is the poor quality of comparative data, even basic data such as population. Population censuses are of varying degrees of accuracy, are taken at different times, and have differing definitions of city and metropolitan regions. The population figures for most cities around the world are more estimates than precise figures, averages across a wide band of error, and more appropriately used as rough comparisons rather than precise absolute values.

REFERENCE

Short, J. R., Kim, Y.-H., Kuus, M., and Wells, H. (1996) The Dirty Little Secret of World Cities Research. *International Journal of Urban and Regional Research* 20: 697–717.

the centers of political power in Europe, the capital of the vast Austro-Hungarian Empire. Vienna flourished as an imperial capital and the most important city in central Europe, but the loss of empire after the First World War (1914–1918) reduced Austria to a small German-speaking state, and Vienna found itself a large city in a small country. Budapest was also the center of a much larger country and empire. After 1920, Hungary forfeited two-thirds of its area. The end result was a large city in a much-reduced country. Nearly one-fifth of Hungary's population lives in Budapest, which is now more than nine times larger than the nation's second-largest city. London was not just the capital of the United Kingdom but the central node in a worldwide formal and informal empire. The retreat from empire and decline of global economic dominance has left a huge city in a relatively small country. The imperial primacy has persisted in a post-imperial society. The wealthy, the influential, the movers and shakers live in the city; it is home to royalty, the political elites, those who manufacture the dominant forms of representation, and those who control much of the making and moving of money.

At the other end of the continuum of primacy values are the non-primate distributions. Table 17.2 lists all those countries with a recorded primacy value of less than 0.9. The case of Bolivia disproves the easy assumption of Latin American countries always having primate urban distributions. No simple generalizations can be made, as low primacy is found in the Netherlands as well as India, the United States as well as Venezuela, Canada as well as Benin.

The Case of Australia

Using aggregate data sets provides only a first glance at urban national networks. If we look at one example in more detail, more complex patterns are revealed. Consider

TABLE 17.2	Low Urban Primacy
COUNTRY	PRIMACY
Benin	0.58
South Africa	0.59
Venezuela	0.65
Netherlands	0.70
Egypt	0.72
Australia	0.74
China	0.78
USA	0.84
Bolivia	0.84
India	0.86
Canada	0.89

TABLE 17.3	Urban Primacy in Australia
COUNTRY/STATE	PRIMACY
Australia	0.76
New South Wales	5.44
Queensland	5.32
South Australia	33.48
Tasmania	1.13
Victoria	14.66
Western Australia	11.94

Australia, which by 2013 had a national population of 23 million. At first blush, it appears to have a non-primate distribution, because the largest city, Sydney, has a population of 4.6 million, followed by Melbourne and Brisbane (see Figure 17.7) at 4.2 million and 2.1 million, respectively. The primacy value is 0.76. One could deduce that the urban system is evenly distributed. But on closer examination, as shown in Table 17.3, Australia is best depicted as a series of primate states in a federal system.

The case of Australia reminds us of the importance of scale, geography, and history. The primacy value for Australia is suggestive of non-primacy, as if activities were evenly distributed throughout the urban system. Yet when we look at the level of individual states, we see there is a marked concentration of activity, enterprise, and population in large cities. The Australian Federation of 1901

17.7 Brisbane is one of the large cities in Australia

brought together a series of **hyperprimate** economies that the passing of time has done little to change. The major cities continue to dominate their respective states. The geography and (white) history of Australia play an important part. An antipodean gulag for the British soon developed into a colonial enterprise organized and structured through the port cities of the different states. A classic case of colonial spatial organization of a vast country was soon embedded into a series of primate states. The case of Australia shows that reliance solely on national data abstracted from issues of scale, geography, and history has limited explanatory value.

Global Networks

Cities are also part of global networks through which capital, goods, ideas, and people flow. Some of the flows have been identified and measured.

The GAWC Network

A substantial body of material has emerged from the work of Peter Taylor and his colleagues at the **Globalization and World Cities (GAWC)** research network. The website (http://www.lboro.ac.uk/gawc/) lists data sets as well as more than 350 research papers and is an indispensable guide to the metageography of urban networks. In 2000, GAWC collected data on the distribution of 100 global advanced producer service firms, which includes accountancy, advertising, banking/finance, insurance, law, and management consultancy, across 315 cities. They analyzed the resultant data matrix to identify a global urban hierarchy. In 2008, 2010, and again in 2012 they extended the analysis to 175 firms in 526 cities. The result was a fivefold hierarchy that identified cities as alpha, beta, gamma, high sufficiency, and sufficiency. Figure 17.8 is a cartogram of alpha cities in the network of advanced producer services. Note how New York (NY) and London (LON) dominate.

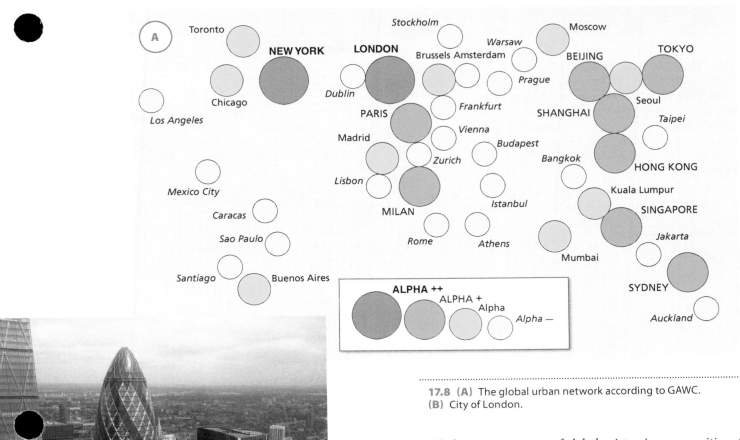

17.8 **(A)** The global urban network according to GAWC.
(B) City of London.

NYLON is an important pivot in the global networks. In 2000, Shanghai was three steps below, in the alpha minus category, while Beijing was only a beta city. In 2008, both Shanghai and Beijing were classified as alpha plus, only one step below NYLON. Both cities are moving into the top tier as China's economic growth, both absolute and relative to the rest of the world, continues apace. The diagram highlights the centers of new metropolitan modernity, the rapidly growing cities of the Far East.

Financial services continue to be concentrated especially in the global cities. At the apex are the supranational centers of London and New York, followed by a second tier of international centers that include Tokyo and Zurich, and host centers that attract foreign financial institutions and include such places as Sydney, Toronto, and Vienna.

The globalizing economy creates lots of information, narrative uncertainty, and economic risk that all have to be produced, managed, narrated, explained, and acted upon.

Global cities are centers of global epistemic communities of surveillance, knowledge production, and storytelling. Trust, contact networks, and social relations play pivotal roles in the smooth functioning of global business. Spatial propinquity allows these relations to be easily maintained, lubricated, and sustained in an efficient means of communication that helps solve incentive problems, facilitates socialization and learning, and provides psychological motivation. There are positive benefits of face-to-face contact. Global cities are sites of dense networks of interpersonal contact and centers of the important business social-capital trust vital to the successful operation of international finance.

Black Holes and Loose Connections

Networks also have holes. Table 14.4 lists eleven cities that met three criteria: they had a population of over 3 million, were not identified by GAWC as a world city, and did not share their national territory with a world city. They ranged from Tehran, Iran, with a population of 10.7 million, to Chittagong, Bangladesh, with a population of 3.1 million. There are a number of reasons behind these very large cities' non–world city status. They are the **black holes** of advanced global capitalism, with many people but not enough affluent consumers or complex industries to support sophisticated producer services. There are also cases of catastrophic decline where there has been an

17.2 Visualizing National Urban Network Flows: The Case of the United States

Xingjian Liu and two colleagues estimated the US national urban network from a study of sixty major cities. Using 2010 data, they map four different types of intercity flows, of which two are noted here: Internet connectivity and recorded business travel flows. Note the patterns they reveal.

REFERENCE

Liu, X., Neal, Z., and Derudder, B. (2012) City Networks in the United States: A Comparison of Four Models. *Environment and Planning A* 44: 255–256. http://www.envplan.com/openaccess/a44496.pdf.

Network flows in the United States: Internet connectivity

Network flows in the United States: recorded business travel flows

almost complete collapse of civil society. In recent years, Khartoum and Kinshasa, for example, have witnessed social anarchy and the decline of the rule of law. War and social unrest have been the norm rather than the exception. These two cities typify those that have internally collapsed for all intents and purposes and have been abandoned or bypassed by global capitalism. Sustained social disruption reinforces the global disconnect.

Poverty and social anarchy do not explain all the cases. Tehran and Pyongyang, North Korea, for example, are cities where national ideologies have not encouraged global economic connections to the advanced capitalist economies.

We can identify three ideal types of large non–world cities: the poor city, the collapsed city, and the excluded city. There are clearly connections between these types, and in reality most of the cities listed in the table have elements of all three in differing proportions. It is just as important to identify the black holes and loose connections as to point out the important nodes and connected cities of global urban networks.

TABLE 17.4	Black Holes
CITY	**COUNTRY**
Tehran	Iran
Dhaka	Bangladesh
Khartoum	Sudan
Kinshasa	Congo
Lahore	Pakistan
Baghdad	Iraq
Rangoon	Myanmar
Algiers	Algeria
Abidjan	Ivory Coast
Pyongyang	North Korea
Chittagong	Bangladesh

Flows of People

Lisa Benton-Short and colleagues also sought to identify a global network, but their work was based on flows of people. They looked at immigration into cities around the world and established an index based on the percentage of foreign-born, the total number of foreign-born, the percentage of foreign-born not from a neighboring country, and the diversity of immigrant home countries. The result was a threefold division into alpha, beta, and gamma cities. There are similarities and differences between the GAWC and the Benton-Short results. Among the similarities are that some cities appear in the same category in both analyses. New York and London are alpha cities and sit atop the apex of both hierarchies. Other shared alpha cities include Toronto, Los Angeles, Sydney, and Amsterdam. Among the differences, while Miami, Melbourne, Vancouver, and Dubai are considered pivotal points of global migration reaching alpha status, in the advanced producer service category they only make it to beta status.

Cities as Space of Flows

Some of the differences relate to issues of data and data availability. But they also refer to the nature of the global urban network. Networks vary according to the flow. Some flows "pool" in some cities rather than others. Advanced producer services, for example, which can be considered command functions of the globalizing economy, are still concentrated in just a few cities because of the need for social interaction in global financial business deals. The flow of people, while matching the connections in command functions, also has slight differences. National regulations concerning immigration, the demand for labor, and the relative openness of societies to foreign migrants all play a part. The sheer need for labor, whether the global talent pool of specialized knowledge experts or cheap unskilled labor, varies throughout the urban network. The result is a similar but not identical match of command centers with immigration hubs. There is no one fixed global urban hierarchy; it is more a global urban network with different configurations of hierarchy depending on the flows.

Money, people, ideas, goods, and practices do not just flow through the urban network; they are transformed in the process. To be more accurate we should use the term "space of transformative flows." Consider flows of people. The movement through the urban network is not a simple geographical movement; it involves cultural exchanges. These can refer to the new work habits and job practices of the temporary worker as well as the complex cultural transformations of long-term migrants as they adapt to a new milieu and in turn transform their surroundings. Rather than mere transfers, flows along the urban networks are transformative experiences, even in the flows of inanimate things such as money, goods, and services. A small amount of money in London or New York, when remitted back to Ghana or El Salvador, can become the source for a land and house purchase, enable a new business, or pay for better schooling. Flows through the urban network change the medium and the networks; as people adapt, ideas are tweaked, money is reimagined, and commodities are reappropriated.

Cited References

Benton-Short, L., Price, M., and Friedman, S. (2005) Globalization from Below: The Ranking of Global Immigrant Cities. *International Journal of Urban and Regional Research* 29: 945–959.

Berry, B. J. L. (1976) *Geography of Market Centers and Retail Distribution.* Englewood Cliffs, NJ: Prentice Hall.

Bromley, R. and Bromley, R. D. F. (1979) Defining Central Place Systems Through the Analysis of Bus Services: The Case of Ecuador. *Geographical Journal* 145: 416–436.

Frantzman, S. (2010) Hitler's Geographer: Walter Christaller and Nazi Academics. *The Jerusalem Post,* August 27. http://www.jpost.com/Magazine/Features/Article.aspx?id=186068

Galpin, C. J. (1915) *The Social Anatomy of an Agricultural Community*. University of Wisconsin Agricultural Experimental Station, Research Bulletin 34.

Hill, P. and Smith, R. H. T. (1972) The Spatial and Temporal Synchronization of Periodic Markets: Evidence from Four Emirates in Northern Nigeria. *Economic Geography* 48: 345–355.

Rössler, M. (1989) Applied Geography and Area Research in Nazi Society: Central Place Theory and Planning, 1933–1945. *Environment and Planning D: Society and Space* 7: 419–431.

Short, J. R. (2004) Black Holes and Loose Connections in a Global Urban Network. *The Professional Geographer* 56: 295–302.

Short, J. R. and Pinet-Peralta, L. M. (2010) Urban Primacy: Reopening the Debate. *Geography Compass* 3: 1245–1266.

Vijver, E. V. D., Derudder, B., Bassens, D., and Witlox, F. (2014) Filling Some Black Holes: Modeling the Connection Between Urbanization, Infrastructure, and Global Service Intensity. *The Professional Geographer* 66: 82–90.

Select Guide to Further Reading

Alderson, A. S., Beckfield, J., and Sprague-Jones, J. (2010) Intercity Relations and Globalisation: The Evolution of the Global Urban Hierarchy, 1981–2007. *Urban Studies* 47: 1899–1923.

Brenner, N. and Keil, R. (eds.) (2006) *The Global Cities Reader*. New York: Routledge.

Castells, M. (1996) *The Rise of Network Society*. Oxford: Blackwell.

Derudder, B. (2008) Mapping Global Urban Networks: A Decade of Empirical World Cities Research. *Geography Compass* 2: 559–574.

Derudder, B., Hoyler, M., Taylor, P. J., and Witlox, F. (eds) (2012) *International Handbook of Globalization and World Cities* Cheltenham: Edward Elgar.

Jefferson, M. (1939) The Law of the Primate City. *Geographical Review* 29: 226–232.

Knox, P. (ed) (2014) *Atlas of Cities*. Princeton: Princeton University Press.

Moomaw, R. and Alwosabi, M. (2004) An Empirical Analysis of Competing Explanations of Urban Primacy Evidence from Asia and the Americas. *Annals of Regional Science* 38: 149–171.

Neal, Z. P. (2010) From Central Places to Network Bases: A Transition in the US Urban Hierarchy, 1900–2000. *City & Community* 10: 49–75.

Taylor, P. J. (2004) *Global City Network*. London: Routledge.

Selected Websites

Comparative urban data is available at
http://www.citypopulation.de/

The Globalization and World Cities (GAWC) website
http://www.lboro.ac.uk/gawc/

Interactive Map Websites

Airline networks
https://www.oneworld.com/flights/where-we-fly/interactive-network-map-instructions

Cities along the Silk Road
http://en.unesco.org/silkroad/network-silk-road-cities-map-app/en

Cyberconnections
http://www.govtech.com/network/Can-Cities-Wait-Until-2084-for-Google-Fiber.html
http://www.businessinsider.com/wi-fi-map-of-largest-cities-in-the-world-2015-9

Learning Outcomes

- Cities exist as nodes in a variety of regional, national, and global networks.
- One of the earliest forms of urban network is the periodic market system, in which villages and towns with low demand density and relatively mobile vendors periodically host open markets where goods are sold.
- However, as demand grows, some traders stay in place, and eventually periodic markets are replaced by fixed markets.

- The basics of central place theory, and in particular the notions of complementary region and of the range and threshold of goods and services, are still important tools for uncovering some of the socio-spatial relations in urban networks as the measurement and definition of complementary regions allow us to identify the effective influence of a city across space.
- Cities are also part of national networks.

- In primate distributions, one city dominates the urban system.
- The Globalization and World Cities (GAWC) research network attempts to provide an understanding of global urban networks by collecting data on the distribution of global advanced producer service firms across cities in order to identify a global urban hierarchy.
- An alternative method looked at immigration into cities around the world and established an index based on the percentage of foreign-born, the total number of foreign-born, the percentage of foreign-born not from a neighboring country, and the diversity of immigrant home countries, broken down into a threefold division.
- The two methods have some similar results, as some cities are in the same categories in both studies, but there are also an array of differences that demonstrate the need for better data as well as the ways in which network flows vary.
- A global system of cities, connected by flows of capital, ideas, information, and skilled personnel, "landscapes" the world into concentrated nodes of heightened connectivity and accumulated control.
- Rather than mere transfers, flows along the urban networks are transformative experiences that change the medium and the networks; as people adapt, ideas are tweaked, money is reimagined, and commodities are reappropriated.

The Internal Structure
of the City

LEARNING OBJECTIVES

Identify how the internal structure of the city is shaped over time.

Understand the city as a site for investment and the role of building cycles.

Relate the internal structure of the city as residence, and how economic, social, and cultural factors form residence patterns.

Outline models that describe patterns of the urban housing market and neighborhood development.

Reflect on how the city's internal structure forms the social and political setting for people's movement, attachment, and interaction.

Consider the social and economic changes that cities in the developing world particularly are experiencing today.

In this chapter, we will focus on the internal structure of the city and, in particular, on four specific themes: the city as investment, as residence, as social context, and as political arena. The chapter ends with a discussion of the major contemporary trends of gentrification, suburban decline, city marketing, and the hosting of global events.

The City as Investment

The city is a site of investment. This investment is cyclical, as capital flows into the building of a city when relative rates of return are high. **Building cycles** occur on average every eighteen to twenty years. There is distinct periodicity to the building cycle. In the *trough*, there is limited investment and little building. During the *upswing*, the lack of supply just as demand picks up leads to higher levels of construction. At the *peak*, building activity is more feverish, land prices increase, and there is more speculative development (see Figure 18.1). But as the supply expands just as the demand falters, there is a *downswing*, and in more extreme cases a *crash* as prices fall and supply outstrips demand. The cycle then begins again.

Different sectors of the property market may experience slightly differently timed cycles in varying places. The commercial office property market shows the largest peak and troughs. What was distinctive about the US housing market crash in 2008 was its national character. Previous US housing cycles had a more regional character.

Building cycles often align with transport improvements and changes. A building boom in the 1950s was associated with mass car ownership, giving a more dispersed and road-connected character to urban development compared to the booms in higher-density development spurred by the spread of trams and railways. It is more accurate, then, to speak of a building-transport cycle.

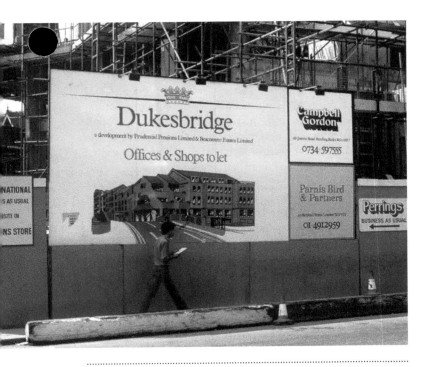

18.1 Office development in Reading, UK. A pension company, eager to place large sums of money in secure, high-return investments, funded the project.

The built form of many cities is a visible legacy of the boom of the past. Each major building cycle is also associated with changes in architectural style. Capital is locked into place, with the architecture of the day acting as the equivalent of a date stamp. The very largest and oldest cities have examples of all the major building cycles. A transect

through a major city is the equivalent of flipping through the pages of an architectural history text. Yet some cities, in contrast, experience massive growth only at specific times. The old part of Tallinn in Estonia is a reminder of the city's growth during the Hanseatic League, Venice is filled with structures built during the Renaissance boom (Figure 18.2), and the Georgian architecture of Edinburgh is a sign of the city's vitality in the eighteenth century. Housing built in the 1920s boom was influenced by Art Deco and Art Moderne sensibilities, while housing and office buildings constructed in the immediate postwar era were more modernist-inspired. More recent postmodern buildings indicate the latest rounds of investment, such as the frenzied growth that produced Shanghai's signature postmodern buildings.

The City as Residence

We can distinguish between production, supply, and demand for accommodation.

The Production of Housing

Housing is produced in a variety of ways. There is the individual contract, whereby a household hires builders to construct a custom-designed dwelling. This is normally very expensive and limited to the very rich. The tiny size of this sector, however, is dwarfed by its architectural importance, as many new residential styles and designs first saw the light of day as individual contracts. Many of the early individual house projects of such famous architects as Frank Gehry, Frank Lloyd Wright, and Le Corbusier influenced changes in overall architectural style. More common is the **institutional contract**, whereby an institution or a government entity builds dwellings and then rents them out or sells them. Institutional contract production was particularly important during the boom in public housing from the 1950s to the 1970s, when local governments hired builders to construct large projects.

Currently, the two most important forms of production are speculative and self-build. **Speculative building** involves builders acquiring land and constructing dwellings for a general demand rather than specific customers (Figure 18.3). This form of production is dependent on macroeconomic conditions, especially the interest rate, as builders use borrowed

18.2 Renaissance architecture in Venice, Italy

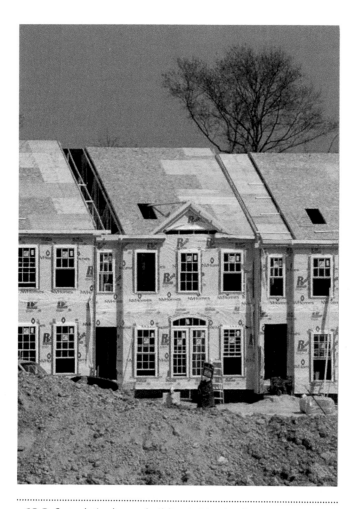

18.3 Speculative house-building in Maryland

Around 1 billion people live in such slums. The figure is estimated to rise to 2 billion by 2030.

The Housing Stock

The rise and fall of housing stock depends on many factors. There is the natural aging of the housing stock, although the aging process can give older housing the allure of the established, much the same way that antiques have value. Building cycles involve not only the building of the new but often the destruction of the past as old buildings are demolished. Urban renewal projects in the United States from 1949 to 1973 demolished 600,000 units, covering 1,000 densely packed urban square miles and ultimately displacing 2 million people. A program begun with the best of intentions, getting rid of substandard housing, ended up demolishing basically sound housing, destroying functioning neighborhoods, laying the seeds of subsequent inner-city decline, and concentrating poor people—especially poor black people—in increasingly segregated urban neighborhoods. A similar process is occurring in the rapidly growing cities of the Chinese coast. Michael Meyer describes life in the vanishing backstreets of Beijing as they are demolished to make way for malls and new high-rise buildings. From 1990 to 2003, more than half a million residents were evicted from central Beijing, a process that accelerated just before the hosting of the 2008 Summer Olympics.

There are also environmental shocks to cities. Just before five in the afternoon, on January 12, 2010, an earthquake measuring a devastating 7 on the Richter scale occurred in Haiti. Over 300,000 people were killed, a similar number injured, and almost a million made homeless; much of the capital city of Port-au-Prince was reduced to rubble. Despite international relief efforts, in 2012 half a million people still lived in temporary shelter, often in appalling conditions. In October and November 2012, Hurricane Sandy etched a destructive path through the Caribbean and along the northeastern Atlantic seaboard. Fifteen thousand homes were destroyed in Cuba, and hundreds of homes were flooded along the New Jersey coast.

The Supply of Housing

Housing supply can be broken down into three main tenure types. First is **public housing**, in which a government agency provides accommodation for citizens. In much of postwar Europe, public housing was an important part of the housing stock. For example, by 1972 more than 80 percent of housing stock in Scotland was public (Figure 18.4). Public housing was, and in many countries still is, an important part of the social contract between the people and the government. In the United States, by contrast, public housing was always a much smaller percentage of the housing stock and often marginalized to house the very poorest. Every major US city had "projects," which tended to be large

money and sophisticated credit facilities that allow households to purchase homes. Because the system is based on long credit lines, any change in the financial structure reverberates in this sector of the housing market. Wider economic changes, such as major interest rate shifts, are reflected in the booms and slumps of the speculative housing sector.

Regarding housing in much of the world, incomes are too low, credit facilities for purchasing homes are lacking, and cash-strapped governments are unable and unwilling to provide accommodation. In many cities in the developing world, **self-build housing**, in which residents simply build their own residences, is an important form of housing production. Areas of self-built housing have a variety of names, including slum, shantytown, informal housing, and squatter housing. Much of this self-build is considered illegal, because the occupiers hold no title to the land, do not pay taxes, and have constructed some form of shelter that does not meet building codes. Because these are illegal structures, they often lack public services such as clean water, sewage connections, and sanitation infrastructure.

modernist units of public housing providing accommodation for the very poor. Many of them have since been demolished.

One public housing trend is privatization. In Britain after 1979, much of the public housing stock was essentially sold off to private residents. A total of 1.6 million units were sold to sitting tenants. The proportion of public housing declined from around 30 percent in 1975, concentrated in the larger urban centers, to around 6 percent by 2016.

Across Western Europe, public housing, sometimes known as **social housing**, is still an important part of the housing stock, at the highest levels ranging from 40 percent in the Netherlands to 20 percent in Sweden. In Sweden, 850,000 social housing units are provided by 300 local housing authorities. There are three allocation models: "universal," in which everyone has access to social housing, as in the Netherlands and Sweden, where rents are fixed at market rates with subsidies for the poorest; "restricted," where access is means-tested, as in France and Germany; and "residual," as in the United States, the United Kingdom, and Ireland, where it is reserved for the poorest.

A second form of tenure is **private renting**, whereby landlords rent out property to tenants. Through most of the nineteenth and early twentieth centuries, this was the dominant tenure form in cities of North America and Europe. Landlords tend to want to raise rents and keep maintenance costs low, while tenants want exactly the opposite. History is studded with landlord–tenant conflicts. Generally, the conflict remains at the level of individual landlord and tenant, but in a few rare cases the conflict can widen. In Glasgow, Scotland, in 1914–1915, a rent strike broadened into industrial stoppages. Because the country was at war, the government quickly intervened with legislation that limited rent increases and laid the basis for tenant protection and controls on rent that have lasted to the present day. In the United States, rent controls are in operation in around 140 cities. In New York City, rent control has been in operation since 1943 and covers around 185,000 residences, while rent stabilization, for apartments constructed after 1945, covers around one million residences.

Third, and in most affluent countries the largest form of housing supply, is **owner occupation**, also known as **homeownership**, whereby households purchase property. Housing is very expensive in relation to income, so immediate house purchase is beyond the reach of all but the

18.4 Public housing in Glasgow, Scotland

wealthiest. Credit in the form of long-term mortgages is a vital prerequisite for mass homeownership. In the United States, for example, there is a steady encouragement of owner occupation through various inducements to lenders to provide mortgages. Government-backed agencies such as Fannie Mae and Freddie Mac provide liquidity to the system by purchasing mortgages from banks. Tax policies also encourage homeownership. In the United States, homeowners get exemption from the capital gains tax on house sales and tax relief on mortgage interest payments. These are massive government subsidies that are mostly regressive, in that the richest groups tend to benefit most.

Financial institutions play a key role in owner occupation. Their willingness to lend and the spatial patterns of their lending influence urban housing markets. In the 1970s, for example, many institutions in the United States issued fewer mortgages in lower-income minority neighborhoods. The process was known as **redlining**, and it resulted in a lack of housing finance to poorer neighborhoods in the inner city. It became both a cause and effect of inner-city decline, because residential areas denied mortgages tended to deteriorate more rapidly. In 1975 Congress passed the Home Mortgage Disclosure Act in response to evidence that banks were not lending in inner-city minority neighborhoods. Two years later, the 1977 Community Reinvestment Act prodded banks to lend in inner-city communities. Promoting homeownership became a national fixation.

From the 1990s on, lending requirements and banking regulations were relaxed, leading to two developments. The first was a relaxation of strict requirements on borrowers. Previously, most mortgages required a 10 percent to

18.1 Home Sweet Home

The home is of huge social significance. We spend much of our lives in the home, and our primary emotional and connections are shaped in its domestic arena; where we live and how we live are important determinants of our social position, physical health, and individual well-being. Home is a central element in our socialization into the world. Despite its huge significance, there has been comparatively little work on the meaning of the home. However, recent work by geographers focuses on the home, where space becomes place and where family relations and gendered and class identities are negotiated, contested, transformed. The home is an active moment in the creation of individual identity, social relations, and collective meaning. The home is a nodal point in a whole series of polarities: journey–arrival, rest–motion, sanctuary–outside, family–community,

space–place, inside–outside, private–public, domestic–social, spare time–worktime, feminine–masculine, heart–mind, Being–Becoming. These are not stable categories; they are both solidified and undermined as they play out their meaning and practice in and through the home. The home is a space riven by ambiguities, a place of paradoxes.

REFERENCES

Brickell, K. (2012) "Mapping" and "Doing" Critical Geographies of Home. *Progress in Human Geography* 36: 225–244.

Blunt, A. and Dowling, R. (2006) *Home*. New York: Routledge.

Cieraad, I. (ed) (1999) *At Home: An Anthropology of Domestic Space*. Syracuse, NY: Syracuse University Press.

20 percent down payment and clear evidence of ability to pay. These were known as **prime mortgages**. By the mid-2000s, more mortgages were **subprime**, given to borrowers making no down payment and with fewer verifiable means of income. Inner-city neighborhoods now became a prime site for predatory lending. These subprime mortgages grew in size and importance. By 2006, there were almost 7 million subprime loans.

The second development was the securitization of mortgages. Previously mortgages had been kept on the books of the lending institutions. With securitization, mortgages were bundled up and sold as a commodity. The emphasis was on generating turnover rather than assessing risk, as the risk was passed onto the next buyer on the chain. It was in the interests of the originating company to allocate as many mortgages as possible and then sell them on. Credit agencies were supposed to assess the risk, but a government inquiry revealed massive abrogation of due diligence by the three main agencies. The result was that toxic mortgages infected the financial system. When the housing bubble burst in 2006–2008, 7 million mortgages went into foreclosure, and banks and other institutions had toxic assets on their books. The net effect was the financial crisis of 2008. The sorry tale reminds us of the importance of the housing finance sector in the overall economy.

The Demand for Housing

Demand for housing varies along a number of dimensions. Perhaps the most important is income and wealth. The private housing market covers a variety of income levels. At the top are luxurious apartments and palatial houses on sprawling estates. At the bottom are the cheapest apartments and

smallest houses. In between are a variety of middle-income areas. The income hierarchy is reflected in the variegated housing market. The more income you have, the more housing you can consume, whether in terms of more space or more prestigious space. Most large cities, for example, have their exclusive neighborhoods where residency is a sign of wealth and achievement. At the other extreme every city has its "bad" neighborhoods, with reputations sometimes undeserved but nevertheless an important part of the social geography of the city. The city's housing market can be considered as a space-packing problem. The richest get to choose the best housing in the best areas. Groups with successively lower income get less choice, and as income declines, the choices narrow (see Figures 18.5 and 18.6).

Demand for housing also varies over the course of the life cycle. Young single adults, for example, need less space than families of older adults with children. At the later stages of life, even families with children may shrink in size and hence space requirements diminish as the children mature and leave to start their own lives. This simple stage in the life cycle model of housing demand is undermined by two trends. First, in the developed world, average family size is decreasing, so the demand for family-style housing is declining while the demand for smaller accommodation for single-person and childless households is increasing. The suburbanization of the US population from 1950 to 1990 was driven in part by the shift of families with children moving to single-family homes in the suburbs, the move in part a decision to get better educational opportunities for their children. Since the 1990s, there has been a move into the city, in part driven by more single-person and childless households for which a house in the suburbs is not as good a choice

18.5 Luxury home in exclusive Palo Alto, California

as a more central location. Second, throughout much of the world the extended family is the norm rather than the nuclear family. Many adults tend to remain with their existing family rather than establishing new independent households. Interestingly, the trend is also reappearing in the richer countries. In the United States, for example, more adult children are remaining longer with their parents than ever before, and sometimes returning home after college; the shift is a function of a difficult job market and expensive housing.

Beyond income and life cycle stage, cultural factors also affect housing demand. Households vary in their ethnic, racial, and other forms of identity. In some cases, certain ethnic and racial groups are discriminated against. To take a prominent example, throughout the first two-thirds of the twentieth century, blacks in the United States faced severe discrimination that limited housing choice in cities. Black concentrations in central cities, initially a sign of discrimination, also emerged through time as a platform for political power as the black population found a voice in majority-minority cities such as Baltimore, where blacks make up 63 percent of the population. In the past fifty years, even after formal discrimination has ended, there still have been marked patterns of segregation. Cities around the world still bear the historical imprint and/or contemporary operation of racial and ethnic segregation. Although apartheid is no longer the formal policy of South African cities, there is still marked residential segregation by race in the social geographies of major South African cities.

Certain groups cluster together in distinct residential areas through choice sometimes as much as constraint. There are many reasons for the clustering, including feelings of security. A neighborhood of similar people can provide a safe platform for entry into a large city. Part choice, part constraint, racial and ethnic clustering is an important part of the social geography of the city. The geography shifts as new migrant streams enter; old neighborhoods are gentrified or commodified, as in the case of Chinatowns becoming places of Chinese restaurants and festivals rather than places where Chinese people live. Some clusters became centers of entertainment as well as source of identity, such as the well-known Little Italy in New York City and Little Havana in Miami. In Washington, DC, there is Little Ethiopia, home to the largest population of Ethiopians outside Addis Ababa, around 250,000. It is not just the formal place of residence of all American Ethiopians but also the setting for

18.6 Modest row houses in inner-city Baltimore

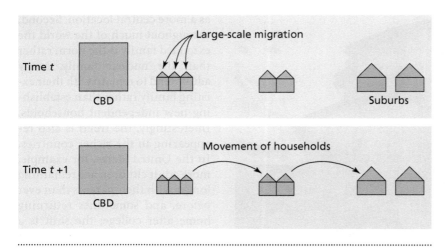

18.7 The process underlying the Burgess model

restaurants and stores that sell Ethiopian food and goods. Little Odessa in Brighton Beach, New York City, is home to more than 350,000 people from Russia, Ukraine, and Georgia.

There is also clustering and segregation along religious lines. The geographer Frederick Boal has long studied the segregation between Catholics and Protestants in Belfast. The city was at the frontier zone between Irish and British realms. The segregation not only gave a sense of safety during the period of conflict; it was also part of the conflict, as it reinforced group stereotyping and the spatial basis and embodiment of community conflicts. Severe ethnic national and religious divisions can result in highly segregated cities with borders that scar and divide. In the most severe cases, the lines of segregation are patrolled and policed.

In recent years, sexual identity has also become a basis for residential clustering. In the Paddington district of Sydney, Australia, from the 1970s onward, an inner-city neighborhood of small family homes became a favored residence for openly gay households who were discriminated against in much of the city. The neighborhood became a gay-friendly district encouraging an open display of gay identity. As cultural attitudes shifted, the previously transgressive nature of the place was soon commodified and then incorporated the city's economy and national and international image. The annual Mardi Gras festival, initially supported only by the LGBT (lesbian, gay, bisexual, and transgender) community as a celebration of sexual diversity, is now one of the city's biggest cultural events, drawing visitors from around the world. In other cities, such as San Francisco,

sexual tolerance is now at the very heart of the city's identity. Elsewhere, in much of the world, where LGBT people face open discrimination, there are far fewer opportunities for neighborhoods of sexual diversity and tolerance to flourish. Homosexuality continues to be categorized as a crime in many societies.

Models of the Residential Mosaic

The housing market produces a residential mosaic of different types of neighborhoods. The modeling of the residential mosaic constitutes a consistent theme in urban geography.

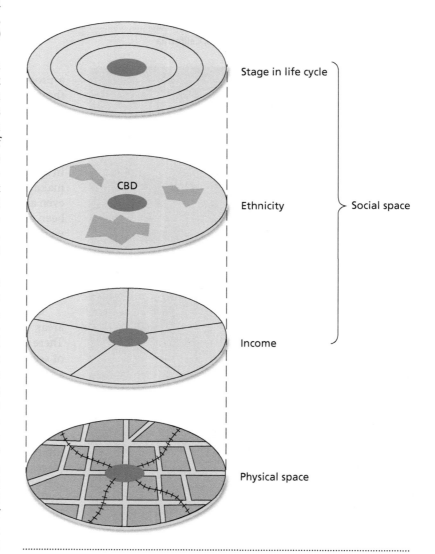

18.8 Murdie's composite model

Different models focus on different processes. Throughout the twentieth century, a number of different models were proposed. The **Burgess model**, named after urban sociologist E. W. Burgess, was based on the experience of Chicago in the first third of the twentieth century, a time of massive immigration from overseas. Burgess saw a process of new migrants moving into the cheaper central city areas and more established residents moving further out (see Figure 18.7). The result was a concentric ring pattern of rising socioeconomic status and increasing family size toward the city's edge. Later, land economist Homer Hoyt pictured socioeconomic groups arranged in different sectors of the city. Robert Murdie combined various observations into a layered model of concentric rings of households at different stages in the life cycle: socioeconomic groups in specific sectors, and ethnic, racial, and other minorities in distinct clusters. The composite **Murdie model** is shown in Figure 18.8. These simple patterns have long been overlain by more complex patterns. An even more recent composite, the Hanlon-Short-Vicino model, is shown in Figure 18.9. Around the central area of this idealized US city, there are pockets of both concentrated poverty

and gentrification, and in the inner ring there are suburbs in crisis, with declining prices and aging housing stock. Toward the city's edge there are "boomburbs" and affluent enclaves. This model reflects the contemporary reality of the heterogeneous US metropolitan areas more accurately than the simplified picture of the traditional models.

These models apply only to the developing world. In the less developed world urban structure takes a different form because some cities may have a colonial legacy, often embodied in the central business district and elite neighborhoods. Most cities have large informal settlements or slums that are found on the periphery of the city as well as distributed throughout it. The city center has the formal CBD as well as traditional markets and areas of slum housing. Some cities also have marked segregation between different ethnic and religious groups. Furthermore, some neighborhoods are undergoing gentrification. In the large cities the route to the airport, often following the direction of elite residences, is also an area of new commercial and industrial building as well as new suburbs, many of them gated communities. Throughout cities in the developing world there is a fuzzier demarcation between rich and poor and

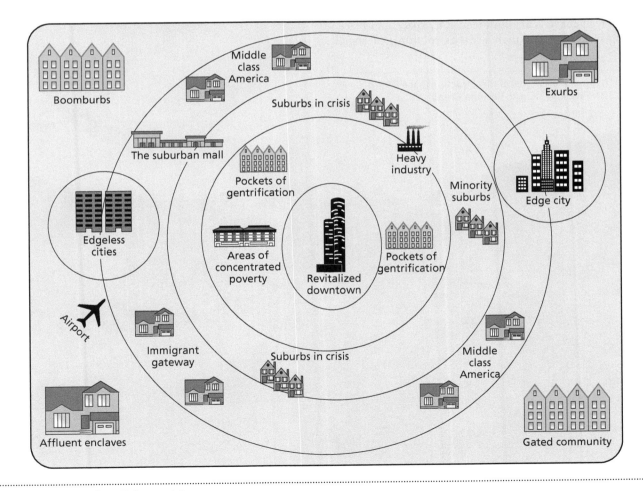

18.9 The Hanlon-Short-Vicino model

between commercial and noncommercial real estate than in the more developed world. Figure 18.10 presents the basic elements of a model for cities in the developing world.

The City as Social Context

Human geographer David Seamon argues that we can consider the relationship between people and urban space in terms of three interlinked processes: *movement*, or how people move through space; *rest*, which is the physical attachment to place; and *encounter*, or how we interact in place with other people.

We can also think of **timed space** and **spaced times**. Certain spaces change through time. Entertainment districts, for example, can remain quiet until the nighttime brings more customers and increased activity. Certain behaviors are more permissible at certain times and places than others. Act rowdily in New Orleans during Mardi Gras, and no big deal. The same activity in Milwaukee on a Sunday afternoon may provoke very different responses.

How we behave is also influenced by space and place. Sociologist Erving Goffman asks us to think of social

activity as a performance and employs spatial metaphors such as "front stage" and "backstage." When we are front stage, in front of strangers, such as a job interview, we are on our best behavior. Backstage, as with friends relaxing, we can let our hair down. The city is a series of stages where different parts of our character are reinforced, highlighted, or suppressed. It provides opportunities to play a multiplicity of roles, allowing both anonymity and

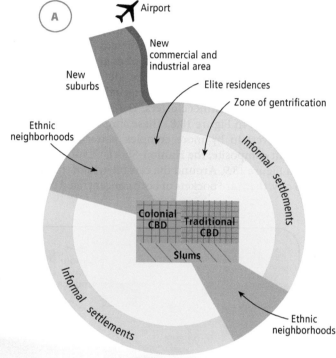

18.10 (A) The Short Model of City Structure in the Global South. **(B)** Favela, or slum, in Rocinha, Brazil.

18.2 Measuring Segregation in Cities

The level of segregation in a city is measured in a variety of ways. The **index of dissimilarity** measures the unevenness in the distribution of two groups across spatial units. The index ranges from 0 to 100, with values closer to 100 indicative of more dissimilarity. Table 18.1 notes the index values for racial groups in counties in the Megalopolis region of the United States in 1960, 1980, and 2000.

The dissimilarity in the spatial spread of blacks and whites remains consistently high. In fact, there was an increase in the index for 2000 compared to 1960, which is a staggering finding given the decline of many of the formal and explicit practices of racial discrimination. A striking feature is the persistently high indices for all the groups. For only one pair, Asians and Hispanics, did the index fall below 30, perhaps due to the shared spaces of more recent immigrants. The data reveal persistent segregation between all the groups, increased segregation between blacks and whites from 1960 to 2000, and stable and high levels for whites and Asians, blacks and Asians, and blacks and Hispanics, with a slight decline in segregation levels between whites and Hispanics and Asians and Hispanics.

The **index of segregation** measures the distribution of one group compared to the total population. This index also ranges from 0 to 100, with values closer to 100 indicative of a more segregated distribution.

We can interpret the data in Table 18.2 using the example of the index of 43.3 percent for whites in 1960; this implies that 43.3 percent of whites would have to move to another area to achieve the same distribution as the total population. This is a high value, indicating marked segregation. The high values shown in this table indicate consistently pronounced levels of segregation for all the groups.

These data can be interpreted as reflecting the differential nature of suburbanization until 1980: as blacks were moving into the central cities, whites were moving out to the suburbs. The 1980 values indicate the high-water mark of this trend. The 2000 values reflect the greater suburbanization of blacks and consequently slightly less segregation. Asians remained approximately the same, while Hispanics became slightly less segregated in 2000 compared to 1980—in part a function of the wider suburban spread of Hispanics through Megalopolis.

The **location quotient (LQ)** measures the degree of concentration of one group compared to the total population; it is found by dividing the percentage of the subgroup by the total population of each and every spatial unit. Values greater than 1 suggest a higher concentration of the subgroup in that

TABLE 18.1 Index of Dissimilarity in Megalopolis

	YEAR		
	1960	**1980**	**2000**
Whites/Blacks	43.9	49.9	47.1
Whites/Asians	39.5	38.2	39.2
Whites/Hispanics	NA	49.3	42.3
Blacks/Asians	40.2	42.2	41.2
Blacks/Hispanics	NA	37.5	36.2
Asians/Hispanics	NA	32.9	29.6

TABLE 18.2 Index of Segregation in Megalopolis

	WHITE	**BLACK**	**ASIAN**	**HISPANIC**
1960	43.3	43.8	37.8	NA
1980	51.2	48.4	35.1	48.4
2000	40.8	42.6	34.2	38.0

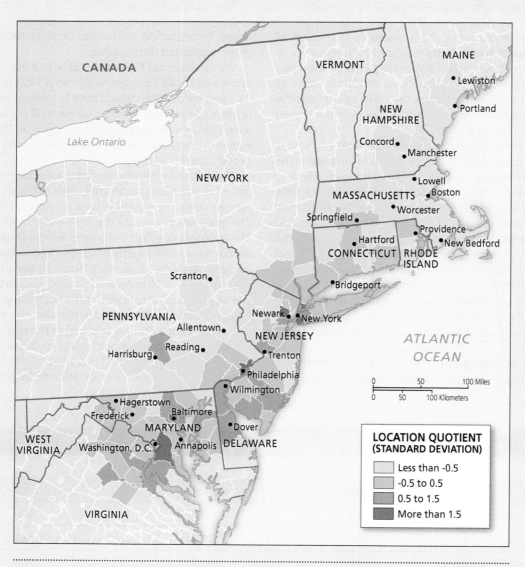

Location quotient for blacks in Megalopolis, 2000

area. Values of less than 1 indicate lower concentrations. LQ values in 1960 were wide, from 0.002 in Perry, PA, to 5.48 in the District of Columbia. By 2000, the range of values had shrunk somewhat, from 0.028 in Perry to 4.16 in Baltimore City. There was a thin spread of blacks throughout the region and increased concentration in selected cities and suburbs. The map shows the LQ values for blacks in 2000, by standard deviation, and reveals increased concentration in the urban centers and the inner suburban areas around the largest cities.

Each of the different measures provides a different picture of residential segregation and concentration. All of them are very sensitive to the scale of the analysis. Larger observation units, such as the county level, may mask patterns of segregation that are revealed only at the level of census tracts.

REFERENCES

Johnston, R. J. and Jones, K. (2010) Measuring Segregation: A Cautionary Tale. *Environment and Planning A* 42: 1264–1270.

Short, J. R. (2007) *Liquid City: Megalopolis and the Contemporary Northeast*. Washington, DC: Resources for the Future.

identification, letting us be both performers and audience in a myriad of fleeting and more sustained social interactions. The characterization of the city as stage has unlocked a whole host of useful metaphors, including plot, script, roles, back/front stage, and so forth that have been enormously useful to understanding life in the city.

Other scholars explore the relationships between the official city and the everyday lives we lead. "Strategy" refers to spatial ordering by powerful interests, "tactics" to appropriations and transgressions of ordinary citizens. Through tactics, the strategies of power can be undermined and appropriated. Life in the city is conceptualized as reinforcing both the spatial strategies of the official city and the tactical appropriation and resistance of everyday dissent. To understand the city in all its complexities, we need to connect the official and the everyday, the compliance and dissent embodied in the space-time paths we make across the urban built form and the routes we weave across urban social space.

The third element that David Seamon outlined was encounter, the interaction between people in space. Much of city living is spent with people who are similar to us. Elijah Anderson looks at how people share urban space with people different from themselves. His ethnographic research examines street strategies adopted by different racial-ethnic groups in Philadelphia neighborhoods. He identifies what he terms "the **cosmopolitan canopy**," islands of civility in the segregated city. These islands are a space where people of diverse backgrounds come together and interact. In much of the city, people interact with people much like themselves in terms of race and class. The more successful cities have an extensive and variegated cosmopolitan canopy.

Anderson's work is also part of the long tradition of urban ethnography fieldwork that moves through the city with care, recording and noting, making sense of the kaleidoscope of urban experience. This more textured qualitative examination of the city complements the quantitative approach.

The City as Political Arena

The city is a political arena where different groups and stakeholders compete and cooperate. Let us consider the case of households.

Households

Households have diverse interests. As residents, they want a full range of public goods and services, including such **public goods** as clean water, efficient transport systems, and good schools. We can also identify public bads as well as public goods. **Public bads** are all those unwanted facilities such as trash disposal sites, motorways, and halfway houses for convicted criminals. In many cases, the process is referred to as NIMBY—"not in my backyard." Households individually and collectively seek to attract public goods and deflect or relocate public bads. The process reflects the political power and economic strength of residents' organizations. The rich and powerful live in neighborhoods without toxic waste sites.

It is a consistent finding that toxic facilities are predominantly concentrated in lower-income and minority-dominated areas of the city, and major infrastructure projects with negative environmental impacts, such as urban motorways, are more commonly found in poor and minority neighborhoods. Study after study from around the world reveals a correlation between negative environmental impacts and the presence of racial/ethnic minorities. Studies in the United States highlight that race is the most significant variable associated with the location of hazardous waste sites and that the greatest number of hazardous commercial facilities are located in areas with the highest proportion of racial and ethnic minorities. Three out of every five black and Hispanic Americans live in communities with one or more toxic waste sites.

In cities around the world, the poor and the marginal more often than not live in the areas with the worst environmental quality. Social inequalities are expressed and embodied in urban environmental conditions. In Europe, waste facilities are disproportionately located in lower-income areas and in places where ethnic minorities live. Urban environmental inequality is large and pervasive.

Poorer communities have less pleasant urban environments and often bear the brunt of **negative externalities**. It is through their neighborhoods that motorways are constructed, and it is in their neighborhoods that heavy vehicular traffic can cause elevated lead levels in the local soil and water. There is a direct correlation between socioeconomic status and the quality of the urban environment.

Households are also taxpayers, and as such they want to minimize their taxes while maximizing their benefits. Attitudes to taxation vary, with Scandinavian taxpayers more willing to bear a large load than those in the United States. In all cities in all countries, however, there are limits to the acceptable amount of taxation. Proposition 13, which passed in California in 1978, limited the tax based on property values. Prior to the legislation, increasing property values meant increasing taxes. In a period of rapid house price inflation, taxes skyrocketed. Households organized to place the proposition on the ballot. The effects were felt immediately in reduced public spending and public services.

Households are also users of services. They want police and fire protection, libraries, and schools. And they want them to be of high quality. So residents have conflicting goals: they want low taxes and quality services. They will fight against increasing taxes but also resist a decline in services; they will work with local governments to improve services but fight to resist increasing taxes.

18.3 Population Change in US Cities, 1900–2010

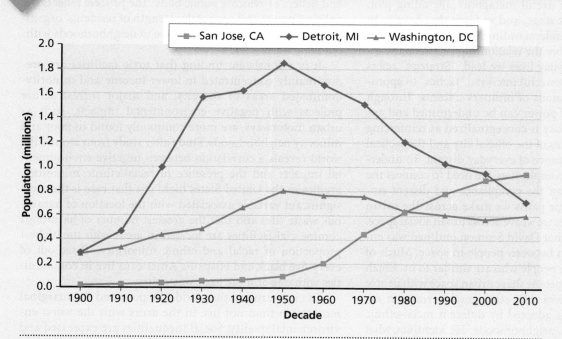

Population change in three US cities

After plotting the individual trajectory of the population size at each successive census of the 100 largest cities in the United States from 1900 to 2010, we identified four model types. The first type of city is *steady decline*. A typical city in this category is Detroit, which experienced a peak of 1.84 million around the mid-twentieth century and then continuous decline; its 2012 population was 701,475 (see first figure). The city embodies the rise and rapid fall of the older, industrial city. Other cities in this category include Akron, Baltimore, Birmingham, Buffalo, Cincinnati, Cleveland, New Orleans, Pittsburgh, Rochester, and Toledo. These are cities that have borne the brunt of an urban fiscal crisis as the steady loss of population and tax base eroded their revenues.

The second model type is *continuous increase*. Here the story is of rising economic and population growth and ease of annexation of surrounding suburban districts. A typical case is San Jose, CA, a Sunbelt city with an expanding economy based largely on information technology. In 1950 the city population was only 95,280, but by 2012 it was 982,765. Decades of spectacular growth, fueled in particular by the

Silicon Valley boom in high technology and computer-related industries, have made San Jose one of the most prosperous and economically dynamic cities in the country. The city was also able to annex territory, increasing its area from 17 square miles in 1950 to 177 in 2010. Other examples of this type include San Diego, Las Vegas, and Orlando.

The third model type is what we term *growth interrupted*. Here examples include New York City, Atlanta, San Francisco, and Seattle. These cities' upward population trajectories saw some decline before returning to growth, and eventually surpassed their previous population peak. The second figure plots the trajectory of New York City, where the population was relatively flat from 1950 to 1970 before a twenty-year decline and then resurgence after 1990. The city's area remained constant at 303 square miles from 1950 to 2010. In this category, some annexations did occur, but, especially for the large cities of Atlanta, New York, San Francisco, and Seattle,

Population change in New York City

population growth within stable boundaries was the most important process.

Finally, there is the *slowly resurgent* city, where previous peaks are not reached but there is a slow and steady return of population. Examples include Boston, Philadelphia, and Washington, DC. In all three cases, the area of the city has remained roughly the same since 1950. Washington, DC's population peaked in 1950, then saw continuous decline until 2000, at which point the city's population began a slow upward resurgence. Cities of this type are big urban areas that did not reach the free fall of continuous decline or the pronounced return to an upward trajectory of the growth-interrupted cities. These cities show signs of population recovery—if not quite to past peaks, at least a bending of the curve from decline to upswing.

REFERENCE

Short, J. R. and Mussman, M. (2014) Population Change in US Cities: Estimating and Explaining the Extent of Decline and Level of Resurgence. *Professional Geographer* 66: 112–123. http://www.tandfonline.com/doi/abs/10.1080/00330124.2013.765297

Exit and Voice

If there is a perception of declining services or increasing taxes, households have a variety of options. The most cited pair of options have been referred to as "exit" and "voice." Choosing the first, households can leave and move to another city or separate municipality. This is more common in the United States, where large metro areas are often balkanized into separate municipalities, often with different tax regimes. People move, if they can, to higher-taxed areas if they believe that the result is better schools and other public services. If households stay, they can choose the second option, to voice their concern, and if they are not heeded they can protest. There are three sources of protest: the call for goods and services, the expression of cultural identity (Figure 18.11), and the demand for political power. Collective action can range from the persuasive (petitions) and the collaborative (lobbying elected officials) to the confrontational (including allies, marches, and in extreme cases forms of civil disobedience).

When residents stage a protest, it represents their failure to influence the political process without direct action.

The more powerful groups do not need to protest, because their interests are reflected in the political process. That is why we rarely see the very rich and the very powerful taking to the streets; they do not have to. It tends to be the poorer and the more marginalized.

Forms of civil disobedience were evident in the Arab Spring that swept through the cities of the Middle East in 2010–2011 and in the Occupy Wall Street movement that occurred in selected cities in Europe and North America the following year (Figure 18.12).

Urban Regimes

Protests and uprisings in cities are rare. Much of the political life of the city consists of the more banal and everyday operation of political agreements, deal making, and compromise. Clarence Stone identifies what he calls "**urban regimes**," which he defines as the "informal arrangements that surround and complement the formal workings of governmental authority." Urban regimes consist of informal governing coalitions that make decisions and get things done in a city. Stone developed the notion of urban regime to refer to the informal partnership in Atlanta between city hall and downtown business elites. Political questions of maintaining and extending political support and leadership dominate City Hall, while economic issues of profit and loss concern the business elites. The combination of political and economic logic, with all the ensuing tensions, conflicts, and ambiguities, constitutes the local urban regime. The options and concerns of urban regimes vary over time and space—they may be inclusionary or exclusionary and will vary throughout the

18.11 Political protest in Barcelona for Catalan autonomy

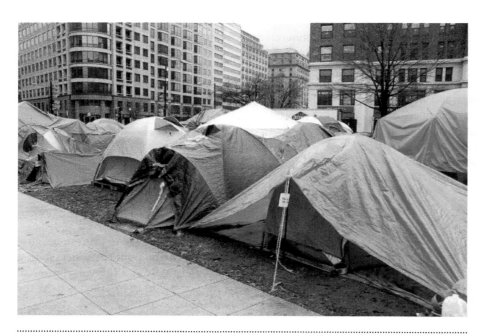

18.12 "Occupy" tents in McPherson Square in Washington, DC. They first went up in October 2011 but were cleared by June 2012.

Gentrification

It was urban scholar Ruth Glass who in 1964 coined the word **"gentrification"** to refer to a certain form of urban change. A strict definition of the term is the replacement of lower-income households by higher-income households. She employed the term to describe how middle- and upper-class residents moved into working-class areas in London. The term is now used to refer to the movement into city neighborhoods of higher-income households, whether through the replacement of lower-income households in the existing housing stock or in the form of movement into custom-designed buildings. The process is now found in cities across the world as the more affluent move into central and inner-city neighborhoods previously neglected or inhabited by the poor. It is promoted by developers and downtown elites eager to generate business, and by city governments eager to lure higher-income households into the city to provide a stronger tax base and a population with higher disposable incomes. Gentrification is often trumpeted as revival, bringing life back into the city. Advocates of poor and lower-income people, in contrast, see gentrification as a form of class warfare over space. Gentrifiers tend to be richer, younger people, while the displaced tend to be older, poorer, and from racial minorities.

In Harlem, home to many poor African Americans in New York City, major construction projects such as Towers on the Park, which opened in 1988, and individual households buying up brownstone dwellings have led to a distinct form of gentrification along the neighborhood's western corridor. Just across the Hudson River from lower Manhattan, a quick ferry ride from New York City's financial district, sits the small, formerly working-class city of Hoboken. The process of gentrification began there around 1980. The housing stock was effectively emptied of lower-income residents in favor of higher-income households, especially those working in the financial services sector in lower Manhattan, turning what was long a working-class town into a yuppie suburb. The increased employment opportunities in San Francisco from the early 1990s generated an extraordinarily tight housing market. The Mission District, close to its downtown, was particularly vulnerable to gentrification, as it had relatively cheap housing stock. This area of working-class families and recent Latino immigrants quickly transformed as housing values, rents, and evictions all increased.

The Olde Towne East district, adjacent to downtown Columbus, Ohio, was like many late nineteenth- and early

metropolitan regions. Suburban regimes, for example, are more concerned with preserving property values.

In the United States, three regime types have been identified: organic, instrumental, and symbolic. **Organic regimes** occur in small towns and suburban districts with a homogeneous population and a strong sense of place; their chief aim is to maintain the status quo. **Instrumental regimes** focus upon specific targets identified in the political partnership between urban governments and business interests. An urban growth machine regime is concerned with securing urban business interests. **Symbolic regimes** occur in cities undergoing rapid changes, including large-scale revitalization, major political change, and image campaigns that try to shift the wider public perception of the city. We should see these three categories as ideal types, with any one city's regime exhibiting characteristics, albeit in different proportions, of each type.

Changes in the Contemporary City

In the rapidly changing cities of the contemporary world, a number of transnational themes can be noted. Among the many, let us consider four: gentrification, suburban decline, **city branding**, and the hosting of global events.

often unable to afford the costs of meeting historic preservation codes. The property cost increases forced out many of the older owners and low-income renters. The recent history of this area encapsulates a number of different cleavages, including race, class, income, and sexual orientation, as two marginalized communities fought a turf war over the price and meaning of the residential space.

Gentrification tends to occur when there is recentralization in central city areas. In the case of Sydney, Australia, the immediate postwar years saw a steady suburbanization. The city extended far into distant suburbs, increasing commuting times. Starting in the 1970s, more affluent younger and gay households began buying up the high-density terraced housing in inner-city neighborhoods such as Paddington. The central location reduced commuting times for those working in the city and provided very cheap housing, and the neighborhoods became safe for openly gay people. Paddington is now one of the more fashionable areas of the city (see Figures 18.13 and 18.14).

Gentrification can also be part of a more fundamental rewriting of urban space associated with more aggressive policing, in what is referred to as the punitive turn. This involves, according to proponents, making the streets safe; for critics, it is an aggressive form of policing directed

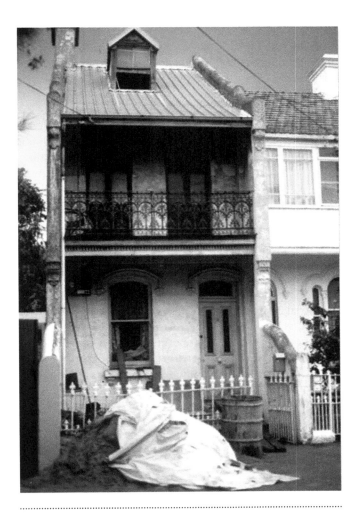

18.13 Gentrification in Paddington: before

twentieth-century residential areas. Built close to the city, it contained both high-income and middle-income residents in a housing stock that contained some large dwellings. Post–Second World War suburbanization drew off the wealthier residents, and then a mixture of urban renewal and highway construction ripped the heart out of the neighborhood. Originally a stable residential area throughout most of the early twentieth century, by the 1960s the dwelling stock was in decline and most of the residents were low-income African Americans living in rented accommodations. However, being so close to the downtown and with some substantial housing remaining, the area soon attracted the attention of gay white men seeking to buy and improve housing. In many cities, gentrification has been associated with the coming out of a gay community. In Olde Towne East, the struggle, crystallized between the gay white newcomers and the longer-term residents—mainly low-income African Americans—took a particular form as the new residents sought to have the area designated as historic. For the longer-term low-income residents, this was a financial penalty, as they were

18.14 Gentrification in Paddington: after

specifically against the poor, minorities, and marginalized people in areas of the city undergoing gentrification. In selected neighborhoods, the gentrified city embodies all these active political moments in a socio-spatial urban transformation that marginalizes the poor and the homeless. Urban policies and market forces combine to promote gentrification, marginalize the poor, and make the homeless disappear from public view.

Suburban Decline

An image persists, especially in the United States, of suburbs as bucolic idylls, different from the grit of the cities. Yet many older postwar suburbs struggle to survive, caught in the stagnant middle between the dynamic processes of urban decentralization, as more distant suburbs are developed, and recentralization, as investment shifts back into the city. The deindustrialization of older suburbs also leaves these places with a smaller economic base for opportunity and mobility. Once prized as the ideal location for families, many older inner suburbs now exhibit symptoms of aging. Those built during the postwar period of mass suburbanization are particularly outdated, as the housing stock now lacks the size and amenities to compete with newer housing on the outer fringe of the metropolitan area or in the redeveloped heart of gentrifying cities. Caught between city gentrification and outer sprawl, many postwar suburbs are currently losing population and the battle for investment resources. Bernadette Hanlon and her colleagues write of the emerging feature of "suburban gothic" as numerous suburbs in crisis appear on the urban scene.

City Branding

The marketing of cities, also referred to as city branding, has a long history. The frontier town, the resort, the suburb, and the industrial city were all marketed with appropriate images to attract investment and people. In the contemporary world, three new forms can be identified.

The first is the marketing of the postindustrial city. In the wake of the deindustrialization of large swathes of urban North America, Australia, and Europe, formerly industrial cities were placed in the unenviable position of losing investment and jobs. The global shift of manufacturing creates new industrial cities in the developing world but leaves behind industrial cities in the developed world. These industrial cities, or, to be more accurate, these newly postindustrial cities, became associated with the old, the polluted, the past, and the failed. A major theme of marketing these postindustrial cities is to distance them from their recent industrial past. In the case of Syracuse, New York, this involved a new logo for the city that replaced images of smoking factories with a postmodern building skyline and a transformation of a local lake from a dump site to a scene of regeneration. Formerly industrial cities are rebranded

in more attractive packages that emphasize the new rather than the old, the fashionably postmodern rather than the merely modern, the postindustrial rather than the industrial, consumption rather than production, spectacle and fun rather than pollution and work.

Then there is Wollongong, an industrial city on the New South Wales coast of Australia. A massive steelworks dominated the urban economy. The steelworks shed 15,000 jobs from the early 1980s to the mid-1990s. The rate of job loss was only one major strand in negative imagery associated with the city in the national imagination and among foreign investors. City leaders decided to rebrand the city in the public imagination. An image campaign was built around the idea of "innovation, creativity, and excellence." An important part of the campaign involved planting half a million trees on the site of the steelworks while the council found funds to clean up the beaches and construct cycle ways. Such "greening" is now an integral part of a city's attempt to shed its hard industrial image for a softer postindustrial feel.

The second form is the marketing of a city as global. Because of the growing competition, there is constant need to upgrade and improve a city's image. At the top of the global hierarchy, as measured in terms of the concentration of advanced producer services, sit London and New York. Both cities actively position themselves as global cities. In 1965 New York hosted the World's Fair, and after a period of fiscal crisis and economic decline the city was rebranded in 1977 with the successful I❤NY campaign. In 2012 London hosted the Summer Olympic Games, presenting itself as a multicultural, cosmopolitan, cool city.

The rhetoric of global city branding is often associated with attempts to justify controversial and costly urban redevelopment projects, tax incentives, and business-friendly economic policies.

Darel Paul explores the politics of "global imagineering" in Montreal, Canada, where business interests triumphed, and Minneapolis–St. Paul, Minnesota, where similar interests were trumped by populist politics. In St. Petersburg, Russia, not only was the old Soviet name of Leningrad dropped, the city was actively promoted as an international hub of circulatory capital, host to corporate power, and stage for globalist megaprojects. St. Petersburg was rebranded as an entrepreneurial, economically competitive, globally connected city.

The third form of marketing is presentation as a business-friendly city. As capital becomes ever more mobile and more cities enter the competition, cities need to aggressively promote themselves as good places to do business. Here are some examples from advertisements of US cities in the business press:

Dallas	*The city of choice for business*
Milwaukee	*The city that works for your business*
New York	*The business city that never sleeps*
Phoenix	*Moving business in the right direction*

More specific themes around this general message include emphasizing a pro-business political climate, a skilled labor force, good infrastructure, and accessible location.

There is a growing body of critical work that highlights the silences of city marketing. The emphases of most marketing campaigns are on the city as a place of business, less on social inclusivity; more on the profitable city, less on the fair and just city. The dominant themes represent the power of the dominant groups. The more marginal are excluded. In an interesting twist on this finding, however, Jaime Hernandez and Celia Lopez suggest how slums can be used to positively brand a city, in this case Bogotá. They suggest that highlighting rather than ignoring the city's slums can invoke positive responses around ideas of genuine difference, cultural authenticity, and vernacular architecture. The poor and the marginal become less a political threat and more a marketing opportunity. In this case, the ethos of marketing is so pervasive that even the silences are uttered and the problems exposed, not as issues to be solved or addressed but as opportunities to sell and rebrand.

Signature Buildings and Global Events

Cities seeking a global reputation often employ big-name architects to design buildings and campaign to host global events.

The marketing of cities can involve the construction of buildings by prominent architects, so-called starchitects, showing the centrality of "**starchitecture**" to the symbolic capital of cities. To have a new building by a starchitect denotes cultural heft and global intentionality. Individual buildings and assemblages of certain built forms—museums, art galleries, airports, and the like—built in the latest style by the hottest architects is an important way for a city to position itself. Perhaps the best-known and most-cited case study is Bilbao in northern Spain. It was a major banking and industrial center that from the 1970s experienced deindustrialization and massive job loss. The old industrial area along the river was a site of abandoned factories. The city authorities began an ambitious urban renewal policy. A centerpiece was a branch of the Guggenheim Museum designed by the famous American architect Frank Gehry. The museum, built along the river, is now an iconic image of the city and a classic case study of having a starchitect design a major cultural center to successfully rebrand a city (Figure 18.15).

Hosting events is now a significant part of marketing a city. So-called spotlight events highlight the city for a wider audience: the larger the event, the greater the potential marketing opportunity. The hosting of events such as film festivals, expos, and Olympic Games provides opportunities for a city's elites to stage a widely accessible and globally pervasive marketing campaign. In a detailed case study, Claire Colomb explores the marketing of Berlin between 1989 and 2011. She argues that the city was marketed through rebuilding and the staging of a new Berlin. In the inner city, sites were turned into tourist attractions. Place marketing and urban reinvention were intertwined in a symbolic politics of representation.

One of the most important mega-events is the Summer Olympic Games. From humble beginnings in Athens in 1896, when only 200 athletes from fourteen countries competed, with limited press coverage, the modern Olympic Games have now grown to a truly global spectacle involving most countries in the world (Table 18.3).

Hosting the Games involves providing venues and improving urban infrastructure. Since Barcelona in 1992 and especially since Sydney in 2000, emphasis has also shifted to using the Games as way to undertake urban renewal of industrial and abandoned sites (Figure 18.16). For the 1992 Games, Barcelona built a new waterfront and upgraded a declining area of the city, as well as making numerous improvements throughout the metro area, including new roads, a new sewer system, and the creation or improvement of over 200 parks, plazas, and streets. For the 2000 Games, Sydney built a new road linking the airport to the downtown and constructed the main venue on a previously contaminated inner-city

18.15 Guggenheim Museum in Bilbao

TABLE 18.3	The Modern Summer Olympic Games		
DATE	**HOST CITY**	**PARTICIPANTS (WOMEN)**	**COUNTRIES/NATIONAL COMMITTEES**
1896	Athens	200 (0)	14
1900	Paris	1,205 (19)	26
1904	St. Louis	687 (6)	13
1908	London	2,035 (36)	22
1912	Stockholm	2,547 (57)	28
1920	Antwerp	2,668 (77)	29
1924	Paris	3,092 (136)	44
1928	Amsterdam	3,014 (290)	46
1932	Los Angeles	1,408 (127)	37
1936	Berlin	4,066 (328)	49
1948	London	4,099 (385)	59
1952	Helsinki	4,925 (518)	69
1956	Melbourne	3,184 (371)	67
1960	Rome	5,346 (610)	83
1964	Tokyo	5,140 (683)	93
1968	Mexico City	5,530 (781)	112
1972	Munich	7,123 (1,058)	121
1976	Montreal	6,028 (1,247)	92
1980	Moscow	5,217 (1,124)	80
1984	Los Angeles	5,330 (1,567)	140
1988	Seoul	8,465 (2,186)	159
1992	Barcelona	9,634 (2,707)	169
1996	Atlanta	10,310 (3,513)	197
2000	Sydney	10,651 (4,069)	199
2004	Athens	10,625 (4,329)	201
2008	Beijing	11,028 (4,746)	204
2012	London	10,960 (4,676)	204
2016	Rio	11,303 (5,086)	206

site. Beijing undertook a building frenzy with an estimated $40 billion of Olympic-related buildings and infrastructure, including a new expressway and ring roads, miles of rail and subway tracks, and a new, $2.2 billion airport that is the largest in the world. For London 2012, much of the construction of the new sports venues were part of the regeneration of the lower Lea Valley, an area of postindustrial decline.

The Olympics also provide an opportunity to make a modern city. Athens spent close to $16 billion in large-scale

18.16 Olympic ring fountain in downtown Atlanta, built for the 1996 Games

coherent planning. But modernity can come at a cost. The changes often come with increased costs for city residents unless there is a specific commitment to redistributional outcomes. The Chinese government used the 2008 Games as an opportunity to modernize Beijing. One plan involved the destruction of the old, high-density neighborhoods of small alleyways in the central part of the city, seen by officials as a remnant of a premodern past. Almost 20 square kilometers were destroyed and almost 580,000 people displaced in this one program.

Hosting the Games provides an opportunity for massive urban renewal, major environmental remediation, dramatic infrastructural improvements, and the creation of a positive image across the globe.

public investments in water supply, mass transit, and airport connections to get ready for the 2004 Games. Hosting the Olympics provided the opportunity for citywide, However, there is rising criticism of the costs of the Games, the impact on marginal groups, and the huge costs compared to the less-than-predicted benefits.

Cited References

Anderson, E. (1990) *Streetwise: Race, Class, and Change in an Urban Community*. Chicago: University of Chicago Press.

Anderson, E. (2012) *The Cosmopolitan Canopy: Race and Civility in Everyday Life*. New York: W. W. Norton.

Boal, F. (2002) Belfast: Walls Within. *Political Geography* 21: 687–694.

Colomb, C. (2011) *Staging the New Berlin*. London: Routledge.

Goffman, E. (1959) *The Presentation of Self in Everyday Life*. Garden City, NY: Doubleday.

Golubchikov, O. (2010) World-City-Entrepreneurialism: Globalist Imaginaries, Neoliberal Geographies, and the Production of New St. Petersburg. *Environment and Planning A* 42: 626–643.

Hägerstrand, T. (1970) What About People in Regional Science? *Papers of the Regional Science Association* 24: 7–21.

Hanlon, B., Short, J. R., and Vicino, T. J. (2010) *Cities and Suburbs*. New York: Routledge.

Harris, J. (2016, January 4) The End of Council Housing. The Guardian, https://www.theguardian.com/society/2016/jan/04/end-of-council-housing-bill-secure-tenancies-pay-to-stay.

Hernandez, J. and Lopez, C. (2011) Is There a Role for Informal Settlements in Branding Cities? *Journal of Place Management and Development* 4: 93–109.

Knox, P. (2012) Starchitects, Starchitecture and the Symbolic Capital of World Cities. In Derudder, B., Hoyler, M., Taylor, P. J., and Witlox, F. (eds), *International Handbook of Globalization and World Cities*. Northampton, MA: Edward Elgar, pp. 275–283.

Meyer, M. (2008) *The Last Days of Old Beijing*. New York: Walker.

Murdie, R. A. (1969) Factorial Ecology of Metropolitan Toronto. Research Paper 116, Department of Geography, University of Chicago, Chicago.

Paul, D. E. (2004) World Cities as Hegemonic Projects: The Politics of Global Imagineering in Montreal. *Political Geography* 23: 571–596.

Seamon, D. (1979) *A Geography of the Lifeworld*. London: Croom Helm.

Short, J. R. (2008) Globalization, Cities and the Summer Olympics. *City* 12: 321–340.

Stone, C. N. (1989) *Regime Politics: Governing Atlanta, 1946–1988*. Lawrence: University of Kansas.

Stone, C. N. (2008) Urban Regimes and the Capacity to Govern: A Political Economy Approach. *Journal of Urban Affairs* 15: 1–28.

Select Guide to Further Reading

Barras, R. (2009) *Building Cycles: Growth and Instability.* Oxford: Wiley-Blackwell.

Bieri, D., Knox, P. L., and Wei, F. (2012) Changing Social Ecologies of U.S. Suburban Areas, 1960–2000. http://www.david-bieri.com/docs/SocialEcologies_BKW_Jun12.pdf. Accessed December 12, 2012.

Bridge, G. and Watson, S. (eds) (2011) *The New Blackwell Companion to the City.* Chichester: Wiley-Blackwell.

Calame, J. and Charlesworth, E. R. (2009) *Divided Cities: Beirut, Belfast, Jerusalem, Mostar, and Nicosia.* Philadelphia: University of Pennsylvania Press.

Castells, M. (1989) *The City and Grassroots.* London: Edward Arnold.

De Certeau, M. (1984) *The Practice of Everyday Life.* Berkeley and Los Angeles: University of California Press.

Gold, J. and Gold, M. (2012) *The Making of Modern Olympic Cities: Critical Concepts in Urban Studies.* New York: Routledge.

Gottlieb, M. (1976) *Long Swings in Urban Development.* New York: National Bureau of Economic Research.

Hall, T. and Barrett, H. (2012) *Urban Geography.* 4th ed. Oxford: Routledge.

Hall, T., Hubbard, P., and Short, J. R. (eds) (2008) *The Sage Companion to the City.* London: Sage.

Harvey, D. (2012) *Rebel Cities: From the Right to the City to the Urban Revolution.* London: Verso.

Immergluck, D. (2011) The Local Wreckage of Global Capital: The Subprime Crisis, Federal Policy and High-Foreclosure Neighborhoods in the US. *International Journal of Urban and Regional Research* 35: 130–146.

Knox, P. (ed) (2014) *Atlas of Cities.* Princeton, NJ: Princeton University Press.

Kramer, K. and Short, J. R. (2011) Flânerie and the Globalizing City. *City* 15: 322–342.

Logan, J. and Molotch, H. (2007) *Urban Fortunes: The Political Economy of Place.* Berkeley and Los Angeles: University of California Press. First published 1987.

Martuzzi, M., Mitis, F., and Forastiere, F. (2010) Inequalities, Inequities, Environmental Justice in Waste Management and Health. *European Journal of Public Health* 20: 21–26.

Mikelbank, B. A. (2011) Neighborhood Déjà Vu: Classification in Metropolitan Cleveland, 1970–2000. *Urban Geography* 32: 317–333.

Mohai, P. and Saha, R. (2007) Racial Inequality in the Distribution of Hazardous Waste: A National-Level Reassessment. *Social Problems* 54: 343–370.

Parker, S. (2011) *Cities, Politics and Power.* Oxford: Routledge.

Short, J. R. (ed) (2017) *A Research Agenda for Cities.* Cheltenham: Edward Elgar.

Short, J. R. and Kim, Y.-H. (1998) Urban Crises/Urban Representations: Selling the City in Difficult Times. In Hubbard, P. and Hall, T. (eds), *The Entrepreneurial City.* Chichester: Wiley, pp. 55–75.

United Church of Christ Commission for Racial Justice. (1987) *Toxic Wastes and Race in the United States: A National Report on the Racial and Socio-economic Characteristics of Communities with Hazardous Waste Sites.* New York.

Vicino, T. J., Hanlon, B., and Short, J. J. (2011) A Typology of Urban Immigrant Neighborhoods. *Urban Geography* 32: 383–405.

On gentrification:

Duman, A. (2012) Dispatches from "The Frontline of Gentrification." *City* 16: 672–685.

Lees, L., Slater, T., and Wyly, E. (2010) *The Gentrification Reader.* New York: Routledge.

Short, J. R. (2006) *Alabaster Cities.* Syracuse, NY: Syracuse University Press.

Wyly, E., Newman, K., Schafran, A., and Lee, E. (2010) Displacing New York. *Environment and Planning A* 42: 2602–2623.

On suburban decline:

Hanlon, B. (2009) *Once the American Dream: Inner Ring Suburbs of the Metropolitan United States.* Philadelphia: Temple University Press.

Short, J. R., Hanlon, B. and Vicino, T. (2007) The Decline of Inner Suburbs: The New Suburban Gothic in the United States. *Geography Compass* 1: 641–656.

On city branding:

Dinnie, K. (ed) (2011) *City Branding: Theory and Cases.* New York: Palgrave.

Short, J. R., Benton, L. M., Luce, W. B., and Walton, J. (1993) Reconstructing the Image of an Industrial City. *Annals of Association of American Geographers* 83: 207–224.

Selected Websites

Cities around the world
http://www.fourmilab.ch/earthview/cities.html
Iconic houses
http://www.iconichouses.org/

Resilient cities
https://www.rockefellerfoundation.org/our-work/initiatives/100-resilient-cities/

Interactive Map Websites

Housing markets
 http://www.jchs.harvard.edu/research/interactive-maps
 http://www.nytimes.com/interactive/2008/05/28/business/
 20071031_HOUSING_GRAPHIC.html
 https://www.washingtonpost.com/graphics/business/wonk/
 housing/overview/
 http://www.zillow.com
Population
 http://www.census.gov/2010census/popmap/
 http://usa.zoom-maps.com

Segregation
 http://www.nytimes.com/interactive/2015/07/08/us/census-
 race-map.html
US cities
 http://www.nationsonline.org/oneworld/most_pop_cities_
 usa.htm

Learning Outcomes

- The four specific themes of the internal structure of a city are the city as an investment, as a residence, as social context, and as a political arena.
- As a site of investment, capital flows into the built environment when relative rates of return are high and experience a distinct cycle of troughs and peaks, known as the building cycle.
- Building cycles often align with transport improvements and changes in architectural style.
- The supply of housing can be broken down into public housing, private renting, and owner occupation.
- The demand for housing varies by income and wealth, the stages of the life cycle, and other cultural factors that lead to clustering and segregation.
- There are various models of the residential mosaic from the Burgess and Murdie models to the Hanlon-Short-Vicino model.
- The Short Model of City Structure in the Global South highlights some of the distinctive features of the city in the developing world.

- The social context of cities can be viewed through various processes based on how we move, how and where we rest, and how we interact with others.
- Cities are political arenas where different groups and stakeholders compete and cooperate.
- In the United States, three types of informal urban governing regimes have been identified: organic, instrumental, and symbolic.
- Contemporary urban change includes gentrification, suburban decline, and city branding.
- Gentrification tends to occur when there is recentralization in central city areas.
- There is also decline in many postwar suburbs caught between city gentrification and outer sprawl.
- Cities are marketed and branded. Among the many types are the marketing of the postindustrial city, the marketing of a global city, and the presentation of a business-friendly city.

Glossary

Abrahamic religions   Common designation for the religions of Judaism, Christianity, and Islam, which share Abraham as a central figure.

acid rain   Rain that is contaminated by nitric and sulfuric acid. Caused by volcanic eruption and burning fossil fuels.

adopted ethnicity   An expressed and chosen ethnic identity.

advanced producer services   Includes business consultancies and accounting, advertising, financial, and legal services used by large firms and corporations.

affect   In contemporary geographical discussions, refers to forces that make us feel, think, or act.

age of distraction   The current period of time marked by people becoming increasingly distracted by communication via mobile devices.

age of imperialism   The period from 1880 to 1918 in which there was a scramble for territory as old and new European powers struggled to incorporate more of the periphery. Africa was important in this race; hence the name of one of the main processes, the scramble for Africa.

agrarian transition   Occurs as agriculture shifts from subsistence to commercial and moves from meeting local market demands to provisioning national and global food supply chains.

agricultural productivity   The amount of food that can be grown given a set of limited resources such as arable land and labor.

anarcho-communism   Theory of societal organization that favors eradicating government in favor of communal ownership of society's resources.

animism (animist religions)   The belief that the Earth is a living entity filled with sacred space and spirits. It is prominent among hunter-gatherers.

Anthropocene   The current geological period in which human activity is deemed responsible for the profound and fundamental restructuring of the environment.

Arab Spring   The independent democratic uprisings across the Arab world in 2011.

arable agriculture   The practice of growing crops in soils.

Asian Tigers   The first batch of newly industrializing counties in East Asia, after Japan. Generally taken to include Hong Kong, Singapore, South Korea, and Taiwan. Also describes those countries that developed later, including Malaysia and Thailand.

assimilationist model   Robert Park's description of how new immigrants are socialized to a new society and in turn provide the dynamic raw material for the creation of a new urban society.

asteroids   Rocky objects that orbit the Sun and occasionally hit the Earth. An asteroid strike is probably responsible for a mass extinction during the Mesozoic era, as its impact triggered volcanoes and earthquakes.

atlas   A collection of printed maps.

authoritarian regimes   A system of government in which power is concentrated amongst a few, but not so deeply entrenched as in a totalitarian regime.

axial age   A period that lasted from 900 to 200 BCE, during which the great religious traditions of the world came into being across the globe—Confucianism and Daoism in China, Hinduism and Buddhism in India, monotheism in Israel, and philosophical rationalism in Greece.

Baumol's disease   Named after economist William Baumol, who first noted that while innovation and efficiency occur rapidly in the manufacturing sector, leading to price reduction, this trend is less pronounced in the service sector.

Bennett's Law   The phenomenon of how consumption of starch staples declines as household income increases.

Bible Belt   A region encompassing all or parts of the southeastern and south central USA marked by greater churchgoing and formal religious belief than the rest of the country. Protestant and Evangelical religions are particularly strong in the region.

Big Bang theory   Describes the origin of the universe as emanating from an explosion of energy that sent matter expanding outward.

bilateral oligopolies   A food supply chain in which large-scale producers combine with large-scale retailers to supply food goods.

biodiversity   An indicator that describes and measures the number of distinct plant and animal species in a given geographical area.

biofuels   Fuels that are made from living matter, such as ethanol and biodiesel.

Black Death   A recurrence of the bubonic plague in the fourteenth century that killed up to 60 percent of the population of Europe.

black holes   Cities with a population of over 3 million that are not identified by GAWC as a world city, and do not share a national territory with a world city.

body mass index (BMI)   A measurement defined by a person's weight in kilograms divided by the square of their height in meters. Useful in comparing levels of obesity, defined as having a BMI over 30, across the world and over time.

BRICS   An acronym for the countries Brazil, Russia, India, China, and South Africa that are grouped together because they are all large countries with developing economies.

bubonic plague   A pandemic that spread the disease in the sixth, seventh, and eighth centuries across modern-day Europe and the Middle East, killing more than 25 million people.

budding   Migration that involves small subgroups breaking away to settle new places.

building cycle   Cyclical explanation of how cities are developed and grow, as they go through troughs and peaks of investment.

Burgess model   Depicts a concentric ring of the residential mosaic, with the poor at the center and the rich on the periphery. Named after Chicago sociologist E. W. Burgess.

buyer-driven chain   A food supply chain comprising a small number of retailers who consolidate their supply around a few suppliers.

bystander effect   The phenomenon of individuals not responding to help those in distress.

capital   In standard economics and finance, describes real assets such as factories, property, and machinery, or financial assets such as bonds, securities, and stocks. However, human geography seeks to examine the social relationship behind capital.

carbon dioxide   A gas produced in the burning of carbon. One of the main greenhouse gases.

carbon economy   Since the nineteenth century, economic growth and much of our transportation has been fueled by the growing use of carbon-based energy resources, especially coal and oil.

carbon footprint   The amount of carbon dioxide produced by human activities.

carrying capacity   A theory that the world can support only a specific level of human population as well as sustain a limited amount of its growth.

caste system   The social hierarchy in Hindu culture that divides the population into different social categories known as castes.

casual racism   Often found in early geographical works, reflecting the belief of the time that certain races were viewed as inferior to those more dominant in the world.

central place theory   A theory attempting to describe the organization of regional networks in which market towns are distributed in a hierarchy, with higher-order goods and services located in the larger towns.

centrifugal forces   The influences that disrupt a state's power across space, including political and economic inequality.

centripetal forces   The influences that unify a state's power across space, including through external aggression, which may stimulate national bonding against a common enemy; federal structures, which allow the safe expression of regional and other subnational differences; and national mass media and education, which create a shared culture and language.

chain migration   Migration in which people tend to move along channels and to destinations followed by friends and families.

Chicago School   A body of work in the field of urban sociology created during the first half of the twentieth century and centered upon the University of Chicago.

city branding   The marketing of cities with distinct messages such as postindustrial cities, global cities, business-friendly cities, or other forms of branding in an attempt to shift a city's previous image.

climate change   Significant long-term change in regional and/or global climates.

code switching   How people can change the language they speak according to where they are, whom they are with, and the topic of conversation.

cold war   A form of hostility between countries that is marked by threats, military posturing, and other acts short of actual warfare. More specifically, the Cold War was the conflict between the USA and the USSR between 1945 and 1989.

Columbian Encounter   The arrival of Europeans to the Americas, resulting in vast demographic changes to indigenous populations, environmental transformation, and an exchange of goods between the Old and New World.

commodification   Describes how material becomes a resource by being valued and traded, assigned a price equivalent, and entered into the arena of things bought and sold.

commodified agriculture   The use of agriculture as a means of profitability rather than only a source of local sustenance.

commodity cartels   Organizations of primary producers who band together to ensure specific price levels and market shares for their resource.

communal economy   Sector of the economy that involves the cashless exchange of goods and services.

comparative advantage   The ability of a region or country to produce goods more efficiently than others.

conic projection map   A map created by charting the Earth on a cone, then flattening it with lines of latitude as concentric circles. Ptolemy was the first to implement this map-making technique.

contact language   A language that develops as a form of communication between speakers of two or more other languages.

core of believers   Most faiths have a hard core of true believers and a wider periphery of occasional practitioners and more casual believers.

core-periphery model   A dynamic economic model that classifies countries as either core, periphery, or semi-periphery by the economic interactions between them.

corporate industrialized food production system   A system of industrial food production dominated by large corporations.

cosmography   Study of the Earth through the understanding of the interconnected cosmos, incorporating other disciplines such as astronomy, astrology, and even magic.

cosmopolitan canopy   Space in the city where people of different types can meet in a relaxed manner.

creole   A pidgin that has native speakers.

crime maps   Maps of local areas used to depict data on crime statistics, including the spacing and timing of criminal activity.

crisis mapping   The mapping of humanitarian crises such as environmental disasters and wars.

cultural geography   Looks at the spatial expression of culture. It is, for example, concerned with landscape as a socially contested, materially produced cultural artifact.

cultural globalization   The flow of ideas and cultural practices, such as speaking and writing English, around the world.

cultural industry   Also known as the cultural products sector or cultural creative sector. It is made up of the artistic industries of music, dance, theater, literature, the visual arts, crafts, and many newer forms of practice such as video art, performance art, and computer and multimedia art.

cultural region   A region sharing a similar cultural feature such as language, ideas, religion, or cultural practices. Often divided into a core and a periphery.

culture   The collective materials, beliefs, ideas, and practices represented by a particular people group, or in a particular location, but can also be globally representative.

cumulative causation   Describes how growth in a region can spread out into other lagging regions through the diffusion of innovations and by providing markets, which leads growth to feed on itself.

dark energy   The mysterious energy force that accelerates the expansion of the universe, overcoming the effects of gravity.

decolonialization   The process by which former colonial territories achieve independence.

deconstructivism   A postmodernist style of architecture that involves the use of nonrectilinear surfaces.

deep ecology   A belief that humans should alter their lifestyles to be more biocentric and should question the morality of economic growth.

deep state   State organizations and interests that persist despite changes in governments.

deepwater wells   Offshore oil wells that extract oil from 5,000 feet or more below the surface.

deindustrialization   A marked decline in the size and importance of industry in an economy. Often caused by manufacturing job losses.

democracy   A system of government in which political power arises from the majority will of the people.

democracy index   An index produced by the Economist Intelligence Unit to gauge the level of democracy. It uses sixty indicators related to electoral process, civil liberties, functioning of government, political participation, and political culture.

demographic cumulative causation   A process whereby population movement from a declining to a growing region exacerbates these regional differences.

demographic dividend   A relative and absolute increase of younger, more productive workers. Occurs when birth rates fall to a point that requires less investment in the very young, but before more investment in the elderly is required.

demographic transition   A global change in mortality and fertility that occurred around 1800 and began the shift away from higher to lower birth rates and from higher to lower death rates.

dependency ratio   A measure of the number of individuals below 14 and over 65 when compared to the total working aged population.

dependent population   The ratio of retired persons to working population, calculated by dividing the number of people over 65 by those aged 15 to 65, and multiplying the result by 100.

deterritorialization   When referring to culture, describes how cultures can be removed from pertaining only to specific locations.

development state   Describes countries that have successfully crafted their economic policies to move from the periphery of the global economy closer to the core.

dialect maps   Maps that show how the same language varies by dialect.

diaspora (diasporic communities)   Any group of people living outside of their homeland.

digiphrenia   Created by social media, the phenomenon of being in multiple places and more than one self all at once.

disease mapping   The mapping of the incidence, causes, and consequences of disease and ill health.

disglossia   Occurs when two varieties of the same language are used in different contexts.

distance-decay effect   The negative relationship between distance and similarity. Similarity between things tends to decline as distance increases. For example, linguistic similarity declines between peoples as distance between them increases, and early imperial expansion was marked by the decay of influence of effective military outreach and political control.

domestic economy   Sector of the economy that constitutes the amount, type, and division of labor within the home.

double exposure   Harmful impacts on a geographical area caused by both climate change and economic restructuring.

Dutch Disease   When countries rely on nature's bounty to the point where it hampers and restricts the innovation necessary for long-term economic growth.

earthquake   A violent shaking of the ground caused by shifts in the Earth's crust or volcanic eruptions.

Easterlin hypothesis   Named after the economist Richard Easterlin, who suggested that the fortunes of a cohort depend on its size relative to the total population.

Easterlin Paradox   A theory suggesting that once basic needs are met, there is little difference in reported happiness with increasing incomes.

easy oil   The rapid and cheap extraction of oil.

ecofeminism   Highlights the gendered nature of many environmental ideologies and practices, and tries to engage feminist and ecological issues in a shared dialogue.

ecology   A study of the relationships of organisms to each other and their environments.

economic development   The improvement in an economy leading to more and better jobs and higher standards of living.

economic union   An agreement between countries that attempts to extend the size of their economic market.

ecosystem   A group of interconnected organisms that function as a unit.

ecstatic worship   A type of religious activity that involves festive events or ceremonies filled with music and dance.

Enclosure Movement   The period from 1760 to 1820 when much of England's land was privatized, replacing a system of traditional rights and obligations with agrarian capitalism.

environmental change   Change in the type and quality of the environment. May result from natural processes such as volcanic eruptions or human behaviors such as the introduction of farming, burning fossil fuels, or greater car use.

environmental determinism   The idea that certain climatic conditions are "especially favorable to human progress," first argued by a Yale geography professor in the early twentieth century. Often cited as evidence for why certain races and cultures were less advanced than others.

Environmental Kuznets Curve   A graphical representation depicting the relationship between economic growth and environmental quality.

environmental legislation   Legislation enacted to protect the environment and safeguard environmental standards.

environmental movement   A social movement concerned with protecting and conserving the environmental quality and diversity.

environmental racism   The practices and processes whereby minorities live and work in the poorest environmental conditions, especially low-income minority groups.

environmental stressors   Changes that put pressure on the livability and sustainability of the environment.

epistemological break   Epistemology is the theory of knowledge; an epistemological break implies a fundamental reordering of what constitutes knowledge and how knowledge is understood and used.

eutrophication   The enrichment of ecosystems with chemical nutrients, which creates algal blooms that result in dead zones.

evolution   The process by which living organisms are thought to have developed and diversified from earlier forms of living organisms.

faith communities   A group of people who share the same religious beliefs and principles.

famine   Widespread extreme hunger caused by food shortage.

female participation rate   Measures the amount of women, usually those aged 15 to 64, employed in the labor force.

feminist critique   An argument that drew attention to the bias in academia and the dominance of men in geography.

fertility   The birthrate of a population.

first past the post   An electoral system in which the politician with the majority of votes in a district wins that district.

First Urban Revolution   A period of time between 11,000 and 5,000 years ago that was marked by the development of the world's first cities.

first wave of globalization   A distinct period of globalization that followed the Columbian Encounter of 1492 and inaugurated a truly global economy.

fiscal crisis   Occurs when a state has more expenditures than revenue and lacks the means to reverse the trend.

fixed market system   A system in which permanent marketplaces are established to replace period markets.

"flat" world   A condition of relatively reduced economic distance and lesser friction between different national economies.

flight capital   The movement of money to a location that is considered safer and less transparent.

food desert   Parts of a city or region with very limited access to affordable, nutritious food.

food security   When people have relatively easy access to safe and nutritious food. When they don't, either on a permanent or transitory basis, food insecurity occurs, leading to hunger and undernourishment and making people more susceptible to disease and illness.

food supply chain   A system of connected suppliers that distribute food throughout the world.

forced migrations   Movements of people forced to relocate, often caused by war, political conflict, or social unrest.

Fordism   A system of production, generally controlled by a small number of very large companies operating under oligopolistic conditions, which employs mechanized and repetitive manufacturing processes.

formal economy   Sector of the economy that involves the sale of labor in the marketplace and is recorded in government and official statistics.

fracking   Short for hydraulic fracturing, a process for removing natural gas in which high pressure liquid is injected into the ground to crack rocks and release gases and oil.

functioning core   In a theory describing the Pentagon's view of the world, describes the group of nations marked by relative stability, rule of law, and integration into the global community.

gateway cities   A city through which international migration channels are routed, such as Amsterdam, New York, or Singapore.

gender inequalities   The unequal treatment of individuals based on gender.

gendering of space and place   The processes by which gender differences are created and reinforced in spatial arrangements and in the character of places.

genetic variation   The diversity of the genetic materials found in DNA.

genetically modified organism (GMO)   When the genes of plants raised for crops are engineered to make them more resistant to disease and predators in order to produce higher yields.

gentrification   The replacement of lower-income households by higher-income households.

geographic profiling   Using crime maps to determine the probable location of criminals and future criminal activity, based on the assumptions that criminal activity is place-specific and that criminals like to use known areas that are not too close to their home location. The technique has been expanded to analyze invasive species, disease outbreaks, and even to identify well-known but anonymous artists.

geographical imagination   A way of thinking about the world that focuses on space, place, location, and their interaction.

Geographic Information Systems (GIS data)   A computer storage and retrieval system for displaying and analyzing geographically coded data.

gerrymandering   The manipulation of voting district boundaries to engineer specific political outcomes.

Gini index   A measure of income inequality named after the Italian statistician Corrado Gini, who proposed the measure in a 1912 paper. It is generally expressed from 1 to 100, with 1 representing maximum equality and 100 maximum inequality.

global climate change   Long-term worldwide climate change.

global epistemic community   The transnational network of knowledge-based workers.

global languages   Languages used across the world.

global shift   The movement of employment from the industrial regions and cities of the developed world to the cities and regions of the developing world.

global warming   An increase in the Earth's surface or atmospheric temperatures, often associated with an increase in atmospheric greenhouse gases.

Globalization and World Cities (GAWC)   A research network that compiles data on the distribution of global advanced producer service firms across 315 cities to identify a global urban hierarchy and network.

glocalization   A business strategy to create worldwide operations attuned to local markets and conditions.

Great Migration   The movement of US African Americans between 1916 and 1970 from the rural south for the urban industrial north.

Green Revolution A series of innovations and transfers, especially marked from the 1960s to the 1970s, that increased agricultural production through new high-yielding varieties of crops (especially maize, rice, and wheat) and liberal helpings of pesticides and fertilizers.

greenhouse gas A gas emitted into the atmosphere that produces a greenhouse effect by trapping heat in the atmosphere.

gross national happiness A measure of national collective happiness.

gross national income A way to measure a country's economy, frequently used by the World Bank.

ground truthing The identification of finer-grained processes in order to see the effects and impact on households and their vulnerabilities and coping strategies.

hard border A country's nonporous boundary.

hard power The ability to coerce others through the superiority of military forces.

heartland theory Sir Halford Mackinder's theory that whoever ruled the central part of the Eurasian landmass ruled the world.

hegemony Dominance or leadership of one or more state by another either directly or indirectly through cultural, economic, or political control.

highway to India Sea route that reduced travel times between Western Europe and East Asia by linking the Mediterranean and Red Seas through the Suez Canal.

homeownership When households purchase residential property to own.

hominids Primates, including humans, which first appeared on Earth around 50 million years ago.

homogenization thesis A theory about consumer activity that assumes the same things are consumed in the same ways in very different parts of the world.

Hotelling's Law Refers to the fact that is it often rational for producers to make goods similar to their competitors. When applied to location, it leads firms to cluster together.

Hubbert Curve Named after a geophysicist who employed a theory to predict the decline of US oil production, suggesting that nonrenewable resource use rises quickly from zero to peak production levels, then tapers off forming a bell-shaped curve.

human geography A branch of geography dealing with the interaction between human activity and the spaces around them.

humanist critique A critique that developed in the 1960s, arguing that there was a lack of concern with personal geographies that ignored feeling, perception, and intersubjectivity.

hunting-gathering societies Groups of people more dependent on hunting fauna and gathering seeds and berries than on settled agriculture.

hybrid identities A state of multiple identities. Often occurs in immigrant communities, as when Turkish immigrants to Germany have children, and their children are German citizens but are brought up in a Turkish home—their hybrid identity is German-Turkish.

hyperprimate An urban system marked by extreme primacy (i.e., the dominance of one city).

illegal economy Sector of the informal economy that includes the shadowy world of the semi-legal and illicit.

immigration regimes The set of rules that governments implement to govern the entry, settlement, and assimilation of foreign migrants.

imperial incorporation The annexation of areas and peoples into an empire.

imperial overstretch The tendency for empires to expand beyond their ability to maintain economic dominance and military power.

imperialism A doctrine or policy that, to improve a home country's economy, argued for protection of home industries and the possession of overseas colonies to ensure cheap raw materials and a secure market to which competitors' access would be denied.

import substitution An economic development strategy for bolstering local growth marked by protectionist policies.

imposed ethnicity When a group of people are named and treated as ethnically different.

index of dissimilarity Measures the unevenness in the distribution of two groups across spatial units. Ranges from 0 to 100; values closer to 100 indicate greater dissimilarity.

index of segregation Measures the distribution of one group compared to the total population. Ranges from 0 to 100; values closer to 100 indicate a more segregated distribution.

Indic region One of the main hearths of the axial age, where Jainism, Buddhism, and Sikhism were founded. Centered in northern India's Indus Valley.

industrial city Cities developed from industrial capitalism and factory growth.

Industrial Revolution The development of new mechanized manufacturing processes, occurring first in Britain between 1760 and 1850 and also involving the creation of industrial and urban societies.

industrialized agricultural system Agriculture that relies heavily on machinery, technology, and chemicals.

infant mortality rate A measurement of the number of deaths of children under one year of age in a specific geographic location.

informal economy Sector of the economy that is not recorded in government and official statistics, where few, if any, taxes are paid.

institutional contract When an institution or a government entity builds dwellings and then rents them out or sells them.

instrumental regimes Focus upon specific targets identified in the political partnership between urban governments and business interests.

intergenerational inequities The differences in wealth, power, status, etc., that can exist between different generations of a society.

intervening opportunities The opportunities between an origin and possible destination that can disrupt the flow of movement between them.

isothermal map A map composed of lines connecting points of similar temperature.

just-in-time production A system of producing goods in which they are made to meet demand.

Keeling Curve A graphical representation of the changes in the concentration of atmospheric gases since 1958.

kleptocracies States geared toward the enrichment of a tiny political elite.

Kondratieff cycles   A series of 50-year cycles identified by Nikolai Kondratieff, each of which is associated with key innovations that structure society and space.

Kuznets Curve   A diagram that depicts rising inequality, with early economic growth turning into increased equality as growth continues. A modified Kuznets Curve depicts an uptick in inequality. Named after economist Simon Kuznets.

labor   Describes organizations of workers in their dealings with government and employers.

land bridge   Dry land that allowed humans to pass between continents that rising sea levels have since separated.

land deficit   Difference between the amount of land needed to feed the world's population and the amount of land available.

land grab   When land is taken by force or by illegal or semi-legal methods.

language domain   A context for language use defined by place, role, relationships, and topic.

language shift   Change in the degree of use from one language to another.

latitude   Lines of latitude that run parallel to the equator, depicting positions north or south of it. Expressed in degrees, minutes, and seconds.

legitimation crisis   When the state loses its popular appeal and its ability to govern.

Lewis Turning Point   When the supply of cheap labor runs out within a specific geographic location. Named after economist William Arthur Lewis.

life expectancy   The length of time a person is likely to live.

light year   Equivalent to 5.8 trillion miles, the distance light travels in an Earth year, calculated from the speed of light in a vacuum (186,000 miles/second).

limits to growth   An idea that gained prominence following the publication of an influential book in the 1970s that suggested the Earth and its resources could support only a finite amount of continued growth.

linguistic landscape   Refers to the display of language in the physical world around us. Examples include street signs, advertising slogans, and words on billboards.

Little Ice Age   From 1640 to 1690 world temperatures cooled, leading to declining agricultural yields and increased hunger.

liturgical language   A language used in religious worship and ceremonies, or in their sacred texts.

location quotient (LQ)   Measures the degree of concentration of one group versus the total population.

longitude   Also known as meridians, lines of longitude depict positions east and west on the Earth from a defined point. Expressed in degrees, minutes, and seconds.

lunar cycle   The 27.3 days it takes for the Moon to revolve around the Earth. The basis for our division of time into months. Responsible for the movement of tides.

malapportionment   The imbalance between the numbers of voters in similarly represented constituencies. An example: The five largest states in the USA hold more than a third of the entire US population but have only 10 percent of seats in the US Senate. The five least populous states have only 1 percent of the nation's population but hold 10 percent of Senate seats.

male gaze   The way things are seen and understood from an overtly masculine perspective.

Malthusian checks   Thomas Malthus suggested that when population increases beyond needed food supplies, checks such as misery, poverty, and famine bring the population back into alignment with available resources.

Marxist political economy   An approach that draws on the work of Karl Marx to look at issues in capitalist economies such as the endemic crises of capitalism and the conflict between capital and labor.

Mason-Dixon Line   A line drawn across America in the 1760s by two surveyors, Mason and Dixon, to settle a dispute between Maryland, Pennsylvania, and Delaware. It took on a wider significance as a division between North-leaning and South-leaning states during the Civil War and as a marker of different racial attitudes and practices.

mass extinction   Unusually high levels of species extinctions. For example, rising sea levels caused the extinction of many species of shallow-water fauna during the Cambrian period.

megacity   Large urban agglomerations with more than 10 million inhabitants.

megafauna   Large animals weighing more than 220 pounds, which were slower than other animals and therefore especially vulnerable to human hunters immediately at the end of the last Ice Age, leading to the eventual extinction of many.

Megalopolis   Region spanning 600 miles from north of Richmond, Virginia, to just north of Portland, Maine, and from the shores of the Northern Atlantic to the Appalachians, which includes the consolidated metropolitan areas of Washington-Baltimore, Philadelphia, New York, and Boston.

megapolitan regions   The extended metro areas of very large cities.

melting pot   The metaphor that portrayed different immigrant groups blending into a more homogenous society.

Mercator projection   A map projection of the Earth depicting the surface area of degrees of latitude as equal in size.

merchant city   Cities that developed based on a money economy in association with the extension of trade and the emergence of the merchant class.

metropolitan statistical area   Urban areas with a core area of at least 50,000 residents and economic links to surrounding counties.

metropolitanization   Dispersal of people and activities away from the tight urban cores of preindustrial and industrial cities.

Metz model   Identifies four phases of mobility: human travel; settlement; mass mobility; and an era in which travel time, trip rate, and distance traveled hold steady.

micropolitan statistical area   Urban areas with a core area of at least 10,000 and less than 50,000 residents.

modernist architecture   Architectural style dominant from the 1950s to the 1980s. Concerned with form following function, truth to material, and a revulsion against bourgeois decoration. Led to designs that incorporated clean-cut lines and soon became serialized into tall, flat-topped buildings worldwide.

modernity   Of or relating to anything modern.

monotheism   The belief that there is only one god.

moral statistics   The tabulation of crime, pauperism, and a wide range of other social phenomena. First appeared in an 1833 essay by André-Michel Guerry; such statistics were an important part of nineteenth-century thematic mapping.

mortality rates   The measurement of the number of deaths.

mountaintop removal (MTR)   A process for extracting coal by removing mountain peaks to gain access to their coal seams and then dumping into neighboring valleys.

Murdie model   Depicts a concentric ring pattern for stages in the life cycle, a sectoral pattern for socioeconomic states, and a clustered distribution of ethnicity. Named after geographer R. A. Murdie.

nation   Community of people with a common identity, shared cultural values, and a commitment and attachment to a particular area.

nation-state   Describes the territory under the control of a state. Using both terms, nation and state, implies that there is subtle and complex relationship between the two.

national imaginaries   A distinct form of geographical imaginary that refers to the spatial nature and representation of the nation-state.

negative externalities   The cost that a party not participating in an economic transaction must bear as a result of the transaction. For example, the cost borne by residents of a particular neighborhood due to air pollution emitted from a nearby factory.

neighborhood effect   A local effect found in voting outcomes in which neighborhood allegiances can trump sectional ones.

neoclassical architecture   A type of architecture that echoes the classical architecture of Rome and Greece.

neoliberalism   An economic ideology that promotes deregulation, minimal or small government, low taxation, and free trade.

nongovernment agencies   Organizations that are not sanctioned or affiliated with any particular government or state, but may be funded by particular countries. Examples include the International Monetary Fund, World Bank, and World Trade Organization.

non-integrating gap   According to a theory describing the Pentagon's view of the world, this is a group of nations marked by political instability, social unrest, and lack of integration into the global economy.

nonporous boundaries   Partitions that do not allow for the easy movement of people and goods. The border between the former East and West Germany is one example.

nonrenewable resources   Resources such as coal and oil that have a finite global supply.

obesity   Describes the state of individuals with a body mass index greater than 30.

obesogenic   Conditions that tend to increase obesity.

Occidentalism   The writing and description of the world from the viewpoint of those from or who live in the East.

oil cartel   A group of oil suppliers that band together to regulate the supply of their oil on the world market in an attempt to control oil prices and member countries' market shares.

oil crunch   The large and growing gap between the supply of oil and its demand.

oil shock   An unexpected shift in the supply or demand of oil, which leads to a significant price change.

organic regimes   Occur in small towns and suburban districts with a homogeneous population and a strong sense of place with a chief aim of maintaining the status quo.

Organization of Petroleum Exporting Countries (OPEC)   An oil cartel dominated by Middle East oil producers.

Orientalism   Writing and describing the world from the viewpoint of those from or who live in the West.

Ornamentalism   The trumping of class and status over ethnicity in world views. A wrinkle to the Orientalist thesis.

Out of Africa hypothesis   A theory that all humankind are descendants of early humans originally found in Africa. The idea is substantiated by the finding that genetic variation decreases in humans found farther away from Africa.

overfishing   Exploitation of the supply of fish in a certain geographic area that depletes its fish stocks.

overkill   The extinction of megafauna, such as the mammoth and the mastodon, that occurred on every continent and was experienced by every community of large terrestrial vertebrates due to humans' increased technological sophistication in hunting.

overpopulation   A state in which there are more individuals that resources available to sustain them at healthy levels.

overweight   Describes the state of individuals with a body mass index greater than 25.

owner occupation   A type of housing supply that results when households purchase their residential property.

pandemic   An outbreak of an infectious disease more widespread than an epidemic.

paracme   Technically, the point beyond a peak; in diffusion studies, the abandonment of an innovation.

participation rate   The percent of those in a given demographic, such as women, who are working in formal employment.

pastoral farming   Raising livestock to produce meat, wool, and dairy products.

patriarchal society   A society in which males are dominant over females.

patriarchy   The belief system and practice that men are society's authority figures and women should be restricted to home, hearth, and the bearing and rearing of children.

peak oil   A hypothetical point in time at which global oil production reaches its maximum, or peak, declining thereafter.

Peasants' Revolt   A social upheaval in England in 1381 against a proposed reduction of wages.

periodic market system   System of non-permanent marketplaces established in villages and towns by mobile vendors.

permafrost   A subsurface layer of soil that usually remains frozen all year round. Common in the higher latitudes of the Northern Hemisphere.

pidgin   A language created by contact between two or more other languages.

plate tectonics   A theory describing the origin and movement of the Earth's crust and its numerous plates, which float and slowly move atop a molten rock layer (the mantle).

Pleistocene Overkill   The extinction of many large mammals around 11,000 years ago. The main reasons are thought to be human overhunting and climate change.

political ecology   The study of the struggle over resources, who controls them, and how the costs and benefits are apportioned.

polyculturalism   How various groups interact and influence each other, with both immigrant groups and host societies metamorphosing. Distinct from multiculturalism.

polytheism   The belief that there are many gods and deities.

porous boundaries Partitions that allow for the relatively easy movement of people and goods. The US-Canada border is a good example.

postmodernism The rejection of universal, stable and objective ways of knowing.

Potato Blight The Irish famine that began in 1845, leading to millions of deaths and substantial migration out of Ireland.

preindustrial city The form of cities before the onset of the Industrial Revolution. Sometimes also used with reference to cities before 1800 and especially before 1700.

primacy ratio The ratio of the population of the largest city in a country divided by the combined population of the next two largest cities.

primary sector The sector of the economy that includes agriculture, forestry, mining, and fishing.

primate distribution A national urban network in which one city dominates.

prime mortgage Residential mortgages that require a 10 percent to 20 percent down payment and clear evidence of an ability to pay according to the loan terms.

pristine myth The mistaken belief of a vast wilderness before the first European explorers and conquerors arrived. Instead, it is theorized that a new wilderness developed as indigenous populations were depleted.

private renting A type of housing supply whereby landlords rent out property to tenants.

processed food Foods that are frequently distributed in cans, boxes, bags, or other containers that are higher in sugars and fats than more natural foods. Processed foods have been found to increase obesity rates.

producer services The knowledge-based segment of the service industry made up of advertising, banking services, financial services, business consultancies, and information technology. An example is when a household employs an attorney or agency to complete their tax returns.

producer-driven chain A food supply chain in which producers tend to establish cartels.

product cycle The trajectory of a brand new product.

profit cycle The process of how sales of new products impact how profits grow, mature, and decline over the life of the product.

proportional representation An electoral system in which representatives are elected in proportion to the number of votes cast.

protectionism Any form of economic policy that seeks to support local companies at the expense of foreign corporations, usually through quotas, tariffs, or some combination thereof.

Ptolemy One of the earliest geographers, he mapped both the universe and the terrestrial world, using a conic projection technique to represent the habitable world.

public bads Necessary but unwanted civic facilities and institutions such as trash disposal sites, motorways, and halfway houses for convicted criminals.

public goods Collectively provided goods and services.

public housing A tenure type of housing in which a government agency provides accommodation for citizens.

quantitative geography An approach that emphasizes measurement and calibration.

quaternary sector The knowledge-based part of the economy that includes business advisory services, consultancies, information technology and research and development.

quinary sector Those in the information and knowledge-based economies who make the final executive decisions.

racial/ethnic clustering Occurs when different groups gather together due to similar race or ethnicity.

racialized groups The social construction of groups of people as racial categories.

range When referring to a good or service, the distance consumers are willing to travel to make a purchase.

rank-size distribution The regular relationship between a city's population and its rank in the urban hierarchy. Also known as *Zipf's Law*.

rationality crisis When the state makes enough poor decisions that other crises emerge.

redlining A process by which US institutions restricted their mortgage allocations in lower-income minority neighborhoods.

relative space The social dimension of space, such as the different income of different regions, often contrasted with the space of latitude and longitude

religio-ecological The combination of environmental and religious beliefs that sees environmental protection as God's work. The human occupancy of the Earth has spiritual responsibilities and ethical obligations.

remittances The money that temporary and permanent migrants send back to their home countries.

remote sensing Gaining information about the Earth using aerial sensor technologies such as satellite recordings and monitoring.

renewable energy Energy from effectively limitless sources such as wind, tide and the sun.

replacement rate The number of births required to keep a given population stable.

resource curse Describes how valuable resources can create short-term booms but may depress long-term economic growth. Also known as the "paradox of plenty" or the Dutch Disease.

reterritorialization Referring to culture, describes how cultures are being reformed in new places away from from their traditional hearths.

rewilding A movement that aims to create landscapes and ecosystems similar to those that existed in Paleolithic times.

Ring of Fire The horseshoe-shaped distribution of intense seismic activity around the Pacific Ocean. This region of active plate-tectonic activity is responsible for 90 percent of the world's earthquakes and 75 percent of volcanic eruptions.

Romanticism Grew as a response to the commodification of land and views nature as a source of profound aesthetic experiences and a vehicle for a personal connection to the infinite.

Rostow model A model of economic development proposed by Walt Rostow, describing the five stages of development: traditional society, pre-takeoff, takeoff, drive to maturity, and high mass consumption.

rural-to-urban migration The movement from rural areas to cities.

Rust Belt A region in the upper Midwest and Northeast of the USA that used to be dominated by manufacturing and heavy industry. The term denotes the deindustrialization and shrinking of industrial labor force in the region since the mid-1970s.

salinization   An increase in the amount of salt in soils that reduces their fertility and, at high levels, makes them toxic. Affects 10 percent to 20 percent of world soils.

seasonal affective disorder (SAD)   The tendency for more negative moods among normally healthy people during distinct seasons.

Second Urban Revolution   A period that began in the late eighteenth century with the creation of the industrial city and unparalleled rates of urban growth.

second wave of globalization   A distinct period of globalization that occurred from 1865 to 1970, facilitated by lower tariffs, an international labor market, relatively free capital mobility, expanded transportation, and the effects of two world wars.

secondary sector   The sector of the economy that mainly consists of the manufacturing of goods.

self-build housing   Housing that is built by residents themselves.

Semitic region   One of the main hearths of the axial age. Centered in the Middle East, it is the location where Judaism, Christianity, and Islam were founded.

service sector   A segment of the economy that is made up of workers who provide services as opposed to producing goods.

settled agriculture   Describes when a group of people remain in a specific location and survive through means of growing foods, rather than hunting and gathering.

shadow economy   A reference to the informal economy in which individuals pay no taxes.

sharing economy   A private market hybrid that relies on collaborative consumption. Owners rent assets such as cars and rooms to users, the transaction often facilitated by social media. Examples include Uber and Airbnb.

shipping container   The creation and worldwide usage of a standardized container. Its usage resulted in dramatically reduced shipping times and transportation costs.

slave trade   The forced migration of millions of Africans to the New World from the sixteenth to the nineteenth century.

slum   Unplanned, informal, and often illegal housing in cities that arises due to the inability of formal markets and public authorities to provide enough affordable and accessible housing. Slums are also referred to as "shantytowns," "informal housing," and "squatter housing."

slums of despair   Slum areas where people are on a downward spiral.

slums of hope   Slum areas where people are improving their living conditions and those of their children.

social conflict   An increase in social disorder, including violent conflict that may occur during a significant youth bulge.

social cost   Costs imposed upon society.

Social Darwinism   Theory that argues that people are also subject to the survival of the fittest. First proposed by Herbert Spencer in the late nineteenth century.

social economy   Sector of the economy that consists of not-for-profit activities. Sometimes referred to as the "third sector."

social geography   A branch of human geography concerned with determining the relationship between social phenomena and geography.

social housing   Housing provided by a government agency.

social statistics   The quantification of social problems. Used by mapmakers to track urban inequality and conduct important research on urban environments.

soft border   A porous boundary.

soft power   The ability to co-opt others to your point of view without the use of force.

Soviet bloc   Countries, mostly in Eastern Europe, that were closely allied or controlled by the Soviet Union.

space-time convergence   The decrease in time taken to cover a specified distance, reducing the cost of transporting goods and people, and seemingly pulling places closer together.

spaced time   Changes in places with time, as when a restaurant district become busy at mealtimes.

spatial diffusion   The movement across space and through time of an innovation, idea, or practice. Swedish geographer Torsten Hagerstand did early work on the topic.

speculative building   Involves builders preparing land and constructing dwellings for general demand rather than specific customers.

speech community   People who share a similar language.

starchitecture   The construction of buildings by big-name architects. Often used to market or brand a city.

state   A separate and distinct spatial unit of political authority.

statecraft   Geopolitical strategies used by a nation-state. According to early twentieth-century writer Sir Halford Mackinder, geography was used to create statecraft.

subaltern geographies   Human geography by and about more marginal groups, often ignored in the history of human geography.

subprime mortgages   Residential mortgages that require little or no down payment with few verifiable means of income.

suburbanization   The movement of people from their permanent residences in the central city to the suburbs.

sulfur dioxide   A gaseous oxide of sulfur (SO_2). In prehistory, thought to have poisoned the air following colossal volcanic eruptions, contributing to mass extinctions of animal life in the Mesozoic era. Its main modern source is burning fossil fuels. Affects the human respiratory system.

surveillance   International Monetary Fund monitoring of a member's economic policies and evaluation of their economy.

symbolic ethnicity   See *adopted ethnicity*.

symbolic regimes   Occur in cities undergoing rapid changes, including large-scale revitalization, major political change, and image campaigns that try to shift the wider public perception of the city.

technopoles   Contemporary industrial complexes established around high-technology industry.

tectonic plates   The thin plates that comprise the Earth's crust and float and slowly move on a molten layer (the mantle).

territorialization   When referring to culture, signifies the connection between culture and specific places.

tertiary sector   The sector of the economy that comprises the service industry.

thick market   A location that contains a large pool of labor and specialized firms.

Third Urban Revolution   The global phenomenon in which we currently reside. Marked by five distinct characteristics:

rapid urbanization, increases in the size of individual cities, increased metropolitanization, increased urban sprawl, and the proliferation of slums.

third wave of globalization   A distinct period of globalization that emerged after 1970 and is associated with an even more marked space-time-cost compression, global production chains, and a deregulated global financial system, creating a "flatter" world with easier and hence quicker international economic links.

threshold   Referring to a good or service, the necessary minimum population to support its continued supply.

tides   The ceaseless and regular vertical motion of our oceans and seas created by the lunar cycle and the gravitational pull of the Moon.

time-cost compression   As economies mature, the importance of the exchange of information leads to innovations that lower the costs of communications via telephone, Internet, and the like.

timed space   Behaviors and events that occur in certain space at certain times.

Tobler's Law   A geographical law stating that "Everything is related to everything else, but near things are more related than distant things."

toponymic colonialism   When colonial powers named features of acquired territory and erased preexisting local names.

totalitarian regime   A system of government in which the state has control over wide and deep swathes of social, political, and economic life.

tough oil   When the extraction of oil becomes more expensive and risky, often producing an increased interest in alternative fuel sources.

tourist gaze   The set of expectations that people have and the way they look when they visit places as tourists.

traditional markets   A type of food supply chain that is dominated by smallholders who cater to domestic and local markets.

transhumance   The traditional movement of mountain peoples with their animals from summer to winter pastures.

transnational architecture   A type of architecture that incorporates styles that transcend specific places, particularly those that cross national boundaries.

transnational corporations (TNCs)   Describes different types of large companies that operate in more than one country and can be further classified as multinational, international, global, or integrated networks.

triangular trade   A trading circuit of sailing ships that linked West Europe, West Africa, and the New World. Manufactured goods were shipped to Africa, slaves transported to the Americas, and commodities shipped back to Europe.

tropical zone   Geographical location that covers the equator and extends to the lines of latitude known as the Tropics of Cancer and Capricorn.

tsunami   A large sea wave resulting from a volcanic explosion, or other movement of the Earth's plates.

unequal exchange   The trade between economies producing primary commodities and manufacturing economies, which tends to provide more wealth to the latter.

urban regimes   Informal arrangements that surround and complement the formal workings of governmental authority, which consist of informal governing coalitions that make decisions and get things done in a city.

urban sprawl   The dispersed form of urban development toward suburban areas and farther from central cities.

urbanization   The geographical redistribution of population from rural to urban areas; also marked by social reorganization.

Uruk expansion   A broadening of the influence of Uruk, one of the major cities in early Mesopotamia.

value-added chain   The position of goods and services on a chain of increasing price and value. For instance, making shoes has a lower value than making computer chips, so when a region shifts from shoe to chip manufacturing, it is considered a move up the value-added chain.

vernacular architecture   A style of architecture that uses local resources and traditions in the design and construction of buildings.

von Thünen model   Describes land use around cities; suggests a concentric ring pattern with more expensive land and thus more intensive agriculture closer to the city. Proposed by Johann Heinrich von Thünen in the early nineteenth century.

wheat belt   Countries located in climates amenable to the production of wheat, including Canada, Russia, Argentina, and Australia.

workfare   The welfare-to-work initiatives that have developed since the 1990s.

world population   The number of people inhabiting the world, currently at close to 7.5 billion.

yield gap   The difference between the possible food supply productivity and actual productivity.

youth bulge   A period of time marked by a rapid increase in the number of people between the ages of 15 and 24. It occurs in middle stages of a demographic transition and is the result of a rapid reduction in child mortality before a rapid falloff in fertility.

Zelinsky model   Describes three phases of mobility: premodern traditional society with limited mobility, advanced society of increased mobility and circular movement, and the super advanced society in which there is increased acceleration of movement, but a potential peak.

Zionism   A movement, formally established 1896/97 as a political organization promoting the creation of a Jewish state in the Middle East. Now the term refers to the support of Israel as a Jewish state.

Zipf's Law   The regular relationship between a city's population and its rank in the urban hierarchy.

Credits

Part 1   Opener Image: iStock.com/michaeljung.

Chapter 1   Opener Image: NASA. Fig. 1.2: Adapted from Project Ocean Vision (www.projectoceanvision.com). Fig. 1.6: Adapted from *Science Advances* 19 Jun 2015: Vol. 1, no. 5, e1400253. This work is licensed under CC BY-NC (http://creativecommons.org/licenses/by-nc/4.0/).

Chapter 2   Fig. 2.3: Sheila Terry / Science Source. Fig. 2.5: Library of Congress, Moffett. Fig. 2.6: Courtesy of the Newberry Library, Chicago, Illinois. Map 6F G4104.C6E2 1895 .G7 sheet 1. Fig. 2.7: Adapted from Wikipedia. Fig. 2.8: Astronaut Photograph/NASA/JSC.

Part 2   Opener Image: iStock.com/Jukkisjupi.

Chapter 3   Fig. 3.2: Adapted from Max Roser and Esteban Ortiz-Ospina (2017) – 'World Population Growth.' *Published online at OurWorldInData.org.* Retrieved from: https://ourworldindata.org/world-population-growth/ [Online Resource]. Licensed under a Creative Commons Attribution-ShareAlike 4.0 International License.

Chapter 4   Box 4.1 Figure (top): After Swartz W, Sala E, Tracey S, Watson R, Pauly D (2010) The Spatial Expansion and Ecological Footprint of Fisheries (1950 to Present). PLoS ONE 5(12): e15143. doi:10.1371/journal.pone.0015143. Fig. 4.3: Adapted from "2015 Global Hunger Index by Severity." Map (von Grebmer et al. 2015). From von Grebmer, K., J. Bernstein, A. de Waal, N. Prasai, S. Yin, and Y. Yohannes. 2015. *2015 Global Hunger Index: Armed Conflict and the Challenge of Hunger.* "2015 Global Hunger Index by Severity" Map. Bonn, Washington, DC, and Dublin: Welthungerhilfe, International Food Policy Research Institute, and Concern Worldwide. Fig. 4.5 Adapted from source: Food and Agriculture Organization of the United Nations, FAO Food Price Index, 2016. http://www.fao.org/worldfoodsituation/foodpricesindex/en/. Reproduced with permission. Fig. 4.6: Reprinted from World Health Organization, *Global Status Report on Noncommunicable Diseases 2014.* http://apps.who.int/iris/bitstream/10665/148114/1/9789241564854_eng.pdf ISBN 9789241564854. Fig. 7.2 Age-standardized prevalence of obesity in women aged 18 years and over (BMI ≥30 kg/m2), 2014. Page 100. © World Health Organization 2014.

Chapter 5   Box 5.1 Figure: Adapted from ASPO International. Fig. 5.1: US Department of Energy. Fig. 5.3: Mauna Loa Observatory, National Oceanic and Atmospheric Administration, Earth System Research Laboratory, Global Monitoring Division. Fig. 5.8a: Adapted from National Oceanic & Atmospheric Administration, Earth System Research Laboratory. Fig. 5.8b: Inga Spence / Science Source. Fig. 5.9: Adapted from US Geological Survey.

Chapter 6   Fig. 6.6: Adapted from Etter, Andres, Clive McAlpine, and Hugh Possingham. "Historical Patterns and Drivers of Landscape Change in Colombia Since 1500: A Regionalized Spatial Approach." *Annals of the Association of American Geographers* 98, 1 (2008): 2–23, Figure 4. DOI: 10.1080/00045600701733911. Fig. 6.7: Adapted from Wilbanks, T. J., & Kates, R. W. (2010). "Beyond adapting to climate change: embedding adaptation in responses to multiple threats and stresses." *Annals of the Association of American Geographers*, 100 (4), 719–728, Figure 1. Ed. Richard Aspinall. DOI: 10.1080/00045608.2010.502408.

Part 3   Opener Image: iStock.com/George-Standen.

Chapter 7   Fig. 7.8b: Benny Marty/Shutterstock. Fig. 7.8c: Associated Press Photo/Paul Sakuma. Fig. 8.3: Adapted from The Economist, "The Wild, Wild East." http://www.economist.com/news/special-report/21663328-booming-middle-class-creating-worlds-most-dynamic-consumer-market-wild-wild. © The Economist Group Limited, London (September 12th, 2015). Fig. 8.4: Diagram courtesy of Barcelona Field Studies Centre (www.geographyfieldwork.com). Fig. 8.8: Adapted from Bureau of Labor statistics.

Part 4   Opener Image: iStock.com/Instants.

Chapter 9   Fig. 9.1: Courtesy of Bill Rankin, www.radicalcartography.net. Fig. 9.6: Adapted from data from The World Bank, International Labour Organization, ILOSTAT database. Box 9.1 Figure: Adapted from Yu, Hongbo. "Spatio-temporal GIS Design for Exploring Interactions of Human Activities." *Cartography and Geographic Information Science.* Vol 33, Issue 1, 2006. Pages 3-19. Figure 3. http://dx.doi.org/10.1559/152304006777323136.

Chapter 10   Fig. 10.9: Jon T. Kilpinen, Valparaiso University. Fig. 10.12: Adapted from Pew Forum on Religion & Public Life, April 2010. Pew Research Center, Washington, DC.

Chapter 11   Fig. 11.7: Adapted from Dialect Map of American English by Robert Delaney (www.robertspage.com/dialects.html), after *Success with Words: A Guide to the American Language* (by Reader's Digest; prepared in association with Peter Davies; David Rattray, project editor. Pleasantville, NY: Reader's Digest Association, 1988).

Part 5   Opener Image: iStock.com/hanohiki.

Chapter 12   Fig. 12.4a: Flik47/Shutterstock. Fig. 12.6: Adapted from Statistics Canada. Fig. 12.11: Courtesy of Periscope (www.periscope.co.uk). Fig. 12.12: Courtesy of The Fund for Peace.

Chapter 13   Fig. 13.1b: iStock.com/maomaotou.

Part 6   Opener Image: Dinodia Photos/Alamy Stock Photo.

Chapter 14   Fig. 14.5: Frank Baum, *The Navy Alphabet.* Fig. 14.10b: Handout/Alamy Stock Photo. Box 14.2 Figure: From Thomas P.M. Barnett's *The Pentagon's New Map: War and Peace in the Twenty-First Century (2004)*, © Barnett Consulting LLC.

Chapter 15   Box 15.1 Figure: © Sasi Group (University of Sheffield) and Mark Newman (University of Michigan). Fig. 15.7b: thomas koch/Shutterstock. Fig. 15.12: Adapted from data from US federal election results, 2016. Fig. 15.13: Adapted from data from US federal election results, 2012 & 2016.

Part 7   Opener Image: iStock.com/capdesign.

Chapter 16   Fig. 16.1: Photo by Lisa Benton-Short.

Chapter 17   Fig. 17.8a: Courtesy of the GaWC Research Network, Loughborough University. Fig. 17.8b: iStock.com/IR_Stone. Box 17.2 Figures: X. Liu *, Z. Neal **, B. Derudder, "City Networks in the United States: A Comparison of Four Models," *Environment and Planning A*, 44 (2), (2012), 255–256, Figures C and D. Pion Ltd, London. (www. pion.co.uk, www.envplan.com)

Chapter 18   Fig. 18.10b: Vadim Ozz/Shutterstock.

Index

Page numbers followed by t indicate tables; b indicate boxes; and italicized page references indicate figures.